In Search of Modernity:

SCIENCE AND TECHNOLOGY IN AFRICA

Edited By

Paul Tiyambe Zeleza
Ibulaimu Kakoma

Africa World Press, Inc.

P.O. Box 1892
Trenton, NJ 08607

P.O. Box 48
Asmara, ERITREA

Africa World Press, Inc.

P.O. Box 1892
Trenton, NJ 08607

P.O. Box 48
Asmara, ERITREA

First Printing 2003

Book Design: Getahun Seyoum Alemayehu
Cover Design: Roger Dormann

Cataloging-in-Publication Data is available from Library of Congress

ISBN 1-59221-111-9 (Hardcover)
ISBN 1-59221-112-7 (Paperback)

Acknowledgements

All the papers in this book, except one, were first presented at the 27th Annual Spring Symposium of the Center for African Studies at the University of Illinois at Urbana-Champaign (UIUC), which was organized jointly with the College of Engineering and in cooperation with the Nairobi-based African Academy of Sciences. The symposium was made possible by the generosity of many colleges, departments, and programs at UIUC which we wish to thank. They include: the African Business Center, Beckman Institute, Biotechnology Center, Center for Advanced Study, College of Liberal Arts and Sciences, College of Veterinary Medicine, Coordinated Science Lab, Department of Chemical Engineering, Department of Civil and Environmental Engineering, Department of Computer Science, Department of Electrical and Computer Engineering, Department of Mechanical and Industrial Engineering, Department of Veterinary Pathobiology, International Programs and Studies, Microelectronics Laboratory, National Center for Supercomputing Applications, and Office of International Student Affairs. Financial support was also provided by the US Department of Education through the Center's Title VI grant as a national Resource Center for African Studies. We also wish to thank the staff of the Center for African Studies who worked hard to organize the symposium, making all the necessary logistical arrangements from travel to accommodation that can make or break an international conference. We are particularly indebted to Sue Swisher, the Center's Administrative Aide, Jamie McGowan, the Coordinator of Special Programs, Sheila Calhoun, the secretary, and two graduate students enrolled in the Center's MA program, Bertin Kouadio and Amanuele Bereket. Above all, we are grateful to Ilesanmi Adesida, Professor of Electrical and Computer Engineering and Director, Microelectronics Laboratory, who was co-organizer of the symposium and without whom this book would not even have been conceived.

Table of Contents

Part 2: Information Technology, Development and Democracy

Part 3: Information Technology, Education and Health

Part 4: Information Technology, Biotechnology, and Agriculture

Notes on Contributors

O. Akin Adubifa has been Program Officer of Carnegie Corporation of New York's Program on Strengthening Human Resources in Developing Countries (HRDC) since 1993. E-mail: aaud3@aol.com

Charles Onyekachi Akalezi is Head of Environment Program, Center for Environmental Resources and Sustainable Ecosystem (CE-RASE), Lagos, Nigeria. E-mail: cerase@yahoo.com

Maurice N. Amutabi is a Ph.D. candidate, Department of History, University of Illinois at Urbana-Champaign, Illinois, USA. E-mail: Amutabi@yahoo.com

Chris Benner is with the Institute of Industrial relations, University of California, Berkeley, California, USA. E-mail: cbenner@socrates.berkeley.edu

E. A. Biakolo is a Senior Lecturer in the Department of English at the University of Botswana, Gaborone, Botswana. E-mail: biakolo@mopipi.ub.bw

Mary F. E. Ebeling is a graduate student at the Medill School of Journalism, Northwestern University, Evanston, Illinois, USA. E-mail: maryebeling@yahoo.com

Juliet Evusa is Minority Graduate Student Coordinator and graduate student in the School of Telecommunication, Ohio University, Athens, Ohio, USA. E-mail: je230591@oak.cats.ohiou.edu

Aaron DeGrassi is with the University of California, Berkeley, International and Area Studies Department. E-mail: degrassi@ofc.berkeley.edu

Nancy Hafkin is former Coordinator, African Information Society Initiative, United Nations Economic Commission for Africa (UNECA). E-mail: nhafkin@hotmail.com

Mohamed H. A. Hassan is President of the African Academy of Sciences and Executive Director, Third World Academy of Sciences, Trieste, Italy. E-mail: twas@ictp.trieste.it

Khaled Ismail is Vice President and General Manager, Ellipsis Digital Systems, Cairo Engineering Center, Cairo, Egypt. E-mail: kismail@ellipsisdigital.com

Ibulaimu Kakoma is Associate Professor of Veterinary Pathobiology at the University of Illinois at Urbana-Champaign, Illinois, USA. E-mail: i-kakoma@uiuc.edu

James Katzenstein is Project Director, Mikocheni Health and Education Network, Lucinda, Mission Viejo, California, USA. E-mail: jimkatzenstein@compuserve.com

J. P. Legg is a researcher at the International Institute of Tropical Agriculture- Eastern and Southern Africa Regional Center, Kampala, Uganada. E-mail: iitanam@infocom.co.ug

Sebastian Mathews works for Umbono Management Consultants (Pty) Ltd, Johannesburg, South Africa. E-mail: smathews@umc.co.za

Muadi Mukenge is a program officer at the Pacific Institute for Women's Health, Los Angeles, California, USA. mmukenge@piwh.org

Nora Kasirye Mulira is the Director, Institute of Computer Science, Makerere University, Kampala, Uganda. E-mail: nora@dicts.mak.ac.ug

James C. O. Nyankori is professor of Agricultural and Applied Economics, Clemson University. E-mail: jnynkr@clemson.edu

R. Obonyo is a researcher at the International Institute of Tropical Agriculture, Eastern and Southern Africa Regional Center, Kampala, Uganda. E-mail: iitanam@infocom.co.ug

O. Galman Omitogun is Associate Professor of Agrobiotechnology,

Faculty of Agriculture, Obafemi Awolowo University , Ile-Ife, Osun State, Nigeria

Deborah Akoth Osiro is a graduate student at the Amsterdam Law School, Amsterdam, The Netherlands. E-mail: deb_Osiro@yahoo.com

Rubin Patterson is an Associate Professor Department of Sociology, University of Toledo, Toledo, Ohio, USA. E-Mail: rpatter@utnet.utoledo.edu

E. Babafunso Sonaiya is Dean of Faculty and Member, Nigerian Technical Subcommittee on Agricultural Biotechnology, Obafemi Awolowo University, Ile-Ife, Ogun State, Nigeria. E-mail: fsonaiya@oauife.edu.ng

P. Sseruwagi is a researcher at the International Institute of Tropical Agriculture, Eastern and Southern Africa Regional Center, Kampala, Uganda. E-mail: iitanam@infocom.co.ug

G. Okao-Okuja is a researcher at the International Institute of Tropical Agriculture, Eastern and Southern Africa Regional Center, Kampala, Uganda. E-mail: iitanam@infocom.co.ug

Ian Kumwenda is a researcher in the Ministry of Agriculture and Irrigation, MASIP Secretariat, Lilongwe, Malawi. E-mail: inkumwenda@malawi.net

Denis Rugege is the Program Director of the Land Information Management Project, Center for Environment and Development, University of Natal, Pietermaritzburg, South Africa. E-mail: rugeged@nu.ac.za

Muta A. Tiamiyu is a researcher at the Africa Regional Centre for Information Science, University of Ibadan, Nigeria. E-mail: muta@cear.org.ng

Paul Tiyambe Zeleza is Professor of African Studies and History at the Pennsylvania State University. He is the author of more than a dozen books, most recently, *Rethinking Africa's Globalization* (2 volumes), *Leisure in Urban Africa*, the Routledge *Encyclopedia of Twentieth Century African History*, and *African Universities in the Twenty-First Century* (2 volumes), as well as three works of fiction including

Smouldering Charcoal (a novel) and *The Joys of Exile* (a collection of short stories). He is the winner of the 1994 Noma Award for his book *A Modern Economic History of Africa* and the 1998 Special Commendation of the Noma Award for *Manufacturing African Studies and Crises*.

INTRODUCTION

AFRICA'S SEARCH FOR MODERNITY IN THE INTERNET AGE

Paul Tiyambe Zeleza

Africa has been grappling with the question of modernity and modernization ever since its encounter with modern Europe in the fifteenth century, when the continent, facing increasingly industrial and imperial societies, was forced to reckon with the state of its own economic, technological, and social development. Africa's encounter with Europe made the question of the modern an African one, a practical and intellectual question. The themes of "African regeneration" or "African renaissance" have engaged African intellectuals in debate since the middle of the nineteenth century. In practical terms, African modernization has been seen in terms of continental societies acquiring tools and instituting efficient techniques of European provenance for an amelioration of their conditions. However, insofar as the encounter revealed to African societies how their deficiencies in science and technology and rationalist social organization made them candidates for European conquest, the dream of African modernity has never been simply this. Africa's modern regeneration has consistently been envisioned in terms of an exponential rise of African power. Modernizing development, that is, has been inscribed, as Kwaku

Korang (2002) argues, in an African desire for parity: how is (conquered) Africa to be like (her conqueror) Europe? But this question has quickly begged another: can Africa be like Europe, i.e., absorb European rationality into herself, without ceasing to be Africa and if so how? Can Africa be similar without being assimilated? It is in this tension that African modernity ceases to be a mere practical question of modernizing along European lines and takes on loaded ideological significance — whether one can separate the spirit of the modern from the substance.

In the nineteenth century, the question of reforming African societies preoccupied African leaders and intellectuals, from those rooted in the old Islamic scholarly traditions and states of north, west, and coastal east Africa, to intellectuals newly exposed to western education in west and southern Africa. This quest for modernity, one that was at once global and distinctly African, intensified during the twentieth century, a period that was marked by colonialism, decolonization, and independence. At the center of it all was education, science, and technology seen as vehicles for intellectual enlightenment, social engineering, cultural production, political participation, and above all, economic development. Thus, education and science as processes and projects through which social, cultural, technological, and economic capital is acquired and reproduced, and the nation and the world are productively imagined, have dominated African discourses about development, nationalism, and globalization for a long time.

Clearly, Africa has long sought to fashion its own modernity or modernities. Indeed, modernity, as a process or a project, condition or discourse, has always carried multiple meanings and trajectories of societal modernization — accompanied by cognitive and social transformations as manifested in the ascendancy of scientific consciousness and instrumental rationality and improving material conditions of life — and of cultural modernity — marked by ideas and experiences of novelty and routine, presentness and progress, creativity and coercion, freedom and conformity. Its embodied character shifts as it is creatively adapted in different sites and refashioned by changing historical moments. The progressive spread of modernity from its European heartland over the last few centuries of capitalist globalization and the contingent regeneration and renegotiations of its institutions and attitudes in places that are simultaneously similar and different, divergent and convergent, generate ever-changing combinations and recombinations of modernity so that it is best to talk of multiple, new, or alternative modernities (Ong 1996; Hefner 1998; Robotham 2000; Gaonkar 2001)

Ultimately, Africa's modernity revolves around the question of developing, to quote Abiola Irele (1999: 7-8) "a workable and effective social and political organization of our national communities, and the productive management of our physical environment and material resources, all this in a world dominated by 'instrumental reason' and controlled from ever fewer centers of power and decision." Africa's contemporary crisis of development is both internal and external, as much inward psychologically and morally as it is outward structurally and objectively so that the struggle for reconstruction must "be undertaken at two levels: that of mind, in the first place, as the necessary foundation for the second level, that of action, the two correlated in the reconstitution of our world, so as to create within it our own unique space of life and expression."

The project of African modernity, then, poses challenges that are both political and philosophical, concrete and conceptual, about renewal in economic and epistemic terms and of social and structural conditions, about the development and democratization of African institutions in a world that rewards technological and scientific progress and punishes those lagging behind; a world, in short wedded to the cultures of modernity, despite all the convoluted fulminations among some western scholars about postmodernity. Only when Africa's modernity can withstand the corrosive pressure of modernities elsewhere will the continent's fate and future be secured. But this is to beg the question: what is African modernity? Who are its architects, its guardians, its beneficiaries?

These are fundamental questions that demand serious reflection by African thinkers and policy makers. Modernity, so essential to Africa's renaissance, is a profoundly intellectual, cultural, political and economic project. It entails the revitalization of the continent's educational, material, and governance infrastructures. African scholars have had much to say about these questions, especially concerning the misconceived dichotomies between tradition and modernity and the equation between modernization with westernization. The eminent Kenyan scholar, Ali Mazrui (1999), contends, "modernization is change which is compatible with the present state of human knowledge, which seeks to comprehend the legacy of the past, which is sensitive to the needs of the future, and which is increasingly aware of its global context." In his formulation, modernization minus dependency equals development. He argues that African modernization needs three major revolutions in skills, values, and gender relations. In an increasingly knowledge based global economy, the power of skills is self-evident. Skills often need science and technology at different levels. Science and technology, in turn, are conditioned by culture. He also suggests several strategies for reducing and transcending dependency. Among them is indigenization, domestication, and diversification.

Indigenization requires greater utilization of indigenous techniques, personnel, and approaches to purposeful change, while domestication involves making imported institutions more relevant to Africa in their organization and functions. As for diversification, he recommends that Africa must diversify the foreign cultures from which it seeks to learn and stop its "excessive reliance on the West as the only Other." Another strategy entails what he calls "horizontal inter-penetration and regional integration" which involves not only continental integration within Africa but also South-South integration as well. The final strategy is that of counter-penetration – infiltrating the western citadels of power through trade, migration, and the globalization of African culture. However, Africa's reconstruction or renaissance will not go far, he concludes, without androgenizing modernity.

African philosophers have provided us with some of the most insightful analyses about Africa's problems, possibilities, and prospects of modernity and modernization. On the question of tradition and modernity, Kwame Gyeke (1997), the distinguished Ghanaian philosopher, argues that both the traditionality of a traditional society and the modernity of a modern society tend to be overstated for no human society is absolutely unchanging, that is completely "traditional," and societies that are called "modern" are based on long cherished traditions or recognize certain traditions as relevant, not as obstacles, to their development. If tradition is understood as a practice or belief which has lasted several generations – Gyeke suggests three – and is inherited from the past and becomes an enduring element in the cultural life of a people, its preservation, in part or in whole, depends very much on the attitude of the new generation toward it. In other words, previous generations do not transmit their cultural creations as such for they have no way of knowing which of them will survive and be accepted by subsequent generations. It is the latter who may, based on their normative perspectives or other rational grounds, either accept, refine, and preserve the traditions or spurn, deprecate, or abandon them.

Traditions may be refined or abandoned either as a result of internal criticism or the adoption of alien ideas, values, and practices. But for the adopted elements to be most beneficial the society must shape, assimilate, and appropriate them in line with its ideals, purposes, and aspirations so that they function effectively. External imposition is likely to lead to little commitment, appreciation or understanding of the alien cultural elements and may engender confusion and self-doubt and even provoke resistance. In so far as cultural change and borrowing are universal, the question that African thinkers must pose is not whether African cultures ought to change or borrow from other cultures, but how they can do it most effectively, productively, and progressively. This requires critical inquiries that eschew the fanciful wishes of the cultural revivalists and cultural antirevivalists, whose polarized positions generate unnecessary confusions because they fail to distinguish between what may be regarded as negative and positive features of African cultures upon which African modernity must surely rest, equipping African societies with the scientific, intellectual, and technological resources to function adequately in an ever more complex and globalizing world.

Africa's modernity has to be continuously reformulated out of a history that has been both unique and an integral part of world history. It is a modernity that is imbricated with the modernities of other societies, especially western modernity, which was itself a product of elements derived from both European traditions and Europe's encounters with and appropriations from the rest of the world, including Africa. This not only suggests the complexity of modernity, it also implies that the view of modernity as a European invention, which other societies need only import as a turnkey project, is silly. While many aspects of modernity in various parts of the world reflect certain features of western modernity, the latter does not exhaust all possibilities of being modern. Indeed, even if Africans wanted to copy every aspect of western modernity, they would not be able to because of the sheer weight of their own histories and cultures. The modernization of Asian countries such as Japan and China shows, for example, that it is possible to modernize without becoming westernized. In contrast, many African countries have acquired western values and institutions through their long contacts with Europe, much of it involuntarily, but they are far from being modernized. Thus a country can become modernized without being westernized or become westernized without being modernized. There is, therefore, no logical equivalence between modernization and westernization.

African intellectuals have the responsibility to imagine different modernities for their societies, not simply for the subjective reasons of nationalism, but because our own historical experiences and cultural traditions objectively demand that we do so. It seems to me that in so far as the project of modernity is quintessentially an intellectual one priority ought to be placed on expanding education at all levels, including the universities, which have fallen upon hard times and need to be restored and re-energized so that they can become powerhouses of ideas, rich reservoirs of critical and creative thought, and springs of the endless innovations that come from fundamental and applied research. Greater emphasis needs to be placed on science and technology, which are crucial purveyors of modernity and can reshape Africa's future if they become fully integrated into the continent's cultural fabric and consciousness. We have to create a modernity that combines high economic productivity with humanism and environmentalism, in which economic growth and privatization are accompanied by sustainable distribution and development, the future is not sacrificed for the present, the market is tamed by the state, and rapacious globalization is tempered by new democratic structures of global governance. A modernity in which personhood combines respect for individuality and communality based on the clear understanding that the individual is both autonomous and constituted by society so that she has individual rights and social responsibilities.

One of the distinguishing characteristics of science is its systematicity in terms of how it describes and explains phenomena, establishes knowledge claims, and expands and represents knowledge. But this very systematicity leads to extravagant neo-positivist claims about science's explanatory and predictive power, which post-positivists, especially postmodernists, with their anti-foundationalist and constructionist views, have done much to debunk (Galavotti, 2000). While these critiques are pertinent, we must be careful not to throw the baby out with the bathwater. As I have argued elsewhere, African intellectuals cannot afford the indulgence or angst of antirationalist postmodernist thinking (Zeleza 2002a). Irele (1999: 7-8) sees postmodernism as the "most insidious threat to the contemporary African mind," despite the fact that its radical questioning of the historical and philosophical legacy of the Enlightenment "is one that our historical experience predisposes us to understand and to rally to." He insists that while we indeed have to recognize the limits of science, there is no reason to jettison its real triumphs, and the keys it holds for "advancing our interests in the modern world...." The systematicity of science also breeds specialization and fragmentation, which can undermine communication between science and the public, and between scientists and policy makers. These communication problems can have real consequences on both science and society (Hoyningen-Huene, 2000).

Science is ultimately a social and political enterprise. It needs political will for funding and structured support, effective regulation and application, which entails, in turn, that it must respond to the needs of society, find practical solutions to pressing problems and the inspired innovations to realize society's collective dreams of the future. As Frederico Mayor (2000: 26-7), then UNESCO Secretary-General put it at the opening of the World Conference on Science in June 1999, "science is too important to be left to the markets." Neil Harrison (1998) argues that in the United States federal government funding, policies, and regulations played a critical role in research and development from which emerged many of the new consumer, not to mention military, technologies. To expect market-produced technology to sustain development through the production of eco-efficient technologies without judicious political intervention by the state is equivalent to hoping that "five thousand monkeys may eventually type Hamlet" (Harrison, 1998: 185; also see Levine, 1998). Public support for science is particularly critical with the rapid expansion of proprietary science that is spreading as private profit seeking corporations assume a greater share of scientific research.

If Africa is to become an important producer, not just consumer, of scientific products massively increased investment in basic or fundamental science is imperative, for there are no technological products – applied science – without basic science. Ever since the industrial revolution in the nineteenth century, the links between science and technology have become ever more intimate: There has hardly been any significant technological advancement since the beginning of the twentieth century that has not been the byproduct of scientific research. The relationship between science and technology is of course not unilinear; there are multiple feedback loops between the two and between them and markets and national economic and social well-being (Gibbons 1995)

Science is of course not a panacea. Its capacities to bring progress and destruction are equally awesome, whether we consider the powers of nuclear energy, the impact of the automobile, the effects of ubiquitous information and communication technologies, or the potential of genetically modified foods and organisms, so are its capacities to influence, for better or worse, the spatial and social divides between and within countries (Sagan 1999). By itself of course science will not solve Africa's stubborn legacies of underdevelopment, but without it they cannot be solved. Africa's scientific and technological literacy must be raised. In the developed world, science has become a major, in some cases the leading productive force: all the talk about the new knowledge economy is essentially a tribute to the pervasive role of science and technology in contemporary life. Africa must appropriate, for its own uses, the existing huge stock of scientific and technological innovations that already exists[1] — which should be considered an international or global public good — and participate in the emerging scientific revolutions — the gene revolution (genomics, bioinformatics, molecular breeding, diagnostics, vaccine technology, etc.), the ecotechnology revolution (in the fields of economics, ecology, equity, employment, and energy), the nanotechnology revolution (involves miniaturization of physical and biological objects), and the information and communication revolution (internet, computing, software, and remote sensing) (Swaminathan 2000). The challenges should not be underestimated in a world where scientific knowledge is increasing so rapidly that some say – with poetic license but fallacious quantification – it is doubling every fifteen years (Murphey 1998).

In short, African countries must pursue a multi-pronged agenda: strengthen their national systems of science and technology, which depend on building strong education systems and establishing productive scientific networks with both the South and the North, including with their own scientific diasporas, as part of a systematic and sustainable drive to generate, import, adapt, and disseminate scientific knowledge and new technologies. Enabling conditions must be created not only for Africa's own scientists but also to attract foreign research and development. Even in the scientifically and technologically well-endowed United States, foreign R&D has increasingly come to play an important role, rising from $700 million in 1987 to more than $17 billion in 1995 (which provoked bitter debates between the techno-nationalists and techno-glabalists, see Florida, 1995, 1998; Mahoney, et al., 1995). Africa needs a judicious combination of techno-nationalism and techno-globalism.

The gap between prescription and implementation is of course filled with political calculations, struggles, and uncertainties. African leaders are known for their flowery rhetoric about the importance of science and technology for development.[2] But where is the political will to translate rhetoric into reality, for Africa to forge a new modernity for the twenty-first century, a modernity that prizes science and technology for their instrumental value in the quest for human material fulfilment, but not as intrinsic values to be pursued for their own sake? An economic modernity based on productive agricultural, industrial and service sectors, one that constructs and maintains the basic infrastructures of roads, hospitals, schools, and energy supply and pursues the new frontiers of ecotechnology, information technology, nanotechnology and biotechnology? A modernity in which urbanization does not entail the fragmentation and atomization of community life. A modernity that is grounded in democratic politics if it is to release the participatory and productive energies of African peoples for so long suppressed by colonial and post-colonial tyrannies. A political modernity that seeks to create new nation-states, in which ethnic, linguistic, and cultural pluralities are stripped of their consuming polarities and remolded into a multicultural nationhood or what Gyeke calls metanationality, a state that is constituted, not by ethnic or cultural groups, but primarily by individual human beings who happen to share certain cultural and historical experiences.

Conference Deliberations

Many of these issues were broached in the discussions that accompanied the formal presentations of the original papers that constitute this volume. Deliberations focused on the constraints and challenges, opportunities and developments, and the vision and strategies required to promote the advancement of information technology and biotechnology in Africa in this so-called new age of knowledge economies and globalization. The bulk of the papers focused on information technology and a smaller number were on biotechnology. It was agreed that African governments have a responsibility to identify the problems, harness the progress, and explore the possibilities of science and technology, in general, and information technology and biotechnology, in particular. This can be achieved through the development of appropriate policies and integrating them in the relevant sectors, mobilizing the required investment, especially in education, and creating an enabling environment for entrepreneurship in the technology sectors. There is also need for greater coordination between governments, civil society, the intelligentsia, business, and entrepreneurs to promote the development of science and technology culture and enterprise.

Particular emphasis was put on the question of mobilizing African scientists and entrepreneurs in the diaspora. The presence at the conference of prominent African scientists based at leading American universities gave added resonance to the discussions. As I have argued elsewhere, instead of bemoaning or ignoring the problem, it is imperative for Africa to develop specific strategies to turn the "brain drain" into a "brain gain" or "brain mobility" by establishing and promoting mutually productive and profitable linkages between the African scientific diaspora and their counterparts on the continent (Zeleza 2002b). In order to do so, it was suggested at the conference, we need detailed studies on the dynamics of African intellectual migrations to the North and inventories of diasporic African scientists. Also useful would be investigations of the patterns of the diaspora's (both the historic African American diaspora and the contemporary African diaspora) engagement with African institutions, the sectoral dimensions of these linkages — with the universities, governments, NGOs, and business — and the institutional and technical mechanisms for implementing the linkages and using the diaspora in the development of science and technology in Africa.

This combination of concerns for scientific research and policy relevance, academic discourse and development practice, as well as the interdisciplinary nature of the presentations and participants, made for animated deliberations. It is of course not easy to capture the flavor of the oral conversations in book form. Nonetheless, the papers selected for this volume offer a sampling of the wide range of issues that were discussed. Collectively, we believe, they help advance our understanding of the challenges, conditions, and contexts — infrastructural, political, economic, legal, cultural, and social — of science and technology development in Africa.

We have divided the book into four broad sections: First, chapters dealing with issues of IT policy and infrastructure; second, the developmental and democratic implications of IT; third, the application of IT in the education and health sectors; and finally, the development and debates about biotechnology and African agriculture.

Policy and Infrastructure

The first section contains six chapters. The opening chapter by Mohamed Hassan makes an impassioned call for Africa to take science and technology seriously if it wants to achieve sustainable development. It notes African countries' gross under-investment in science and technology, even in comparison to other developing regions. But the bulk of the chapter stresses the opportunities: how Africa can benefit from revitalizing its once reputable scientific capacities in the universities and mobilizing its scientific diaspora; from exploiting its comparative advantages in biodiversity; from the leapfrogging possibilities offered by the new information technologies; and from collaborating with countries in the South that have made scientific and technological advancement.

As befitting the head of both the African Academy of Sciences and Third World Academy of Sciences, Hassan offers concrete proposals, stressing the need to develop, sustain, and utilize local capacities and leadership in efforts to advance science and technology; to mobilize the best and most relevant science and technology in Africa and elsewhere to address critical social and economic problems; to build a strong case for supporting the development of science and technology; to establish and strengthen centers of excellence in science and technology among African universities and research institutions; to establish institutional networks to address critical problems facing Africa and other parts of the developing world; to create interdisciplinary panels of African experts through close cooperation with academies of science; to establish a trust fund to promote science and technology; to develop strong political will at national and regional levels to support science and technology; to allocate more official development assistance (ODA) to science and technology development; and to incorporate science and technology into the debt process.

This is followed by Okin Adubifa's chapter, which seeks to give a broad overview of science and technology building capacity in Africa. It begins by emphasizing the difficulties of getting accurate data on science and technology, which makes comparisons and effective planning difficult. Nevertheless, it is quite clear that Africa's science and technology capacity lags behind the continent's economic potential and is awfully inadequate to meet the challenges of globalization. But rising levels of literacy, entrepreneurial activities, democratization, and new wireless technologies offer new possibilities for the development of science and technology in Africa and other developing regions. He outlines several successful and not so-successful regional initiatives and national projects in various fields of scientific research and development and what can be learned from them. The point is made that besides developing the capacities of the universities and research centers, African countries need to fully explore the full potential of industrial manufacturing as a vehicle for building scientific and technological capacity. Indeed, there is need for greater coordination between the different stakeholders – government, industry, donors, and universities – in building capacity that is relevant to local needs and is sustainable in the long-term. For this to be achieved, public awareness and advocacy about the importance of science and technology in modernizing national economies, improving national security, and participating competitively in international trade have to be promoted. Science and technology research capabilities and coordinated systems need to be developed by establishing academies of science and encouraging the production and dissemination of scientific knowledge.

Globalization, both as a process in which there are rapid flows of capital, labor, communications and culture, and closer integration of markets and economies, and as a project of neo-liberal ideology, is transforming the importance and contexts for science and technology development. This is explored in the chapter by Rubin Patterson, which examines the implications for Africa in general, the SADC region, and for Botswana and

Zimbabwe in particular, of the new highly competitive global knowledge-intensive economy that is dominated by transnational corporations, which are prowling the planet for creative and skillful talent to generate new products and production processes. Africa finds itself particularly vulnerable in this brave new world as the considerable scientific infrastructure it had built by the 1970s has been eroded by the structural adjustment programs of the 1980s and 1990s, and its access to sophisticated science and technology curtailed by racist donors who push for so-called "appropriate technology." Like the continent as a whole, the development and competitiveness of the SADC region can no longer simply depend on its richly endowed natural resources, but on its ability to mobilize science and technology to establish new comparative advantages in the global economy. The two authors proceed to examine various science and technology models, before looking at the specific cases of Botswana and Zimbabwe, and recommending a synergistic model that carefully navigates local technological innovation and technological transfer in a context of economic liberalism and active state intervention.

Information technology lies at the heart of much of the frenzy about globalization and the knowledge economy. African countries have become increasingly concerned about the development of information and communication technology. The chapter by Nancy Hafkin assesses the impact of the African Information Society Initiative, adopted in 1996 at the instigation of the United Nations Economic Commission for Africa (ECA), the first such regional initiative in the developing world. As someone who was intimately involved with the initiative at the ECA, Hafkin gives a detailed and fascinating account of the forces that led to its adoption, and its vision and framework of implementation. Central to the vision, she tells us, was the conviction that information technology was not a luxury for the elite but a necessity for development. The initiative called for, among other things, the development of national plans, the elimination of legal and regulatory barriers, establishment of an enabling environment, and development of locally based, low-cost, and accessible Internet services and information content. The rest of the chapter offers a candid assessment of what had been achieved by the beginning of 2000 in the areas of policy changes, levels of connectivity, building national and communications infrastructure, training, and improving access to the public and integrating IT in various economic and social sectors. The record, she concludes, is a mixed one, for while advances have been made, a lot still remains to be done. For example, whereas only five countries had Internet connectivity in 1996, by 2001 the Internet had spread to all African countries and there were 3.11 million users across the continent. However, this only represented less than 1 percent of the 407.1 million users worldwide, and the digital divide in Africa itself between and within countries, among different classes, sectors, and between major cities and small towns, not to mention the rural areas, men and women, remained deep and wide.

The last two chapters focus on the development of IT in two countries, South Africa and Uganda. The two countries not only demonstrate the uneven development of IT among African countries, but also within countries, and how this unevenness often reflects and can reinforce existing structural, spatial, and social inequalities. Further, the two cases underscore the important role that the state can play in promoting the growth of IT in general and in democratizing access. Chris Benner argues that given its relatively high level of industrialization South Africa is better poised than any other African country to take advantage of the growth opportunities in information technology industries. Not only does the country have considerable capacity in the manufacture of telecommunications equipment, the telecommunications industry has been deregulated. In fact, South Africa ranked in the top twenty-five in the world in terms of its Internet connectivity. But the IT sector has not proved a panacea. To begin with, despite capacities in certain sectors of the IT industry, South Africa remains heavily dependent on imports for much of its hardware and software. Also, while the IT sector grew faster than any other sector, the economy as a whole shed jobs, particularly in unskilled and elementary occupations. IT development constituted a critical part of the economic restructuring of post-apartheid South Africa as the country became rapidly integrated into the global market. Moreover, the

IT industry has done little to overcome South Africa's enduring racial and regional inequities. The industry is dominated by white-owned companies and by a professional workforce concentrated in the Gauteng province. The rest of the chapter examines government strategies to balance growth and equity in IT industries and policy by examining two critical policies, the South African Information Technology Industry Strategy and the National Skills Development Strategy, both of which the author finds problematic in their conception and implementation.

The experience of IT development in Uganda is particularly pertinent to the many African countries that are still predominantly agrarian and rural, where the role of IT may not always be self-evident for governments with other pressing needs. Nora Mulira's chapter presents a fascinating account of the strides Uganda has made since the early 1990s, once the state and other key economic players and stakeholders recognized the importance of the IT sector. She chronicles the IT infrastructure that has been established in a few short years in various sectors – from health, to education, to agriculture, to business, and to government, including ministries, the judiciary, and the president's office. Clearly, Uganda's IT infrastructure has expanded thanks to the liberalization of the telecommunications sector, government and donor support, private sector involvement, and commitment from a growing cadre of IT specialists, many associated with or trained at Makerere University, which has taken a lead in establishing and spearheading the electronic revolution. Despite the strides that have been made, serious challenges remain. They center around questions of unequal access (only 10 percent of the population have access to telephones and related telecommunication services); connectivity issues with regards to lack of an Internet Exchange Point (IXP) in the country so that much of the traffic not on the same Internet Service Point (ISP) has to be routed outside Uganda and back again; sustainability issues for technologies that depend on donor support; and inertia among many public and private leaders and ignorance about the benefits of IT within the general population. She suggests, therefore, that besides expanding the IT infrastructure, a national IT policy be developed, one that is owned by all the major stakeholders in the public, private, and civic sectors.

Development and Democracy

The next six chapters grouped in section two focus on the developmental and democratic implications of information technology. They raise questions about the connections and contradictions, if any, between information technology as a tool for development and democracy, and the extent to which the two can be balanced. In his chapter, Muta Tiamiyu supports the view that information technologies can greatly facilitate efficiency and effectiveness in organizations with information-intensive activities such as public agencies. The need for African public sectors and agencies to improve significantly their capacity to initiate, stimulate, and coordinate national development processes cannot be overemphasized. Using the example of Nigeria, the chapter examines, on the one hand, the technical, economic, socio-political, and other factors that facilitate or constrain IT development in African agencies, and on the other, how the visions, culture, and processes of these agencies might be exploited or modified to promote sustainable IT development. Grounded in theoretical models about organizational change through IT transfer, the chapter outlines the socio-economic dimensions that affect IT development in public agencies. This requires, he argues, significant changes in the visions, processes, and technology learning capabilities of African public agencies, so that they can best select, synchronize, and integrate foreign and local IT components into their activities. The economic variables of IT integration include cost-effectiveness, affordability, and fund flow, while the socio-cultural variables encompass not only the behavioral inclinations and bureaucratic culture of the various stakeholders themselves, but also external factors including levels of literacy, development, and political openness. He concludes that IT development in African public agencies will only succeed if the prevailing socio-cultural, institutional, and infrastructural constraints are overcome and offers suggestions how this might be achieved.

The global explosion of the IT industry, the speed at which IT innovations are made, and the convergence of telecommunications with the broadcasting regime, have necessitated the restructuring of old regulatory frameworks designed for a previously monopolistic era. Deborah Asiro's chapter analyzes the legal implications of information technology and telecommunications using the case of Kenya. She foregrounds her analysis in the changing contexts of developments in IT and the deregulation and privatization of the industry, noting that the convergence of the traditional modes of communication — telecommunications, computing, and broadcasting — has created a nomenclature problem, as former definitions of services and markets within the industry erode, which is causing regulatory hurdles. At a global level, this has led to the ascendancy of the profit-motivated World Trade Organization's General Agreement on Trade in Services (GATS) at the expense of the much older International Telecommunications Union (ITU). The rest of the chapter is an extended interrogation of the implications of the new international electronic regulatory regime for African countries in general and Kenya in particular, with judicious comparisons drawn with the European Union framework. African regulators have to define new roles for independent regulators and try to balance the complex issues and the contradictory tendencies between competition and concentration, foreign ownership, and local control, as they draft new regimes for market entry and licensing, and grapple with the issues of interconnection, tariff policy, access, and standards, and user rights protection. Africa, she concludes, must develop competencies in these areas in order to exploit the rapidly changing information technologies and their equally rapidly changing regulatory regimes.

Some of these issues are followed up in the chapter by Sebastian Mathews, which offers an upbeat account of the prospects of the IT revolution for Africa, arguing that Africa has emerged as one of the most dynamic telecommunication markets in the world, thanks to economic liberalization, the introduction of competitive markets, and increased investment in the sector, as shown most graphically by the exponential rise in cellular telephony. Rooted in neoclassical premises, his argument is that IT contributes not only to income generation and income distribution (apparently for every dollar invested in the sector three dollars are added to GDP), not to mention that it attracts high levels of foreign direct investment and enhances general commercial efficiency and a country's productivity, it also helps ameliorate cultural schisms and promotes democracy. After outlining the positive contribution of telecommunications on African economic development, with examples drawn from various parts of the continent, Mathews examines the challenges of telecommunications regulation. The role of government, he insists, must be to provide an enabling environment, and regulatory reform requires a stable macroeconomic economic policy, flexible labor markets, and appropriate regulation of capital markets. He then proceeds to examine the different stages of regulatory reforms undertaken in various African countries, with pertinent examples drawn from other parts of the world. Care must be taken, however, to ensure that public monopolies are not replaced by private monopolies. Thus the real challenge is to promote privatization and maintain competitive markets guided by a progressive, credible, and transparent regulatory environment.

Also positive, but more guarded about the prospects of IT development in Africa, is Khaled Ismail's account of IT development in Egypt. He argues quite forcefully that developing countries, including those in Africa, cannot afford to be left behind in the new IT revolution as they were left behind during the industrial revolution a century earlier. The challenge is two-pronged: they need to become both informed consumers of technology-driven products, such as the Internet, mobile communication, and computers, and technology producers by contributing to the design and/or manufacturing of technology-oriented products such as software and Internet services. To do both requires reorienting the education system to produce graduates who specialize in technology areas. It is important that technological literacy be translated into entrepreneurship. Critical to the development of productive technological skills is the establishment of close, mutually beneficial relations between industry and academia. Ismail discusses these issues using the example of Egypt. He presents detailed statistics on the numbers of Egyptian students who graduate from high school, how many of them proceed into technology related fields of study at university and into industry, both at home and abroad. The data reveals that technology attracts some of the brightest students, but not only are many lost to other countries, especially the Gulf region and for those with Ph.Ds to Western Europe and North America, many of those who remain in the country are not always able to advance the technology sector by setting up technology companies due to lack of resources and entrepreneurial skills. That this can be done is demonstrated by the case of the company he runs, Ellipsis Digital Systems, which has been able to attract Egyptian engineers locally and from abroad and has established close relations with local universities.

The last two chapters focus on the social and political dimensions of IT, first with reference to gender equality, and second, press freedom. Juliet Evusa's richly documented chapter examines the extent to which the IT revolution is serving women's needs and preferences. The first part of the chapter provides a survey of theoretical approaches on the intersections between gender, technology, and development from the modernization perspective (with its optimistic assumptions about the effects and gender-neutrality of technology), to the dependency perspective (with its pessimistic views about the relationship between technology developed in the North and underdevelopment in the South), to postmodernist perspectives (which emphasize the social constructedness of technology). Evusa leans towards the dependency and postmodernist approaches, seeing IT as both socially embedded and engendered, which reflects and reproduces spatial and social hierarchies and inequities. The second part of the chapter presents a case study of the role of IT in providing socio-economic empowerment of African women organized in NGOs. She looks at the positive use African women made of the Internet during the preparation of the Beijing 2000 conference. From the e-mail conversations Evusa examined, it is clear that the women NGO leaders were concerned about issues of cyber-democracy and the feminization of IT policies, demonstrating their clear understanding that women have to take an active role in the development and utilization of IT to accelerate and engender Africa's development. The research also revealed persistent digital divides among the women based on location (South Africans and urban-based women being generally dominant) and language

(Francophones being under-represented).

The final chapter in this section by Mary Ebeling makes a compelling case for the democratizing effects of the Internet for Africa's often beleaguered journalists. Using examples from various countries, including Zimbabwe, which has bucked the democratization trend on the continent, Ebeling shows that African journalists have used the Internet to circumvent repressive and restrictive regulations and publicize abuses of power by governments and violations befalling the journalists themselves. Confronted with such "online activism" and other concerns, governments all over the world from China, to the United States, to France have been trying to control and regulate content on the Internet by developing screening or surveillance systems. As the numbers of Internet users in Africa increase and as journalists' and media-rights organizations become more organized and effective some African governments are also trying to place stricter controls on the new media. There is no doubt that African journalists have been empowered by Internet activism to overcome government censorship and reach beyond their domestic audiences. The Internet, however, Ebeling notes, has a paradoxical impact on African journalism. While it has extensive reach with the global audience, it has clear limits in reaching local audiences because of access issues. But she sees hope in the growing numbers of Internet users, the establishment of more locally based Internet Service Providers (ISPs), and the rapid spread of cell phones, which are increasingly capable of providing Internet services. For example, Econet, a Zimbabwe-based cellular company, provides some media feeds on its phones. Thus, while repressive governments will continue to seek ways of controlling the flow of information, the rapidly changing technologies for disseminating information will prove a formidable obstacle for them and an ally for the defenders of the freedom of expression.

Education and Health

The four chapters in this section dwell on the application of IT in two critical social sectors, education and health. While they all seem to agree that IT is indeed critical to effective delivery of social services, IT does not operate in a socio-economic vacuum, but is articulated with prevailing hierarchies and structures of power and access, which need to be addressed in order to maximize the developmental potential of IT in these sectors.

The opening chapter by Maurice Amutabi offers a critical evaluation of the African Virtual University, an online educational system sponsored by the World Bank and linked to university campuses in nine African countries by the beginning of 2001. In the developed countries the impact of IT on education has been quite profound. IT has been incorporated into the knowledge activities of teaching, research, and publication, although opinions differ sharply on the utility and benefits of online education. The debate has centered on two issues, first, the cost and profitability of online education, and second, its pedagogical benefits. African leaders of higher education generally believe that the adoption of the new technologies is critical for the revitalization of African universities and for national economic and social development (Zeleza 2003: Chapter 3). Amutabi contends that, in principle, the AVU is a good idea, but its effectiveness is undermined by the fact that it is the brainchild of the World Bank, the same institution that has wrecked havoc on African higher education through the imposition of its misguided structural adjustment programs (SAPs). Amutabi begins his chapter with a critical assessment of neo-liberalism, the ideological apparatus behind SAPs, and the disastrous impact SAPs have had on Kenyan universities, which despite their increase in numbers have become pale caricatures of what they used to be and could become, as the state has increasingly divested its investment from the sector. Introduced in this context, the impact of the AVU cannot but be limited. Specifically, Amutabi assesses the operations of the AVU in terms of its rationalization, relevance, access, cost, pedagogy, sustainability, and spillover effects. On many of these scores, especially on cost and access, he finds the current AVU model falling short of its possibilities, although he maintains that distance education, of which online education is the latest incarnation, does have the potential and capacity to contribute to growth and expansion of high quality university education. For the AVU, and similar projects to work, he suggests, there is need to abandon

SAPs and increase investment in education as a whole.

The potential of telemedicine for the provision of health care services in a typically underdeveloped African country, such as Tanzania, is the subject of James Katzenstein's chapter. Telemedicine refers to the provision of interactive health care services, through telecommunications technology, where the patient is at a location remote from the provider. Thus, it depends on an electronic infrastructure at both the patient's and physician's sites, which requires, besides the availability of reliable telephone and electrical systems, the installation of computer hardware and the appropriate software, and raises questions concerning Internet connectivity and bandwidth, all of which are costly propositions. But Katzenstein believes that the benefits far outweigh the costs in countries with limited health budgets and shortages of physicians. The chapter describes a telemedicine project in Tanzania in which Katzenstein has been involved, which started in April 1997. What we get is a captivating reflexive analysis of the initiation, design, and implementation of the project, the processes of finding the appropriate technology, the difficulties encountered, the modifications made, the relationships built, and the lessons learned. The project was guided and it helped reformulate what he calls the KM consulting model, which facilitates the development of the human systems interface that converts the telecommunications technology into a telemedicine system. The model involves, among other things, the establishment of an organization called the client consultant system infrastructure, the process of coupling and interpenetration, and techniques of action research. The model, he states, empowers the local medical establishment and ensures that issues and solutions emerge from within rather than being imposed from without.

Also on the subject of IT and health care, Muadi Mukenge provides an interesting account of an IT project involving African women NGOs. The project, sponsored by Pacific Institute for Women's Health in 1999 under its *Women Connect* initiative, sought to strengthen the communication and advocacy skills of women's rights organization, as part of a broader effort to address both the North-South and male-female digital divides. The project provided small grants to twenty-six women's NGOs in

Zimbabwe, Zambia, and Uganda. By strengthening the women's media campaign and Internet skills, the project hoped to reinforce and expand the reproductive and sexual health and rights outreach of these NGOs. Mukenge reports that the project designers were careful not to impose their own agenda. Rather, they sought to respond to the agenda set by the African women themselves, for whom health issues were not narrowly defined, but encompassed other issues that affect a woman's ability to be productive, including land and inheritance rights. An evaluation of the project reveals that it did indeed succeed in increasing women's online advocacy skills. Several of the NGOs have set up their own websites and conducted effective media campaigns, and even set up Internet cafés. But there have also been some challenges from which the project sponsors have learned some invaluable lessons, including a greater appreciation for phased-in approaches to using the Internet and the need for broad and collaborative communications capacity building strategies in order to improve women's health, empowerment, and participation in the complex social, economic, and political processes of development. The rest of the chapter discusses the project methodology and specific developments in each of the three countries.

IT is often seen as a communications medium that is particularly useful for the economic, political, and social sectors, both private and public, while in educational circles it is viewed as a special preserve of the science-related and professional fields with incidental benefits to the social and cultural disciplines. The chapter by E. A. Biakolo on a cultural database project developed as a potential tool to combat HIV/AIDS, demonstrates the fruitful uses that IT can be put to by cultural scholars engaged in addressing pressing national emergencies. A professor of English, Biakolo was commissioned with two other people to undertake this ambitious project to develop a comprehensive cultural database that could be used to combat Botswana's frightening rate of HIV/AIDS, the highest in the world. The chapter presents a report on the objectives, design, and implementation of the project, which sought to collect and develop a website listing the names and addresses of all core and subsidiary cultural, information, and sports groups that could readily be accessible to individuals and agencies. It was hoped the databank would facilitate access to information, promote communication among people, and encourage cultural production in the overall context of the struggle against HIV/AIDS in so far as the syndrome was as much a cultural as it was an epidemiological phenomenon. The project was in fact designed after the abysmal failure of earlier techno-bureaucratic attempts to address the HIV/AIDS pandemic without incorporating cultural practices and values. He concludes that even if the potential benefits of the database to combat the pandemic do not materialize, its creation will still have enormous benefits for the country's cultural and intellectual development.

Biotechnology and Agriculture

The final section contains seven chapters, all of which deal with issues of IT in African agriculture and biotechnology. The opening chapter by James Nyankori and Ibulaimu Kakoma talks about the transformations in the IT landscape in Africa and its potential beneficial impact on agricultural production and marketing. It is noted that while at the beginning of the 1990s the utilization of information and communication technology for electronic working was confined to expatriates working with international and non-governmental organizations, by the end of the decade it had spread to growing numbers of the local population who were using IT services in their daily personal, academic, and business activities. Among them were farmers. The chapter differentiates the Internet industry into two categories, first, those firms that provide basic services and, second, those that provide value added services. The two authors proceed to analyze the nature of the industry in terms of its infrastructure, service distribution, cost structure, and access utilization. They predict continued growth and consolidation of the Internet service industry and conclude that the expansion of the industry has important implications for African agriculture at the production, processing, and distribution levels. It is argued that the Internet has made it possible for African farmers to access a wide range of information sources to guide their production decisions. They give the example of a website, Africanfoodonline.com, which provides African food products to the global market through an online store. Thus the Internet allows African farmers to access new buyers in domestic and foreign markets. Also, information from the Internet enables African farmers more efficient resource allocation and utilization, which can reduce production costs.

For its part, biotechnology may hold the key to increasing agricultural production in Africa. The subject of biotechnology often raises heated debates, especially over the ethical, environmental, and epidemiological effects of genetically modified organisms and cloning (Fukuyama, 2002; Miller, 2000). But as Babafunso Sonaiya and Galman Omitogun argue in their splendid chapter on regional and international cooperation for sustainable agricultural technology in Africa, some biotechnologies are already widely used in Africa. For example, several traditional practices – from beer brewing to intercropping – reveal biotechnology at work. But the question still remains whether modern biotechnology can be applied sustainably to Africa. They believe it can so long as issues of relative costs, appropriateness, environmental suitability, socio-economic viability, and legal and regulatory frameworks are addressed. Africa cannot afford to be left out of the biotechnology revolution because of the continent's extraordinary biodiversity of crops, plants, and animals, which need to be exploited for Africa's benefit, or else they will be prospected and patented by others, as has already happened in some instances. Unfortunately, Sonaiya and Omitogun note, many African countries not only lack sufficient trained scientists in the basic disciplines required for biotechnology, but also effective legislation and regulatory agencies for monitoring the biotechnology industry to ensure biosafety. The development of human resources through mass training and the creation of awareness are essential if Africa is to reap the benefits and minimize the dangers of agricultural biotechnology. Institutional linkages also need to be developed between research centers, universities, government, and entrepreneurs. The rest of the chapter examines the work of some of the organizations that have been created to promote biotechnology on the continent.

The next two chapters demonstrate the practical applications of biotechnology tools to pressing food and environmental challenges. In their chapter Sseruwagi, Okao-Okuja, Obonyo and Legg discuss the use of molecular diagnostic tools in the diagnosis and monitoring of the cassava mosaic virus disease (CMD), which causes up to 100 percent yield loss. Cassava is a major food staple particularly valued for its drought resistant qualities and ability to produce even in exhausted soils. It is estimated that CMD causes annual losses of up to US$60 million in Uganda and more than US$1,200 million across the continent as a whole. Molecular diagnostic tools, such as the Polymerase Chain Reaction (PCR) and Restriction Fragment Length Polymorphisms (RFLP), enable the detection of different virus strains affecting cassava, help in the monitoring of their spread and distribution, the identification of CMD-worst affected areas, and the development of control strategies using host plant resistance. The chapter describes the PCR method. Clearly, these are powerful and useful technologies, although their use is often hampered by high costs, the need for well-maintained laboratories and highly trained personnel, and the hazardous nature of some of the chemicals used.

Charles Akalezi shows us how clean up techniques – called bioremediation – based on the microbial breakdown of contaminants are being used to enhance ecosystem restoration in the highly degraded Niger Delta. Besides the well-known oil spillage, the Delta receives the bulk of the effluent of 18 coastal countries between Senegal and Angola. This has not only affected the quality of the diverse but fragile delta ecosystem – leading to the defoliation, death, and disappearance of mangrove plants and various organisms – and undermined productivity – from farming to fishing – it has also provoked fierce social conflict and civil unrest which have led to widespread destruction and deaths. It was in response to this that in 1995 the Center for Environmental Resources and Sustainable Ecosystems (CE-RASE) was established to promote environmental conservation and development in the Niger Delta. Guided by the concept of sustainable bioremediation, the Center embarked on a program of environmental restoration involving the use, among other things, of kenaf, an indigenous fast-growing herbaceous plant, as an oil absorbent, and vetiver grass, which has a massive fibrous root system, for vegetative soil and moisture conservation. To encourage local participation, both Kenaf and vetiver grass are produced in nurseries by local people who also use the various byproducts the plant and grass yield. This is a remarkable example of sustainable environmental protection.

The subsequent chapters recommend the need for coordinated regional and national policy for the application of science and technology in agricultural development. The chapter by Ian Kumwenda and Wells Kumwenda examines the case of Malawi in this regard. The picture they present is disquieting. In 1997/98 public spending on research in the country was only 0.49 percent of the government's budget and 0.13 percent of GDP, well below international norms even for the low income countries; and the ratio of R&D scientists/engineers per million of population in 1991 stood at 42 against a world average of 1000 and a recommended average for Africa of 200 by 1980. The agricultural extension system is also overstretched because of understaffing and low morale. With specific reference to agricultural research, they note the existence of three streams – research centers that serve the estate sector, national research centers, and international research centers – which are poorly coordinated. The national research centers are poorly funded, thanks to the lack of clear policy and political commitment, which in turn reflects the existence of a weak national science culture. Notwithstanding these constraints, they argue, science and technology have made some contributions to agricultural development in Malawi in terms of providing improved varieties for both field and horticultural crops, livestock rearing methods, agronomic practices, and soil fertility, irrigation, pest management, and labor saving technologies. It stands to reason that more would be achieved if investments were substantially increased in agricultural research.

The development and application of agro-environmental technologies in Africa are deeply influenced by prevailing perceptions about African agrarian practices, which determine national and international policies and consequently become a material force in people's lives. It is, therefore, critical to unveil the intellectual trajectories and ideological permutations that inform the development of African agro-environmental paradigms and policies. This is the task that Aaron deGrassi's sets out to do in his chapter. He gives us a bold and comprehensive critique of what he calls conventional development studies and policy wisdom on change in African agro-environmental technology, according to which increasing population has led to the breakdown of a previously stable shifting cultivation system that may have been eminently suited to Africa's low population densities in the past, but is now the source of environmental degradation, declining productivity, and deepening poverty. DeGrassi shows that this myth, which was born to justify colonial conquest and high-handed economic and political interventions, was recycled during the eco-hysteria of the 1970s by the ahistorical and functionalist farming system studies, and revived in the late 1980s and 1990s by life-science corporations, academics, and policy makers peddling varieties of development populism, co-opted environmentalisms, and pro-biotechnology public relations campaigns. Following his systematic critique of the dominant narrative, deGrassi offers a tantalizing outline of an alternative framework for understanding African agro-environmental technology. His contention is that it is not just demography, formal markets, and ecology that drive agricultural production and environmental degradation/preservation. Rather, these are social processes embedded in social relations – operating at multiple historical and geographical scales – including but not limited to, the nature and development of gender and labor relations, property rights regimes such as land tenure and distribution, resources and their extraction patterns, state policies, agricultural development and environmental conservation projects, social conflicts and struggles, migration and settlement schemes, and the changing patterns of integration into the world capitalist economy.

From the broad historical and theoretical sweep of deGrassi's paper, we close the book with Denis Rugege's chapter on one of the factors that affects agricultural production, namely, rainfall. He presents a mathematically-based derivation of predictive data on rainfall. Agricultural production is predicated on available rainfall and real-time measurements combined with physical, chemical and biological data obtained via satellite imaging to give reasonable predictive parameters subject to numerous confounding factors which are generally assumed. Rugege believes that these data can assist governments to be better prepared for food shortages attributable to drought. A major limiting factor in Zimbabwe, and perhaps elsewhere in Africa is the paucity of meteorological stations, especially in the so-called communal lands. This makes interpolation of meteorological data to depend largely on estimates rather than absolute and real-time measurements. The inevitable result is that we have to depend largely on simulations that assume similarities of weather patterns. Most government policy makers in these regions are not data-savvy in order to use these predictions in a timely and cost-effective manner. It is reasonable to speculate that food-related calamities can be mitigated or completely averted by use of early warning systems combined with stringent stock-taking, retrieval, analysis, and timely intervention through frugal conservation and strategic importation and food-aid contingency plans. Rugege does not promise complete salvation to the farmer, our most important client-stock-holder. But he contends that science potentially has the answers if technocrats and politicians can only work together. The scope of this chapter does not include any contingency plans to bring the Indian-type green and white revolutions to Africa. Nevertheless, it strongly suggests that Africa has the capacity to transform itself into the breadbasket of the world through modern technological leap-frogging and good political and scientific leadership.

Thus, this book offers a rich sampling of the debates, discourses, and analyses of the enormously complex issues concerning science and technology in African societies and African studies. Ultimately, at issue is Africa's capacity to realize the historical dreams of the struggles for development and self-determination; in short, for its own modernities in a world of multiple, often conflicting, modernities.

Notes

1 Susan Adkins and Dennis Bushnell (1995) report that in the mid-1990s the Japan Information Center of Science and Technology (JICST), with a budget of $160 million from the government, had more than 300 full-time employees who

worked with 5,000 cooperative abstractors, most of them scientists and engineers, to collect scientific and technical information from Japan itself and more than 50 foreign countries. Japanese companies also give their libraries a high priority. She suggests that acquiring research results in this manner is cheaper than conducting the research on-shore.

2 See for example, the reports by UNESCO (1994, 1999a, 1999b, 1999c) and Shabani (1997).

PART I:

Science and Technology: Policy and Infrastructure

CHAPTER 1

SCIENCE AND TECHNOLOGY IN AFRICA: PATHWAYS TO SUCCESS

Mohamed H.A. Hassan

Introduction

Africa is a rich continent. Rich in mineral resources, rich in precious stones, and rich in biodiversity. It is also rich in traditional knowledge, especially knowledge associated with indigenous and medicinal plants. And Africa is rich in culture — the place where Homo sapiens emerged from other species between 5 and 8 million years ago and the home to humankind's earliest civilizations and seats of learning dating back 5000 years or more. But Africa, as we all know, is also a poor continent. With roughly 13 percent of the world's population, it enjoys only 1 percent of the world's wealth. An estimated 50 percent of Africa's population lives in poverty and 40 percent suffer from malnutrition and hunger. Two-thirds of Africa's land base is degraded and over half of Africa's population does not have access to safe drinking water. Malaria poses a serious threat in many regions and HIV/AIDS has devastated the youthful populations of many African nations, including Botswana, South Africa, and Zimbabwe, where an estimated 25 percent of the adult population is now afflicted with this deadly disease.

The truth is that stark disparities exist not only between Africa and the rest of the world, but between Africa and the rest of *the developing world.* What accounts for Africa's impoverished state? There are many political, socioeconomic, and environmental factors: Centuries of colonialism and enslavement; weak governments; corruption; lack of transparency in economic transactions; unsustainable use of natural resources; and marginal participation in the global economy. However, there is another factor that I would like to focus on — a factor that may not be as visible or dramatic as those mentioned above but that may nevertheless play a central role in the continent's inability to participate in the global economy, protect its environment, and devise sustainable strategies for growth. That factor is Africa's serious shortcomings in science and technology.

By almost any measure, African science and technology are in a dismal state: in terms of the miniscule numbers and inadequate skills of its research and technical personnel; the poor and neglected quality of its infrastructure; the low level of student enrollment throughout all levels of education; and the paltry investments in universities and research institutes. Several statistics help tell the story. With 13 percent of the world's population, Africa has only 0.36 percent of the world's scientists and engineers. With so few scientists and engineers, it should come as no surprise that Africa also lags far behind the rest of the world in its output of scientific publications. African scientists, in fact, account for less than 1 percent of the publications published in reputable international refereed journals. Switzerland, on the hand, with just over 7 million people or roughly 1.2 percent of the world's population, accounts for 1 .6 percent of the world's scientific publications.

Meanwhile, expenditures in research and development in Africa amount to much less than one half of one percent of the world's total. The United Kingdom, conversely, invests about 2 percent of its gross national product on research and development. Some developing countries — for example, China and India — have recently embarked on programs to sharply increase their research and development expenditures. In China, annual R&D expenditures account for 0.66 percent of the nation's gross national product, or some US$6.5 billion; in India the percentage has climbed to 0.75 percent, or some US$3.3 billion. There is no doubt that the challenges Africa faces in promoting its science and technology infrastructure and in charting a course for sustainable development are indeed daunting. But the current situation also presents opportunities that offer rays of hope for the future.

Opportunity Rising

Nonetheless, the situation has not always been and need not be so bleak. First, it is important to remember that Africa's dedication to science and technology is not just an ancient occurrence separated by eons of time. In fact, in the late 1960s and early 1970s, science departments in many African universities — including the University of Lagos in Nigeria, Dar-es-Salaam in Tanzania, Accra in Ghana, and Khartoum in Sudan — were considered among the finest in the developing world. Examples of the progress that Africa can make in improving its scientific and technological capacities thus are not relegated to dusty history books but can be culled from the experiences of well-educated, successful Africans who are alive today. Many of them, however, have pursued their careers in countries beyond Africa's borders. In fact, studies show that some 100,000 high-level African-born experts, who roughly equal the number of foreign experts working in Africa, left the continent during the 1990s. Despite their migration abroad, the knowledge and experience of Africa's scientists living abroad offer lessons in strength and determination that could be shared with younger African scientists and technologists today — perhaps transforming what has been a brain drain into a brain gain.

Second, as stated above, Africa is not devoid of competitive advantages even in today's global market place. For example, its richness in plant biodiversity and its remarkable centuries-old foundation in traditional knowledge could well serve as the basis of a lucrative continent-wide enterprise devoted to the sustainable use of indigenous and medicinal plants. To date, however, Africans have failed to profit from the exploitation of these resources. Indeed they have more often than not relinquished control of the cultivation, harvest, and processing of these plants to their Northern "partners." Stronger governmental regulations, particularly for the protection of intellectual property rights, could help Africans receive the financial benefits from the sustainable use of these resources that clearly should be theirs to enjoy.

While biodiversity represents Africa's best comparative advantage in the age of globalization and international competitiveness, advances in biotechnology, if tailored to the circumstances found in Africa, could bring "backyard" benefits that resonate closer to home for many of the continent's poorest residents. In the North, biotechnology has raised fears among a wary public about the potential negative fallout created by "engineered" foods, which critics have depicted as both unnatural and possibly dangerous to public health and the environment. In Africa, issues surrounding biotechnology have surfaced within a vastly different context. Questions surrounding the "pluses and minuses" of biotechnology have been overshadowed by the fact that biotechnology research and development have largely been conducted by Northern scientists and have largely focused on Northern products and processes. Biotechnology holds promise for increasing food production, enhancing environmental protection, and improving the treatment of tropical diseases — all areas of vital concern to Africa. However, if the promise of biotechnology has any chance to be fulfilled on the continent, then African scientists must be trained and put to work on biotechnology issues of importance to Africa itself.

The third point is that information technologies, most notably the Internet and the use of e-mail, represent another "double-edged sword" characterized by promise and pitfalls for Africa today. On the one hand, no other technology reflects the gnawing skills and equipment gap between Africa and the rest of the world more dramatically than the new information technologies. Africa has just 2 percent of the world's telephone lines and just 1 percent of the world's Internet connections. Clearly, the continent remains disconnected from the world of modern electronic communications and, as a result, has been denied access to the wealth of information, including scientific and technological information, that is found on the Internet. But the Internet represents warp speed not only in terms of information access for those who are plugged in but also for those who are plugged out. That is because advances in wireless digital communications allow countries with poor telephone systems to leapfrog into the information age by avoiding the construction of costly wired telephone networks. Several African nations, including Botswana, Mauritius, and Ghana, have embraced new wireless technologies with promising results and other African nations are likely to soon follow in their footsteps. The result could be that Africa becomes part of the electronic communications network faster than anyone would have thought possible.

But it is not the hardware itself that will spark advancements in science and technology in Africa. It is the purposes to which the hardware will be put to use that really counts. Providing equitable and immediate access to the world's storehouse of scientific and technical knowledge is one potential benefit; enabling the creation of virtual networks and research teams of scientists from nations thousands of kilometers apart is another. Thus the new information technologies may give nations, which have been left behind the knowledge revolution, the tools that they need to catch up quickly both on their own and through cooperative research ventures with colleagues from around the world. The dream of South-North and South-South cooperation is more likely to become a reality because of the existence of these new technologies — a trend that carries important positive implications for Africa.

Which brings me to my fourth and final point concerning the opportunities that are now at hand for improving scientific and technological capabilities not just in Africa but throughout the developing world. Since the time of the industrial revolution until recently, science and technology were the exclusive domain of Northern nations. That is no longer the case. Today, a number of developing nations — and once-developing nations that have substantially improved their economic status — have built scientific and technological infrastructures that sometimes approach (and indeed sometimes rival) the infrastructures found in developed countries. Such countries include Brazil, Chile, China, India, and South Korea. This development not only confirms the abiding principle that "science is the heritage of all humankind" but also creates an environment where Southern nations can truly learn from one another's experience. What better way for the nations of Africa to build their scientific and technological capacities than by acquiring knowledge and skills from their counterparts in other developing countries? Brazil's recent announcement that it would establish a fund for nurturing partnerships with Africa's Portuguese-speaking countries in support of projects related to science and technology is a promising move in this direction. I suspect we will be seeing more of these ventures between "stronger" and "weaker" developing countries in the future.

Three Step Program

With these challenges and opportunities before us, what overall strategies should African nations pursue to advance their scientific and technological capabilities? I would like to offer a three-step program.

First, *develop, sustain and utilize local capacities and leadership in efforts to advance science and technology.* The truth is that developing scientific and technical capacity is less difficult than sustaining it, and sustaining it is less difficult than utilizing it. Yet, the ultimate impact of science and technology depends on all three factors — development, sustainability, and utilization — working together. That is why it is important for African nations to invest in the education and training of scientists and technologists, and that is why it is important for each nation to develop an economic strategy that offers scientists and technologists employment opportunities once they obtain their degrees. A single talented scientist can make a difference. That is the good news. The troubling news is that past experience indicates educating and retaining scientists and technically skilled workers is much more difficult than it seems. For these reasons, it has become clear that a prerequisite of sustainability is a vibrant educational system and an enduring, yet flexible, job base.

Second, *mobilize the best and most relevant science and technology in Africa and elsewhere to address critical social and economic problems.* As many observers, have noted: the poor, especially the poor in Africa, face a different set of health and environmental sustainability issues than those who are not poor

— not only because they lack financial resources for addressing these issues but because they live in different ecological zones. For example, recent studies show that 8 of the 10 poorest nations are located in arid and semi-arid regions. At the same time, the most advanced science and technology initiatives have rarely been mobilized to solve Africa's most critical problems, including poverty, food deficits, tropical diseases, and HIV/AIDS. The lessons of this experience are straightforward: If Africa expects to use science and technology to tackle its most pressing problems, it must develop its own scientific and technical capacities. Otherwise, it will forever be beholden to the benevolence of others. That is not to say African nations should turn their backs on the research taking place beyond their borders. Indeed a critical challenge facing Africa and those who would like to see more attention paid to science-based development in Africa is to strengthen poorly funded relevant programs of United Nations organizations — for example, the United Nations Educational, Science and Cultural Organization, the Food and Agricultural Organization, and the World Health Organization — whose roles and impacts have been increasingly marginalized by rich countries. Africa should also continue to pursue cooperative projects with constituencies that have special ties to the continent — for instance, scientists in Africa's Francophone countries should continue to work closely with scientists in France. At the same time, sub-Saharan African scientists should seek to tap the distant yet potentially strong ties that exist between them and African-American scientists in the United States.

Third, *build a strong case for supporting the development of science and technology in Africa.* This is a critical challenge for African scientists given the competing demands that are constantly being exerted on the continent's limited financial resources. African scientists not only have an obligation but also a self-serving interest to convince governments, the public, and the private sector of the value of science and the need to support such endeavors. Such efforts must include a willingness to engage the public in discussions of science-based issues, a desire to lobby the government for support, and, perhaps most importantly, a commitment to pursue research agendas that focus on critical social and economic problems — that is, agendas that allow scientists to use their knowledge and skills to help their nations build stronger and more sustainable societies.

Guideposts for Success

While these three broad objectives — the promotion of local capacity building, the use of science to address everyday problems, and the nurturing of public support for science — present a broad pathway for future success, it will be necessary to develop even greater programmatic details to ensure that the proper guideposts have been put in place. Among these details are, first, the need to *establish and strengthen centers of excellence in science and technology among African universities and research institutions.* I am delighted to detail several initiatives that either have taken place or are under consideration to advance this goal. For example, the Third World Network of Scientific Organizations (TWNSO), in collaboration with the South Center, has published two editions of *Profiles of Institutions for Scientific Exchange and Training in the South.* The first edition, which appeared in 1994, included 208 institutions; the second edition, which appeared in 1998, contained 431 institutions (TWNSO, 1994; 1998). Not only has the number of institutions of excellence increased over time but so too has the quality of their personnel, their infrastructure and, most importantly, their research. These paired volumes provide valuable snapshots of the advances that have taken place among the developing world's best research institutes during the past decade, and the portraits that they bring into focus are both clear and compelling. On another front, the Millennium Research Institute (MRI), an initiative supported by the World Bank among others, has begun taking steps to help finance the development of centers of scientific excellence in the developing world. Until now, MRI's major efforts have been based largely in South America. However, serious discussions have recently been held, including at a meeting in Trieste, Italy, in May 2000, to devise a strategy targeted specifically for Africa. With adequate funding and careful planning, the MRI initiative could give a significant boost to the creation of centers of scientific excellence throughout Africa.

Second, there is also need to share *innovative and successful experiences in science and technology in developing countries, including notable success stories in Africa.* Again I am delighted to report that some progress has been made in this effort. For example, TWNSO has received a series of grants — from the United Nations Development Programme Special Unit for Technical Cooperation among Developing Countries, the United Nations Environment Programme Global Environmental Facility, and the World Meteorological Organization — to examine successful experiences in the conservation and use of indigenous and medicinal plants, the conservation and use of biodiversity in arid and semi-arid land, and the wise use of water resources, respectively. These projects, which include write-ups about many successful experiences in Africa, are designed to accomplish two goals: (1) to highlight the good work that is being done in many developing countries in applying science and appropriate technologies to everyday problems of critical environmental social and economic concern; and (2) to foster a social environment that allows scientists and administrators from developing countries to share information and learn from one another. As TWNSO moves forward in this effort, it hopes to launch other like-minded programs devoted to such issues as the development and use of renewable energies and the advancement of information technologies in the developing world.

Third, we need *to establish institutional networks to address critical problems facing Africa and other parts of the developing world.* One of the major pathways to progress throughout the South, and particularly in Africa, lies in the building of institutional networks to exchange information and experiences. Such initiatives not only provide an efficient means of applying science and technology (after all, one of the primary goals of institutional networks is to avoid duplication of effort), but they also serve as "safe-houses" for discussions of common concerns. Such discussions, particularly if they are more than onetime chats, may help tear down both the technological and psychological barriers to success. That is why I am happy to note that the ultimate goal of the TWNSO projects described above is to create a series of institutional networks devoted to science-based issues of critical importance to the developing world. The first such network, dealing with the conservation and use of indigenous and medicinal plants has been launched in Karachi, Pakistan, with the help of the University of Karachi's HEJ Research Institute of Chemistry. Other networks will soon be created focusing on the issues of dryland biodiversity and water management.

Fourth, effective strategies have to be developed to *mobilize eminent expatriate scientist of African descent.* As the distinguished scientist and founding president of the African Academy of Sciences, Thomas R. Odhiambo recently noted: The most crucial element in Africa's future will be the mobilization of skilled and talented Africans in and outside of Africa. In response to this challenge, such modest initiatives as the Return of Talent Program, sponsored by the United Nations Development Programme and the European Union, which resulted in only a handful of Africans returning home since the program's launch in the mid 1980s, must be reconfigured and reinvigorated. At the same time, Africa must not be hesitant to draw on the talents of African scientists and technologists who have left the continent and have no intention of returning. For example, the Edward Bouchetl Abdus Salam Institute, which was launched in the mid-1980s to promote cooperation among African physicists living and working in their home countries and African-American physicists living and working in the United States, has achieved more modest results than hoped for. A recent executive board meeting of the institute in Trieste, Italy, attended by seven Africans, seven African-Americans, and two representatives of the Abdus Salam International Centre for Theoretical Physics (the hosting facility for the institute), sought to lay the groundwork for nurturing a deeper and longer lasting impact for the institute not just in physics but in mathematics and high technology as well.

Fifth, it is important to *create interdisciplinary panels of African experts through close cooperation with academies of science.* One of the principal roles of the African Academy of Sciences has been to increase the number and to build the capacity of science academies throughout the continent. (Africa, home to 53 nations, currently has only 9 science academies). This initial effort for capacity building has recently received a much needed boost with the relocation of the Inter Academy Panel (IAP) to Trieste and the placing of the panel's secretariat under the umbrella of TWAS. The panel is an organization of 80 or so science academies from around the world. Its main goals are to promote cooperation among academies and to raise the capacity and voice of academies, especially when it comes to science-based issues of social or environmental importance. A fundamental aspect of IAP'S responsibilities is to help improve the capabilities of weak academies and to push for the creation of academies in nations where they do not currently exist. In line with this goal, the Trieste based IAP secretariat held its first major workshop in May 2001. The theme of this event was "capacity building for academies in Africa."

Action for Progress

We should think of these programs — the strengthening of centers of excellence, the establishment of institutional networks, the mobilization of scientists within and outside Africa, and capacity building for African science academies — as the core of a proposed "action plan for progress." Other items in the action plan should include: (1) the launching of additional major fellowship programs to support graduate and post-graduate education at centers of excellence in Africa and elsewhere in the South (such programs currently receive support from TWAS and the Swedish International Development Cooperation Agency (Sida), and other donors); (2) nurturing the participation of the private sector, both within and outside the continent, in science and technology development programs and projects (while private-sector investment in science and technology has increased in most countries in the North as well as in many of the "more developed" developing countries, including Brazil, China, and India, it has virtually disappeared from the accounting ledgers among some nations in Africa); (3) and the organization of pan-African science summits (as the experience of RANDFORUM shows, such summits can play a critical role in pushing science and technology towards the top of the policy agenda by sparking local, national, and international enthusiasm and support for such efforts).

The ability to fulfill the items listed in this action program will, of course, depend in large measure on the commitment of Africa's governments to sustainable policies that promote science and technology. It will also rest in some measure on the willingness of international lending agencies to continue to fund and also reform their assistance programs to meet the evolving needs of the recipient organizations and nations. And finally, it will depend to a degree on the enthusiasm and dedication of scientists and scientific institutions, especially those in Africa, to devise projects that put science to work for the benefit of the people of Africa.

Finding Funds

None of this will be easy and it will require significant funding. Where will the money come from? I do not need to state that funding for development assistance has fallen dramatically since the end of the Cold War. In fact, studies show that assistance aid from Northern donors to the least developed countries, many of which are located in Africa, has dropped nearly 25 percent since 1990 and now averages less than of 5 percent of the gross national product for many African nations. While external aid has declined in developing countries, external debt has increased sharply. In fact, external debt now exceeds 75 percent of the gross national product of many African nations and service payments now consumes some 40 percent of annual government expenditures — if the countries had the financial means to pay it.

In light of these two reinforcing and suffocating trends — reduced assistance funds and heavier debt burdens — I would like to conclude my contribution with some modest proposals, which although not new, nevertheless warrant repeating. These proposals offer nothing less than an opportunity to ensure that adequate financial resources will be available for science- and technology-based initiatives in Africa, which represent the only true pathway to sustainable growth for the continent.

- *Establish a trust fund to promote science and technology in Africa.* African Union (AU), in collaboration with United

Nations Economic Commission for Africa and the United Nations Development Programme, should call for the creation of a special fund to support science and technology initiatives in Africa. The African Development Bank, World Bank and relevant United Nations organizations, and international donor agencies should agree to contribute to such a fund. Indeed the first South Summit, held in Havana, Cuba, in April 2000, passed a resolution urging the creation of such a financial mechanism to serve as the centerpiece of a host of initiatives designed to provide adequate financial resources for the advancement of science-based sustainable development in Africa.

- *Develop strong political will at national and regional levels to support science and technology in Africa.* The 30 year-old dream to increase science and technology investments to 1 percent of each nation's gross domestic product must continue to be pursued with full acknowledgement of how difficult this goal will be to achieve, especially in the short term. At the same time, strategies must be pursued to continue to strengthen South-South cooperation both as a tool for progress and as concrete examples of how developing countries can improve their science and technology infrastructures and research output through sustained government commitment and funding.

- *Allocate more official development assistance (ODA) to science and technology development in Africa.* The first step in fulfilling this goal lies in reversing the nearly 25-percent decline in such assistance that has taken place since the end of the Cold War. As called for in the first South Summit in Havana, developed nations should be encouraged to target 0.7 percent of their gross domestic product to ODA, with 0.2 percent of that allocation going to least developed countries in Africa and elsewhere. Such a small amount of assistance represents not only a moral imperative but a reasonable price to pay to help promote global stability. To ensure such efforts help lay the foundation for sustainable development, at least

50 percent of ODA funding should be invested in scientific capacity building, including the rehabilitation of universities and the strengthening of regional centers of scientific and technological excellence.

- *Incorporate science and technology into the debt process.* As has been proposed for domestic investments in nature and the environment, African governments should be allowed to write-off critical domestic investments that they make in science and technology from the burdensome debt that many now suffer. Such relief, which would have the twin benefits of easing the debt burden and advancing science-based sustainable development, represents a prime example of how to turn a paralyzing problem into an attractive opportunity. Representatives at the United Nations Educational, Scientific and Cultural Organization, at the World Conference on Science in Budapest in spring 1999, expressed their overwhelming support for such a proposal and we should not waste any additional time in putting this initiative in place.

Conclusions

The major problems that have afflicted Africa during the past three decades of the 20th century remain stubbornly in place at the beginning of the 21st century. Africa is still the world's poorest continent with many of its nations becoming even poorer. Its rich and unique biological resources are being rapidly depleted. Disease, food deficits, and energy and water shortages continue to be widespread. Previous efforts by national and international development agencies to stem poverty and stimulate growth have not enjoyed much success. Rapid globalization propelled by large multinational corporations and characterized by free trade have only deepened the continent's problems. To address these problems and avoid even greater marginalization, Africa must devise new long-term visions and strategies that enable it to sustain economic growth and compete in a world whose development is becoming increasingly dominated by scientific knowledge and technical skills.

In short, African nations must build and sustain their own capacities in modern science and technology, particularly in the biological and material sciences, biotechnology, and new information technologies and communication systems —and then direct the knowledge and skills that they acquire to problem-solving strategies. Such strategies, in turn, must put the best of science and technology in Africa and elsewhere to work in ways that will build and sustain local and regional capacities as well as address real-life concerns. The bottom line is this: Science and technology alone cannot save Africa but Africa without science and technology cannot be saved.

CHAPTER 2

AN ASSESSMENT OF SCIENCE AND TECHNOLOGY CAPACITY BUILDING IN AFRICA

O. Okin Adubifa

Introduction

Since the 1979 Vienna Conference on Science and Technology for Developing Countries, international development agencies as well as national governments have attempted several initiatives for the economic development of the African continent. The results are mixed. There have been pockets of outstanding successes, but also many abandoned expectations. Regretfully, it is the catalogue of failed efforts and attendant obstacles that have dominated the presentation of the state of science and technology development in the continent. The negative reporting has been so persistent that it has taken on a life of its own, sometimes crippling the enthusiasm and morale of otherwise talented innovators, daring investors, and budding scholarly communities. The pessimism, however, has often been exaggerated. An array of accomplishments under challenging circumstances can also be seen to produce a more optimistic characterization of the continent's potential, pointing to the prospects and opportunities rather than to the failures and obstacles. Although the technological gap between the developed countries and the developing countries has been considerably reduced in many of the countries of Asia, Latin America, and Eastern Europe, this is not the case in Africa where the gap between them continues to grow.

The African Governors of the World Bank (1998) concluded that Africa is a continent rich in natural resources but lacking the capacity to transform that potential into a standard of living that would enable the African people to become full partners in the global economy. Indeed, the present economic profile of Africa is worrisome: the continent is poorly integrated into the innovation-based global economy; its share of global trade fell from 3 percent in the mid-1950s to 1 percent in 1997 (The Economist Intelligence Unit, 1997), and more recently it is estimated to have fallen further (Adubifa, 2000); growth rates of the African economy are projected by the World Bank to increase modestly to only 3.3 percent by the year 2000; export growth is expected to advance by 6.3 percent in 2000 (4.6 percent if South Africa and Nigeria are excluded); and the continent will continue to be highly vulnerable to commodity prices and global economic developments (Prewitt, 1998).

The chapter is divided into four sections. The first reviews the status of science and technology capacity in Africa. It has been necessary to describe the science and technology systems of the countries through subjective interpretation of scattered, incomplete, and sometimes inaccurate or totally missing data on performance indicators. Comparison of available data was difficult since data sources and quality across the continent are not standardized. Nonetheless, the study provides a synthesis of policy analytical documents on the status of science and technology in Africa, supplementing the sparse published information with discussions with donor representatives and locally based practitioners. The second section reviews experiences in science and technology capacity building in Africa. It discusses the elements that have contributed to the successes of some of the national efforts, as well as the factors that have limited the ability of other initiatives to achieve expected impact. In recommending a regional strategy for the development of science and technology capacity, especially the capacity for undertaking research and improving its social benefits in Africa, the third section is a review of lessons from past efforts. The final section offers a number of specific recommendations, with some addressed to national governments and others to the World Bank and its partner development institutions, as well as to other donors.

Status of Science and Technology in Africa

Researchers on African economic development generally confirm that credible data on science and technology indicators are difficult to collect. Available data give skewed results when aggregated, and are sometimes incompatible with physical evidence. While South Africa, for example, is part of Africa, its science and technology indicators are considerably higher in value than those of the rest of the continent. Major international handbooks that regularly publish comparative data on global development (including UNDP's Human Development Report; UNESCO's Statistical Yearbook; National Science Foundation's Science and Engineering Indicators) are usually unable to provide complete data in reporting on African countries. Yet, the availability of such data and the credibility of their sources would facilitate national planning and assist external donors to make wise investment decisions.

Africa has never had a clear definition or a comprehensive characterization of its state of science and technology development. Several studies have been carried out by locally-based researchers to capture the thrust, scope, implementation, and outcomes of the many initiatives that are internally or externally generated, and to describe their effects in building a science and technology capacity at the national level. The outcomes of many of these studies generally fall short of expectation because researchers are unable to find appropriate statistical data; research objective is poorly defined; scope and coverage of research activities are too wide for the available funding; research methodology and data gathering are not designed to allow comparability across countries; and research efforts do not benefit from rigorous peer review.

General Characteristics of the State of Science and Technology

Africa's internal science and technology capacity lags behind the continent's economic potential. This realization is compelling countries to take the S&T sector more seriously, leading to calls from many countries for urgent international assistance to help them develop the incipient capacities and harness their potential more systematically. A crucial beginning of this new willingness to act is to forge an understanding of the existing strengths and assets, which should form the cornerstone of new and innovative initiatives.

In the 1960s and 1970s the international development assistance community gave a lot of emphasis to deepening Africa's science and technology capacity. It supported both government and academic institutions, but not the private sector. The results fell below expectation: Science and technology sectors and professions are still in their infancy and not strong enough to compete globally; universities have suffered from inadequate funding for long periods and their academic standards have declined in many cases; science itself, as a field, is not fully appreciated since it is not introduced into early education; apart from a few groups and individuals scattered across the continent, the academic communities are largely isolated from their global colleagues. Under these conditions, there have been limited private investments in technological development or scientific applications. African countries cannot be said to have acquired adequate scientific and technological capabilities to meet the challenges of globalization, particularly the attendant international negotiations in the economic, financial, and environmental regimes.

A breakdown of global expenditure on science and technology shows that the high-income OECD countries account for about 85 percent while Africa's share is only about 2 percent (UNESCO 1996). Furthermore, the distribution of the world's scientific and technological output as measured by indicators of scientific publications and registered patents shows considerable disproportionality: 80 percent of publications are owned by the highly industrialized countries plus India, while 96 percent of the world's patents are registered in the United States, Japan, and

Western Europe. While capital flows to developing countries as a whole reportedly averaged 3.2 percent of GDP during the 1990s, African countries averaged only 1.9 percent. Africa's share of the total foreign direct investment flowing to developing countries was a mere 3 percent in 1997. The number of African scientists and engineers in professional practice is also considerably lower than the world average on population basis. The continent, moreover, accounts for only 0.8 percent of the world's listed peer-reviewed publications, and only 0.2 percent of registered patents.

Results of science and technology policy research and promotion of policy dialogues in some of these countries, funded by the IDRC and the Carnegie Corporation of New York over the past 10 years, show that the level of public sector awareness has been greatly elevated, and there is increasing collaboration between government functionaries and academics in addressing science and technology development issues. Some countries are improving and expanding their channels of communication between scientists and technologists, government officials and advisers, parliamentarians, and service sector practitioners. For example, the Zimbabwe Institute of Development Studies has been using funding from Carnegie Corporation of New York to convene working groups of stakeholder representatives for each of the economic sectors to deliberate on sectoral performance, and to formulate strategies for strengthening these sectors and optimizing their potential.

Several opportunities for science and technology development are beginning to appear on the horizon, and many countries are determined not only to use them effectively but also to enlarge these opportunities in order to accelerate national development. Several encouraging trends can be identified. First, literacy levels are rising in all countries, elevating the national capacity to facilitate scientific applications and participate more effectively in national development. Second, skilled entrepreneurs with talent and dedication are increasing in number, and equally increasing their global outreach through the new information and communication technologies. Third, democracy, accountability, and the rule of law are spreading rapidly across the continent, and becoming entrenched in irreversible ways in many countries where such prospects could not be contemplated only a few years ago. Fourth, with democratic dispensation taking root and under the pressures of globalization, many countries are beginning to open their formerly closed markets. Finally, new wireless technologies and mobile computing devices are enabling ambitious institutions and civil society organizations to leapfrog into the digital world, avoiding the high costs previously associated with traditional computer and telecommunications infrastructure.

Building Science and Technology Capacity

This section discusses experiences in building science and technology capacity in Africa. Perhaps the best example is the Green Revolution initiative pioneered by the Ford and Rockefeller Foundations. Their funded research for the production of high-yielding rice and maize varieties led to the establishment of the Consultative Group for International Agricultural Research (CGIAR). Similarly, the African Economic Research Consortium (AERC) was initially launched by International Development Research Centre (IDRC), and later supported by Rockefeller Foundation, to strengthen economic management in Africa. It is now a flourishing capacity-building research network funded by a consortium of donors.

These successful models point clearly to the kind of framework that effectively builds and sustains capacity. The framework is characterized by funding commitment to sustain the initiative over the long term, and beyond threshold; sound institutional management under strong scientific leadership; and elimination of resource uncertainties regarding scientific careers, research facilities, access to international research systems, and conducive academic environment.

Regional Initiatives

Regional initiatives concentrated mostly on institution building, but some operated as networks housed in existing institutions. Apart from the CGIAR institutions and the AERC network, other successful programs include the African Technology Policy Studies Network (ATPS), the Coordinated African Program of

Assistance in Services (CAPAS), the Council on Health Research for Development (COHRED), the International Health Policy Program (IHPP), and the Universities Science, Humanities, and Engineering Partnerships in Africa (USHEPiA). All these initiatives receive their entire funding from donors.

While the initiatives have been largely successful, they have also suffered, to varying degrees, from certain weaknesses, including the fact that capacity building activities did not always emphasize scientific quality; programs took on wider scope than optimal, thereby losing thematic coherence; there was lack of clarity regarding financial and administrative support for the host institution, especially as some donors only provided program support and not core support for the institutions; donor support waned just when it became necessary to consolidate the emerging gains of the initiative.

The African Academy of Sciences (AAS) is a special case of an institution in the forefront of building scientific capacity in Africa, but which continually struggles against odds to accomplish its mission. It enjoyed strong donor support when first founded because the need for such an organization was clearly apparent. Its objectives were laudable and its programs compelling. It achieved outstanding success with its regular publication of a world-class scientific journal *Discovery and Innovation.* However, its programs rely almost totally on external funding, which has usually been below threshold.

Less Successful Regional Initiatives

Three examples of important regional initiatives that did not achieve the expected impact are now described. The first is the USAID's Program of Science and Technology Cooperation, which sought to support research in developing countries in collaboration with US scientists (or scientists from other developed countries who would have to bear their own portion of the research cost). This highly competitive international grant program made no allowances for the weak scientific infrastructure of the institutions of the African applicants, or for their lack of access to current knowledge in their research fields. Very few African scholars benefited from the grant scheme. Others expressed frustration at the disadvantages that they faced in the competition against more advanced developing regions.

The second involved several intergovernmental science and technology institutions, which were established by Member States in the late 1970s when African leaders were full of optimism and had liberal access to donor resources for development. These included the African Regional Centre for Technology (ARCT), the African Regional Centre for Engineering Design and Manufacturing (ARCEDEM), the African Regional Standards Organization (ARSO), the African Regional Industrial Property Organization (ARIPO), and the African Institute for Higher Technical Training and Research (AIHTTR). The respective mandates of these institutions were unrealistic with respect to available funding, staffing, and operating facilities. Beyond initial funding, the member states gave little attention to science and technology development, especially as their national economies took an unexpected downturn. The institution-building efforts also took on political dimensions as the management, staffing, service conditions, and international status of the institutions had to be politically negotiated. Without adequate core funding, most of the institutions could not be sustained, and their skilled professional staff departed. This quickly led to loss of project funding from donors. AIHTTR was closed down within ten years. ARCEDEM's elaborate technical and mechanical facilities are largely idle, and ARCT does not have the resources to play a regional role anymore. Thus, a lot of capacity that was built has hardly been put to productive use. ARSO and ARIPO are still functioning, but barely effective as regional engines for development.

Finally, there is the case of the Association of African Universities (AAU), which has a membership of more than 140 institutions across the continent. It has made efforts in recent years to promote capacity building for scientific research in the universities. It established a program to market top research proposals that are developed and submitted jointly by inter-university collaboration between two or more institutions. Forty-two such proposals were submitted for experts review in 1998, out of which the best seven were presented to the meeting of donors to African higher education. Although the donors acclaimed these proposals to be rigorous and compelling, they did not agree to a proposed pool-funding mechanism. They merely endorsed the proposals as worthy of support, and advised the applicants to seek funding individually from interested donors. This raised concerns among African scholars about the conflicting signals that seemed to be emanating from the donor community.

National Projects

At the national level, research systems continue to grow slowly, mostly with the support and advocacy of donor agencies more than by national initiative. Previous efforts have created isolated layers of successful scientific research, but lacking a broad base of competent mid-level technicians (qualified laboratory assistants, analysts, statisticians and data processing specialists, equipment maintenance specialists etc.) all of whom play important roles in research. The national research systems are generally characterized by instability of support to Research and Development; lack of political commitment at top governmental levels; low public awareness of the need and importance of scientific and technological capacity (mostly due to the absence of a science-based culture); and poor recognition and reward for achievements in science.

It is relevant to note that no single indicator is available yet for measuring science and technology capacity in Africa. Only a detailed assessment in a given country can reveal the real and present level of its research capacity, and properly characterize its structure. National experiences can be reviewed under three of the most common models. First, is the question of support for individual scientists. This has had limited success in countries like Nigeria, Ghana, and Sierra Leone. Although the individual scientists received personal benefits that upgraded their skills or expanded opportunities for international exposure, the impact of the projects was hardly felt beyond their immediate environment. Financial support from donors usually did not anticipate the requirements for institutionalization of project outcomes, nor provide specifically for project outreach that would multiply the effects of capacity strengthening.

Second, is the issue of support to universities and research centers. The number of scientists in African universities has often increased more rapidly than the available funding for research. But, regardless of the amount of resources available, a number of practical considerations continue to limit African efforts to increase research capacities through institution building. To begin with, universities and public research centers lack adequate numbers of technicians, specialists, and qualified managers to support the desirable level of research. Universities often tend towards the European tradition, which pursues scientific knowledge without entrepreneurial enthusiasm; they do not generally emphasize applied research that would interface closely with the private sector. Their academic culture promotes scholarly research and publication, but it is not geared towards business development or industrial applications. Also, the trend of diminishing funds per researcher has intensified in the last decade. The imbalance between human and financial resources has exacerbated another problem: The cost of maintaining research infrastructure (laboratories, equipment, physical plant, etc.) has risen sharply, leading to their stagnation, degradation and obsolescence. Research teams do not always have functioning basic facilities, access to current literature, compatible teaching and administrative load, nor the cooperation of their institutions. Because the key ingredient *—funding—* is often not reliable nor programmed for long term support, research teams easily become isolated; they need to be nourished by external contacts at conferences, in networks, and by exchanging research fellows. Nationally sourced funds (especially government subvention) are applied mainly to staff salaries. It is the external funds that usually support research projects. This leads to the tendency to load excessive numbers of under-qualified workers on to the research fund. Promotion policies and rewards are more favorable for administrative and managerial functions than for research and scientific achievements. This leads inevitably to high rates of turnover in research staff, which in turn undermines efforts to develop a cadre of skilled and experienced researchers.

The third area concerns support for networks and promotion of networking. In an initial attempt to identify success factors in research networks a 1998 colloquium in Bellagio, Italy, identified a number of critical elements. These were reported by Prewitt[5] to include balancing open membership with quality; good governance system; common needs assessment with shared ideas about problems and solutions; financial stability; adequate project management capacity; and institutional ownership of the research program. Examples of strong and successful networks that are coordinated from within the region are ATPS, AERC, and USHEPiA all of which have been described as regional initiatives. Others include the Africa Energy Policy Research Network (AFREPREN). They all enjoy, for the most part, adequate donor funding to sustain their research programs. They have also all managed to keep administrative costs low by remaining as programs within existing institutions. In USHEPiA particularly, the enthusiasm of participating institutions, researchers and their joint supervisors, as well as the effective linkages at the senior management level of participating universities have contributed enormously to the network's outstanding impact. A distinguishing feature of the program is the provisions of re-entry grants to visiting researchers to enable them upgrade their local research environment upon returning home. Other important but less successful networks include the AAS's Network of African Scientific Organizations (NASO), the African Network of Scientific and Technological Institutions (ANSTI), and several discipline-focused regional and sub-regional networks. The most crucial limitation that they all face is sustainable funding.

Capacity Building Through Industrial Development

African countries have not explored the full potential of industrial manufacturing as a vehicle for building scientific and technological capacity. Local industries are mostly still in their infancy and do not generally have the risk capital to engage in basic research and technological innovation. On the other hand, the multinationals normally conduct research and development at their parent companies, leaving African subsidiaries to acquire technology through basic training, production, and quality control functions.

In South Africa which is the most technologically advanced

among the countries of the region, efforts are underway to convert some of the existing technological capacities which were built under military and nuclear energy programs into commercial production capacities for public benefit. This link between the public and private sectors will facilitate rapid upgrading of technological skills in industry.

The countries of the region are beginning to discover what works and what doe not with respect to industrial development as shaped by the national science and technology system. They are finding that apart from financing there are four other parameters that need to be firmly in place to assure successful transition of their economies from rural agrarian to modern industrial. These parameters are policies, legislation, institutions, and training.

Even though case studies would be necessary to determine how well the industrial sector has fared with respect to the national science and technology policies and legislation, there are already a number of success stories that can inspire other countries in similar circumstances. A few examples will suffice. Morocco has demonstrated that careful and targeted planning for the application of science and technology to socio-economic development can reap rich dividends. According to Jugessur (1998) the country has adopted modern technologies like green house technology, tissue culture, and overhead irrigation methods to meet national food requirements and produce other goods and services. Mauritius, a country of only one million people, is taking advantage of its high literacy rate, political stability and the high work ethic of its people to achieve an annual growth rate of about 6 percent. Its industrialization process is being driven by a well-focused set of science and technology policies, which include very attractive incentives. The country seeks to become a leading African center for microinformatics technology. Governmental collaboration with the private sector in Zimbabwe has promoted the development of solar photovoltaic technology, which is now being rapidly commercialized throughout the country. This low cost energy system will find wide applications in homes, schools and farms. Zimbabwe's ATPS chapter is taking a leadership role in preparing the country's science and technology policy and planning framework while its Scientific and Industrial Research and Development Council is mobilizing the country's research potential for effective utilization.

Lessons From Past Efforts

Stakeholders in African development, particularly the donor community, have been increasing their quest for lessons of past experience in order to guide their future investment in the science and technology sector. They are keen to find what impact has been achieved by these past efforts (both public and private) to build capacity during the past several decades. No overall assessments are available. However, there is enough anecdotal evidence to suggest that the impact has been below expectation. Many of the following ten experiences address the performance of external assistance.

- The skills and capacities that have been developed have not been harnessed purposefully to maximize their potential. Opportunities for their direct application were often absent as external assistance that built the capacity did not always extend into such long-term practical applications that would be found in the private sector. Generally, the capacity building also was not a direct result of private sector demand.
- External assistance came in various forms from many sources, usually addressing different aspects of the process for developing a science and technology system. Donor collaboration and national programming of assistance were not given sufficient attention, leading to uncoordinated implementation of similar activities and stretching thin the limited human resources available for specialized functions. Generally, this led to a reduction of the possible impact of the projects as initially conceived.
- The models used by external assistance to address science and technology issues were seldom able to build adequate capacity to tackle the problems confronting the countries concerned. Assistance that is provided consistently over the long-term and targeted at a specific need has been shown to achieve better results. An example is the CGIAR model for agricultural

research, well funded by a consortium of donors, and devoted for more than thirty years to the development in Africa of new and improved varieties of seeds and food crops.

- Centers of excellence were sometimes established to build capacity, but these later required high maintenance costs which the beneficiary countries could not afford amid competing social needs.
- Where building of research capacity has been the objective the final product has often been a research report or journal publication. The results of such research were not transferred to industry or commercialized through patents. A general and recurring finding of policy research across the continent is that much of the previous efforts to build capacity did not link projects to end-users. It can, therefore, be said that capacity-building initiatives sometimes met narrow objectives and did not produce sustainable development.
- Past efforts focused too heavily on studying the economic profiles and their indicators without adequate specificity of science and technology components of those indicators. This is probably why analytical data on scientific and technological performance of individual countries have been so hard to find.
- Foreign firms operating in African manufacturing industry usually source their technological and engineering inputs from abroad, and also carry out research and development in their parent organizations. They therefore built local capacity only for equipment maintenance and quality control.
- Global trends in technological advancement and application have not always allowed opportunities for the infant science and technology systems in Africa to impact on domestic economies. In addition, the few scientists and scholars who acquire critical capacity are often drawn away from local needs towards global issues.
- It is fair to say that, after so many years of training, fellowships and research support, a large number of individuals have been strengthened in their personal skills for scientific endeavor. But the impact of all this is not quite institutionalized. This

is a systemic weakness that is proving difficult to overcome. Governments and their external collaborators would need to form a stronger partnership to implement modalities that make all development assistance more effective through appropriate institutionalization.

- Support for science and technology development is still facing a dilemma of choosing between alternative models of building capacity. External assistance would like to achieve rapid results, but is also concerned that previous investment did not point clearly to a specific model as the strategy of preference. For example, in choosing between support for science versus support for technology in Africa the extent to which a country should devote resources to basic research is always difficult to estimate. Longer-term strategy requires the development of capabilities to assimilate and adapt foreign technologies, to produce engineering designs, and to carry out consulting services. The choice between investing in policy research or in technological research will vary from country to country. Very little technological research appears to be going on now in most countries, partly because its costs far exceed the resources needed to pursue policy research.

Recommendations for a regional Strategy for the Development of Science and Technology in Africa

It is recommended that a Regional Strategy should seek to simultaneously achieve the following goals. First, strengthen internal capacity of government machinery to address science and technology effectively and integrate its system into the national development and planning strategies. Science and technology policies need to be explicit, well articulated, and integrated into the overall national economic policy framework. Second, create awareness among stakeholders about the need to invest in science and technology, the potential benefits to society, and the opportunities that become available for utilizing it to foster social and economic development. Third, promote consensus building for science and technology policy and planning through the establishment of Working Groups of appropriate public and private sector representatives.

Operationally, implementation of the strategy should be complemented by national action. While it is recommended that different approaches be targeted to meet the needs of selected countries, a common need for all of them is the use of structured formal training to nurture a cadre of scientists and young scholars who can keep the countries engaged in global intellectual production. The following three elements are proposed as possible mechanisms of intervention: a comprehensive study to collect and analyze credible science and technology data; a program of the Millennium Science Initiative; and specific programs in groups of countries

Study to Collect and Analyze Science and Technology Data

There is a chronic lack of statistical data to describe the characteristics and performance of Africa's science and technology sub-sector. Building a dynamic system for the development of the sub-sector requires considerable, but not insurmountable, investment; but it should begin with the collection and analysis of baseline data as well as time-series measurement of the indicators that would describe the profile of sectoral progress and development. Where information is scanty and indicators for assessment are difficult to measure, the paucity of data increases the risks for investment decisions. This has been the case in much of Africa.

Provision of factual data and quantifiable information will not only enhance the quality of analysis of science and technology parameters for Africa it would also facilitate a more realistic comparison of development potential and characteristics among its countries, and between them and their counterparts in other developing regions of the world. Such comparisons would also assist African countries to identify and select more appropriate strategies based on similarities (or otherwise) of their science and technology profiles with those of more advanced developing nations whose efforts have yielded fruit.

Most countries of the region do not have the appropriate tools and personnel trained to collect and analyze the required data properly. The World Bank should lead efforts to initiate the appropriate process in its client countries, working with the governments and urging them to involve the private sector and academia in the program.

The Millennium Science Initiative

Representatives of Africa's scholarly community are investigating the efficacy of establishing vibrant research programs under the Millennium Science Initiative (MSI) that is being fiercely promoted in Latin America by the World Bank. Several options are available, each of them carrying different risks and opportunities. In order for the MSI to make a significant impact in Africa, especially in meeting its primary objectives of 'promoting frontier research and educating the best scientific talent in developing countries,' it will be necessary for its programs to involve a sizable number of African research scholars and scientists, to nurture a modest number of research centers or units in specific priority disciplines, and to receive the committed support of the respective national governments. Financing the selected centers for the MSI programs would be a challenge unless there is adequate long-term commitment of research funds from external donors. National governments can be persuaded to give firm political support and recognition. Relatively wealthy countries may find it possible to also provide necessary resources to fund the MSI institutes or centers, but it is doubtful if many of the other countries would be willing to take new loans (as expected by the World Bank) to execute the MSI programs.

It is therefore recommended that the World Bank and the African Development Bank jointly assume the responsibility of providing the resources to initiate model MSI institutes and, further, of mobilizing donor collaboration to assure the sustainability of selected centers over a ten-year period, at the minimum.

Specific Programs at the Country Level

The objectives and suggested modalities for implementing the following six types of activities:

- Promoting public awareness and advocacy. There is need to increase public awareness amongst countries of the importance of science and technology in the context of efforts to modernize national economies, improve national security, participate in international trade, and strengthen the base for economic and social development, employment opportunities, and wealth creation. Countries need strong professional associations to guide, advocate, set standards, and regulate the practice of their professions in the scientific and technological fields. They also need to introduce science into early education.
- Promoting S&T policy research. In general, scientific and technological research in Africa has been directed mostly towards agriculture, particularly the production of food crops. Now that several African governments are beginning to recognize the imperative of technology as an important tool for pursuing their development goals, and are also willing to embrace its potential, they need an improved analytical framework of information to enhance their ability to make critical decisions and investment. Policy studies in science and technology can help to illuminate the gaps that must be addressed and provide a practical road map to guide various stakeholders as they contemplate next steps.
- Building capacity to foster national science and technology system. Countries need to establish science and technology Working Groups as vehicles for responsible national dialogue among key stakeholders. Immediate benefits would include better articulation of ways in which science and technology might be applied effectively. External assistance should collaborate with host governments to create policy dialogue mechanisms. It should also contribute technical expertise to backstop these initiatives, which should facilitate the establishment of effective collaboration between policy makers, scientists and technologists, and the private sector. External assistance may also be able to assist countries to establish a funding mechanism, possibly with development

banks, to be dedicated to national science and technology development. Other vital initiatives would include the strengthening of the policy, supervisory and regulatory institutions. All African countries should aim to enhance their capacity for research and development, transfer of technology, and commercialization of significant research findings. This would require partnerships between private, public and academic leadership of respective countries to examine what has worked elsewhere and why, what has been tried that failed, and how to adapt these experiences to their own settings.

- Strengthening national academies. Only a few African countries have functioning academies of science. These honorific organizations play an important role in showcasing the best of the academic community, and in providing appropriate reward and recognition for outstanding performers. They also help to shape the culture and tradition of intellectual communities, placing appropriate values upon merit and performance. It is recommended that national university systems lead a new effort to rally external support, particularly from foundations, to strengthen national academies where they already exist, or otherwise to catalyze the founding of new ones.
- Arranging technology missions. A useful modality for building capacity around critical issues is to mount technology missions. A mission would arise from a deliberate decision by stakeholders to define a social objective that requires a major technological input, and then to mobilize all necessary resources for achieving that objective within a specified time frame. Such a mission should seek specific collaboration that links all stakeholders and leverages other resources. Priority areas, which might present mission opportunities include renewable energy research; food production and security; environmental sanitation and water resources; telecommunication, electronics, and information technology.
- Promoting scientific production. It is necessary for Africa's scholarly community to greatly increase its contribution to the global pool of knowledge. A significant amount of on-going research as well as indigenous knowledge remains

either unpublished or poorly documented. The current global knowledge revolution powered by new information technologies provides unprecedented opportunity for reducing the isolation of African scholars, and for expanding the knowledge resources of the continent. National research institutions should seek to increase intensity and coverage of their research program. They should conduct critical studies that lead to the determination of priorities for strategic research. They should also identify those research areas that are most likely to benefit society within a short period of time, and reposition themselves to address the priority needs of their countries. African institutions need to carefully seek out and utilize emerging opportunities for this purpose. Reform of the university system in Africa is a necessary step to enhance the research environment. Such reform would include cultivating linkages with the private sector and encouragement of collaboration with non-university research and development. The institutions should also improve the reward system to compensate scientific production and technological innovation. Increased contribution of African science to the global intellectual knowledge base would greatly boost the morale and enthusiasm of African scholars whose relative isolation tends to diminish the appreciation of their scholarship.

CHAPTER 3

LOOKING AT SCIENCE AND TECHNOLOGY IN SOUTHERN AFRICA: AN ENTERPRISE AND REGION IN MOTION

Rubin Patterson

Introduction: The New Global Economic Order and its S&T Engine

All people are entitled to knowledge and the benefits of science and technology (S&T) for social, economic, and cultural development. This observation grows in saliency when one considers that science and technology are the raw materials of the future for advancing beyond abject poverty, social injustice, and political subordination. The laudable vision of mobilizing science and technology innovatively to advance society and its citizens has to be translated into action through policies by governments in the developing world and by the international development community. In today's global arena, the successful nations are those that start with a vision, choose areas of specialization that fit that vision, then tailor all policies to turning that vision into reality (UNESCO 1998b). The opportunities and threats to facilitating this endeavor in the vortex of globalization, monopoly of economic liberalization, and cutting-edge generic technology pose daunting challenges even for the industrialized knowledge-societies, let alone for the SADC countries. This paper seeks to illustrate some of these challenges and opportunities by clarifying, first, the global economic order and its elaborate S&T aspects; second, Africa's status on the world's S&T stage; and third, S&T activities and prospects by the SADC secretariat and two member countries, Botswana and Zimbabwe. The paper concludes with a couple of specific recommendations.

In nearly all economically advanced nations,[1] state intervention was decisive in protecting domestic industries from foreign competition until indigenous industries had secured large sections of the domestic mass market (Bhagavan 1990). Twenty-first century globalization—triggered in part by the end of the Cold War, the monopoly of economic liberalization, and techno-scientific advancements—has abruptly terminated this path to development for today's Third World nations. Today, capital, labor, markets, communications, and culture are all becoming more globalized at a rapid rate. But the multiple processes of globalization are occurring unevenly over space and time (Hay and Marsh 2000). The unyielding high price of participation in global markets is performance benchmarking against world-standards. Meeting superior standards of technical performance is an enormously tall order for most nations that barely have 60% of their people literate, let alone technically savvy.

Benchmarking excellence requires extensive capital investments and continuous innovation. Consequently, governments universally are contradictorily prioritizing the opening of their respective markets as much as they are protecting selected domestic industry. For instance, over the period 1991 – 1999, 94% of the 1,035 policy changes to national FDI laws around the world favored investors (UNCTAD 2000). At the same time, in response to concerns about unemployment and politically powerful local constituencies, governments across the globe seek to improve certain domestic opportunities through targeted protection. Despite only ten countries receiving three-quarters of the $800 billion in FDI in 1999 and just ten of the developing countries receiving 80% of total FDI flows to the developing world, all nations continue to reposition themselves to become more investor-friendly.

One of the greatest threats to benchmarking excellence by developing nations is the pace with which the S&T knowledge generation is increasingly becoming monopolized within the networks of large TNCs (UNCTAD Secretariat 1999a). According to estimates published in UNCTAD's *World Investment Report 2000*, TNCs account for as much as 75% to 80% of global R&D, and as much as between two-thirds and ninety percent of international technology flows are also intra-firm, that is, within TNC systems (UNCTAD 2000:17). In 1995, 70% of patents granted worldwide in biotechnology were granted to applicants from the United States and Japan, approximately 35% each (UNCTAD Secretariat 1999b). Despite the fact that the economic inequality gap today, by some measures, is twice as great as it was in 1965, when decolonization was just getting underway in Africa, the gap in knowledge production is even greater and more consequential (World Bank 1999a). For these reasons, it is quite prudent for governments in developing nations to attempt to gain access to foreign advanced S&T while simultaneously engaging in indigenous innovation capacity building.

Fortunately, the necessity for Third World nations to become more attractive to TNCs for technology investment is matched by the necessity for TNCs to make more far-flung investments. Thus, the rising intensity of competition and innovation has led to shortened product life cycles, not to mention the added pressures on TNCs to continuously and quickly introduce new products into the market, as well as improved production processes and more extensive international partnerships. In order for world-class TNCs to keep pace with their counterparts, they are increasingly forced to go offshore with FDI, licensing, OEM, joint ventures, or some other technology transfer arrangement (Patterson and Bozeman 1999).

TNCs are enhancing their portfolio of investments with "locational assets" as a strategic means of capturing new markets. The fact that sales of foreign affiliates worldwide are now nearly twice as high as global exports illustrates the globalized strategy of international production, as opposed to domestic production and international shipments. International production is economically more efficient and politically more palatable in foreign markets than importation. Yet another impetus behind international production is the search for "created assets," that is, new and cheaper sources of marketable knowledge and technical skills. Rich, knowledge-societies monopolize capital markets and patents, though they do not have a lock on human creativity. In fact, TNCs prowl the planet for creative and skillful talent that may result in new blockbuster products and ingenious production processes, which can conceivably be profitable in local markets or in regional or global ones. So Third World governments are liberalizing their economies and encouraging greater scientific and technological activities not only because they see accruable benefits to their respective nations, but also because they see the necessity of TNCs searching for international technology partnership opportunities. But not all FDI and other technology transfer mechanisms have the same impact on the receiving nation. There are "high quality" technology transfer arrangements and "low quality" ones. High quality technology transfer arrangements have intimate forward and backward linkages in domestic sectors, with strong export orientation, with high and upgrading skills, or with favorable technology spillovers (UNCTAD 2000). Low quality technology transfer mechanisms have none of these benefits.

Africa's Status on the World's S&T Stage

For developing countries, there are three principal overlapping means of preparing to take advantage of the opportunities and minimizing the threats discussed above: first, creating and maintaining a stable macroeconomic policy environment; second, establishing the comprehensive machinery for S&T—from funding to competent staff to promotion and diffusion—in order for an innovation culture to become entrenched in society; and third, S&T capacity building so as to adapt to the challenges of continuous change and to innovate trend-setting technological change themselves. These three principals are hardly new. As early as 1971, the *World Plan of Action for the Application of Science and Technology to Development*, prepared by the United Nations Advisory Committee on the Application of Science and Technology to Development, highlighted the following in its assessment of the global South: the need for indigenous capacity building; the weakness of scientific and technological institutions; the economic, political, and cultural problems of accessing the world's state-of-the-art technologies; and the brain drain.

John Forje (1989) and others have documented the travails and hard-won successes of the development of Africa's S&T infrastructures and activities since independence. Like other Third World nations colonized by the West, African nations have struggled with reconciling Western techniques and indigenous cultures. Organically integrating analytical science into cultures that continue to privilege holism has proven to be a formidable problem for most developing nations. Of course, nations such as Japan, South Korea, and Singapore have done an exemplary job of mastering this challenge. Although some scholarship on comparison between the successes of East and Southeast Asia and the challenges still ahead for many African countries in this endeavor are available (Patterson and Bozeman 1999; Patterson 2001), much more is needed. Given the unchallenged fact that the importance of technological progress is independent of culture and political philosophy, the unresolved objective now is mastering the mechanics of building up a minimum threshold of scientific culture. An established objective is for each nation to promote its own technological progress in its own way and in a manner that preserves its identity and culture (UNESCO 1974), all while making requisite concessions to adapt itself to a world that is fraught with accelerating high-impact innovation.

According to Forje and UNESCO, African leaders were never intellectually or ideologically adverse to attempting to integrate analytical science into the cultures of their newly independent societies. The embrace of science early on by African leaders contributed to making well-known improvements in infant mortality, life expectancy, and access to safe water, and literacy rates. As Forje reports, by 1973, as the decolonization process was still underway, just fewer than one-half of African governments had at least a ministry of science, a general science planning body, or a multisector body for coordinating scientific research. By the early 1970s, UNESCO (1974) declared that a considerable scientific infrastructure had been built up in many

African countries (UNESCO 1974:18). Students of national S&T infrastructure in the Third World today, however, are less inclined to overstate progress. Instead, they pay close attention to distinctions between the mere existence of an S&T office, staff, and a small budget versus comprehensive S&T machinery that is intimately integrated into the full socioeconomic development of the country and is able to ignite measurable impacts. Very few, if any, of the nations identified as having a science planning body in 1973 would truly qualify for the latter category. Nevertheless, some observers believe that the reason Africa's endemic problems persist is that modern science and indigenous African knowledge have yet to be fused.

Many governments of deeply impoverished African nations may have been persuaded at the outset by the critically strategic role of S&T in Africa for the promotion of socioeconomic development; but, perhaps, S&T became de-emphasized in light of the persistent palpable pressures of poverty (e.g., malnutrition, premature death from infectious diseases, illiteracy, unemployment, abrogation of sovereignty, corruption).

Another important reason why African nations have plausibly failed to advance in S&T further than they have to date is due to the restraining influence of economic structural adjustment programs (ESAPs) imposed by the International Monetary Fund and the World Bank. Bearing in mind that the years-long austerity measures dictated by the fund and the bank cut healthcare delivery, primary schools, fuel subsidies, and other immediate necessities, it is patently obvious that S&T activities would be among the first to be summarily eliminated. Enos (1995) reports on the status of incomplete studies on the plausible impact of ESAPs on S&T endeavors. More rigorous and comprehensive studies are needed to determine the extent to which ESAPs are inhibiting the development of more vibrant S&T infrastructures in Africa. Yet another inhibiting factor has been other types of foreign domination of S&T activities in Africa. For instance, the Canadian International Development Agency (CIDA) and other foreign aid bureaus have allegedly exhibited subtle racist views about Africans' "inferior" ability to grasp and make productive use of sophisticated S&T. The technologies that CIDA champions for many African projects do not particularly lend themselves to enhancing industrialization in those countries. Conversely, while the World Bank does champion more advanced technologies, the bank has historically transferred large infrastructure projects in turnkey arrangements. These examples illustrate that Africa's S&T foreign assistance is oftentimes paralyzed in this vise. In short, S&T foreign assistance approaches oftentimes do little to promote indigenous technological capacity (Tiffin and Osotimehin 1992). As Tiffin and Osotimehin contend, Africa's S&T needs are largely viewed through foreign eyes. This is particularly debilitating, given that, "to an increasing extent, the major portion of the funds allocated to increasing the capacity of R&D institutes is coming from abroad (Enos 1995: 183).

Foreign funds, not surprisingly, come with their own agenda, which may not be consistent with the coordinated strategy established by the national leadership. That said, caution is needed to avoid an untenable conclusion that all foreign S&T assistance is suspect at best and summarily destructive at worst, while all African governments' positions on S&T are beyond reproach. Ideally, African nations would be governed by "developmental states" (Leftwich 1995), whereby the societies would be steered by competent, largely uncorrupt, determined politicians and bureaucrats who seek to position their nations favorably in the global economy at this moment of globalization and technological advancement. In this ideal scenario, foreign assistance would only support the indigenously designed plans. Given the chasm between this ideal and existing reality, there is ample need for genuine foreign assistance to engage the governments and civil societal organizations in developing countries around enhancing scientific and technological capacity building and infrastructures.

Despite the critically strategic role of S&T in Africa for the promotion of socioeconomic development, it is not a "magic bullet," for S&T must follow certain antecedent factors and be concurrent with others. For purposes of illustration, consider that Rwanda and Sierra Leone had the distinction of being two of only six African nations that met an important objective of a continental-wide S&T forum (CASTAFRICA I) for a threshold number of scientists and engineers per million inhabitants. Sierra

Leone actually had the highest percentage of scientists and engineers engaged in research (Forje 1989). However, due to the compounding impact of national civil war and regional political instability, the influence of international actors from the world's power centers, and entrenched poverty, Rwanda and Sierra Leone have arguably suffered more than any other African nation since independence. The deplorable events visited upon these nations illustrate that "science and technology" alone can make little contribution without the *will to progress*, and the opportunity and organization to use them (UNESCO 1974: 110). Moreover, some have contended that what exists is a crisis in values rather than in science and technology (UNCTAD Secretariat 1999c). The lack of political will by African leaders to whom these institutions point is in reference to overcoming the redoubtable demands they are under, rather than a desire to engage in S&T-led development under propitious conditions.

S&T Activities and Prospects in SADC

In the future, the SADC region will have to compete not only on the basis of its richly endowed natural resources, but also on its ability to mobilize science and technology resources to establish new comparative advantages for competition in the global economy. The region has considerable ground to cover before reaching this objective. Manufactured goods account for more than 70% of SADC imports but less than 10% of its exports. Manufacturers in the region—who are for the most part themselves consumers and operators of imported technology—are fundamentally restricted to a narrow range of consumer goods production that entails simple value-added processing. SADC secretariat is in the process of enlarging and reinforcing its S&T coordinating unit to help its member nations establish the requisite critical success factors so that innovating institutions in the region can be benchmarked against world standards.

S&T policy is reducible to neither R&D policy nor the commercialization of research findings. The thrust of SADC research, like the rest of African research, will be geared for a considerable period towards appropriating research results from around the world and tailoring them to fit the specific African context in question. Since science and technology at the SADC secretariat are bureaucratically under the human resources development sector, there is a lack of leadership and coordination for S&T policy activities to help focus research in the region for such global knowledge appropriation. Although the S&T base in SADC is more developed than any other region of the continent, it is also too woefully inadequate to attract much of the world's technology transfer that was discussed at the beginning of this paper. SADC is still a region that considers the primary to secondary school progression rate of 50% to 70% as high. This is at a time in world production and competitive history when the most competitive nations are characterized as mass college-exposed societies. South Africa has over 70% of Sub-Saharan Africa's researchers in national research centers (UNESCO 1998b), while Zimbabwe has been noted as having the most diverse, sophisticated, and efficient industrial sector in all of Africa, save South Africa (Tiffin and Osotimehin 1992). The senior government officials of member countries can significantly leverage the insufficient resources that can be allocated for S&T development with appropriate secretariat planning and support. To date, however, a low priority ranking of R&D, marginal political will by SADC governments, an unabating brain drain, insufficient funds, and political instability all account for the arrest of greater S&T development in the region.

From the decades-long civil strife between the government and the UNITA belligerents in Angola to the volatile land redistribution struggles in Zimbabwe to the opposition's unrecognizing the government in Mozambique to the "African World War" played out in the DRC theatre, one can only guess the extent to which the region's political instability is deterring potential technology investment endeavors. In this international production era of globalization, TNCs are investing into nations where there is rising regional income, functioning technical infrastructures, competent government bureaucrats and technical labor, and last, but not least, political stability (Patterson and Bozeman 1997). The S&T staff at the SADC secretariat know that their policy recommendations for producing the requisite infrastructures, labor talent, and rising income are impossible to achieve until the region is politically stable.

Table 1. Short-Term Goals of the S&T Coordinating Committee At SADC's Secretariat

ITEM	SHORT-TERM GOALS OF SADC's S&T COMMITTEE
1	Establish an S&T advice clearinghouse
2	Sensitize governments to the importance of S&T in economic development
3	Help all member states establish a national S&T coordinating body in an influential cabinet office (e.g., Ministry of Finance) or office of head of state
4	Foster the development of international and international specialized centers of excellence that comprise: university-industry-public/ private research center links
5	Encourage separate areas of selective excellence in universities across the region, preventing unaffordable overlapping and costly omissions
6	Support establishment of region-wide Master's degree programs in various specialized S&T areas to produce a competitive workforce
7	Study other multilateral secretariats (e.g., Organization for Economic Cooperation and Development (OECD), Asia Pacific Economic Cooperation, and the Latin American Network of Indicators of Science and Technology) for insight into regional development of indicators; data compilation from national collections; region-wide surveys; and monitoring of activities
8	Determine priorities, inventory needs, strengths, and mutually beneficial research opportunities

Notwithstanding the highest priority of political stability, the S&T staff are still playing a catalytic role by working to identify needs and opportunities for advances in innovation that can be implemented immediately while stability is still being secured. A few of many ongoing projects being prosecuted by SADC's S&T staff are cited in Table 1.

Table 2 shows the chief S&T organizing bureaus of selected SADC countries and their strengths and weaknesses. Observations incorporated in the Table were provided by the heads of science and technology in their respective SADC countries at the meeting, "Science and Technology. Recent Noted Strengths and Weaknesses of Selected SADC Countries in the SADC Region for the 21st Century" held in Pretoria, South Africa, in April 1999 under the aegis of the *World Conference on Science* (UNESCO 1999a).

Table 2. Recent Noted Strengths and Weaknesses of Selected SADC Countries

COUNTRY	S&T ORGANIZING BODY	S&T STRENGTHS	S&T WEAKNESSES
Botswana	Botswana Technology Centre	Recommends 3 new institutions: National Commission on S&T Council for Research, S&T National Center for Industrial Research	Lack of R&D expertise
Lesotho	Minister for Natural Resources, Science and Technology	Gvt commitment to establishing a Dept of S&T; introduce S&T in secondary/tertiary Ed; Dev S&T policy; establish indicators	Convincing Gvt to invest in S&T; predominance of Western S&T vis-à-vis indigenous tech; poor assessment
Malawi	National Research Council of Malawi (Office of President & Cabinet)	Recent Gvt stated commitment to S&T (Vision 2020); new National S&T Policy (assisted by UNESCO/UNDP; reviewed by stakeholders)	NRCM lacks regulatory & funding authority; lacks Gvt direction & coordination; prominence of Int'l Orgs; failure to identify emerging issues
Mozambique	None	Focusing on curriculum Dev and ICT for Ed institutions	Just getting started

Namibia	Ministry of Higher Ed, Vocational Training, Science and Technology	Proposed national S&T policy and implementation; conducted recent survey to identify the science, engineering and technical HR base	Low level human and financial resources for S&T; lack of coordination low priority for R&D in industrial enterprises
Zambia	Ministry of S&T and Vocational Training/ National Science and Technology Council	None provide (though obviously some exists)	Weak local/Int'l collaboration; shortage of Sr. scientific personnel; low capital funding for Equip. And facilities
Swaziland	None	None provided (though obviously some exists)	Lacks coordinating body; lacks implementation of S&T policy; lacks funding
Zimbabwe	Research Council of Zimbabwe	Established SIRDC; SIRDC and RCZ drafted a new S&T policy	Limited funding; poor dissemination of info related to R&D impact; poor coordination

The region as a whole will benefit in the coming years by the secretariat's catalyzing role in helping to elevate the prominence of science and technology, and in creating a critical mass of interest and resources for success. Despite this collective good, there are bound to be instances that call for adjudicating contradictions between long-term regional integration needs *vis-à-vis* short-term national needs. Increased political bargaining power, avoidance of economic distortions, larger FDI inflows, and efficiency gains will be encountered to the extent that the former occurrences overshadow the latter (Kenney 2000).

The key is to link S&T policy to economic policy and social needs. Doing so facilitates shifting the debate from whether pursuits of scientific and technological advancements by SADC's secretariat and member states for global competitiveness are feasible, and towards consideration and adoption of mechanisms and timeframes for progress in light of the innumerable daunting challenges. Catalytic action on the part of the regional S&T coordinating body can only occur and can only be effectual to the extent that national leaders have the political will to commit to the long-range success of science and technology. In supporting S&T, leaders would be asked to allocate resources that may not address immediate pressing needs and will often run counter to vocal and powerful national and international constituencies. Moreover, the fruits of such investments into innovation may not be harvested until after the investing leaders are no longer in power. When considering that illiteracy in the region tops 40%, it is clear that enhancing a climate for innovation is a long-term endeavor. What the coordinating body can do is make accurate assessments of needs and help articulate sober priorities, which can both be translated into targeted capacity building agendas, precise technological research strategies, and measurable innovation objectives.

Education is one of the critical components of national (or regional) successful scientific and technological advancements. While official educational exposure and attainment correlates with national scientific prowess and receiving technological transfers, the connection is not perfect. For instance, although the literacy rate for Sub-Saharan Africa and India is 56%, the former is collectively responsible for less than 1% of world scientific output while India was recently ranked as number eight in the world in terms of scientific output (Bhagavan 1997).[2] Despite India's anomaly that thus far defies persuasive explanation, the strongest social science conclusion to drive policy on this question is that greater national educational attainment enhances national S&T innovation.

Table 3 shows the adult illiteracy rate and the illiteracy rate for the 15-24-year-old age group (UNESCO 1999b); secondary school enrollment ratios; the percentage of college students and graduates in natural sciences, engineering, and agriculture (UNESCO 1998a); and FDI figures (UNCTAD 2000). Comparison of columns two and three illustrates that both SADC nations and the selected Asian nations are making notable improvements in terms of educating their respective youthful citizens. Both groups of nations have 15-24-age-group illiteracy rates significantly below that for citizens 15 and above. Hence, younger citizens, the nations' future workforces, are receiving more and better education. Data in columns four and five are from the UNESCO (1998a) *World Education Report 1998*. Column number four informs us about the rate at which pupils that complete primary school progress to and matriculate in secondary curriculums. Column five shows that the technical fields of production are, on the whole, strongly represented among college students and graduates. The more successful NICs—Singapore, South Korea, and Taiwan—for years had as much as 40% of their students and graduates in these technical categories. In light of SADC's distance from its objective of competing successfully in selected technology production arenas, the S&T coordinating body should consider counseling for higher technical academic preparation.

The final column of Table 3 shows the 1999 foreign direct investment inflows to SADC nations and the selected Asian nations (UNCTD 2000). Since FDI is increasingly associated with knowledge-intensive production and with a declining share directed to primary production—SADC's former comparative advantage—it is not surprising that SADC, and Africa as a whole, is not particularly an attractive destination for international investment today. Singapore, with its 3.5 million people in 1999, received 95% as much FDI as did all of Sub-Saharan Africa's more than 650 million. Africa as a whole received only 1.2% of world total FDI in 1999. Only five countries (i.e., Angola, Egypt, South Africa, Nigeria, and Morocco) received 67% of Africa's total FDI receipts of that year. As for SADC, Angola and South Africa received 71% of the community's total. There are two problems particularly striking with respect to FDI in SADC and the rest of Africa. First, the investments are unusually lumpy in comparison with other areas of the world. In the cases of Africa and SADC, a little over 10% of the nations in each region are attracting most of the international investments. Investments tend to be larger when FDI is dispersed with greater equity across an

Table 3: Key Education Statistics and FDI Inflow for SADC and selected Asian Countries

Region/ Country	200 Illiteracy rate Aged 15 and Over (%)	200 Illiteracy Rate Aged 15-24 (%)	1995 Gross secondary enrollment ratio (%)	1995 Students (and graduate) in natural sciences, engineering & agriculture	1999 FDI inflows (US$million)
SADC					
Angola	N/A	N/A	N/A	N/A	1,814
Botswana	22.8	11.8	56	24 (N/A)	112
DRC	N/A	N/A	26	N/A (N/A)	1
Lesotho	16.1	8.8	28	25 (30)	136
Malawi	39.7	28.7	6	18 (12)	60
Mauritius	15.7	6.7	62	N/A (N/A)	49
Mozambique	56.2	40.0	7	50 (49)	384
Namibia	17.9	8.2	62	5 (3)	114
Seychelles	N/A	N/A	N/A	N/A	60
South Africa	14.9	9.1	82	18 (14)	1,376
Swaziland	22.0	12.2	28	N/A (N/A)	-4
Tanzania	24.8	9.3	5	39(26)	183
Zambia	22.0	12.2	28	N/A (N/A)	163
Zimbabwe	7.3	0.6	44	23 (N/A)	59
Total SADC FDI					4,513
Selected Asia					
China	15.0	2.0	67	37 (35)	40,400
India	44.2	30.3	49	N/A (N/A)	2,168
Indonesia	13.0	2.2	48	28 (27)	-3,270
Malaysia	12.5	2.5	57	N/A (N/A)	3,532
Thailand	4.4	0.9	55	19 (24)	6,078

entire region. When this occurs in the developing world context, investors see the region emerging as a mass market. These existing lumpy regional investments are more archetypal of strategic mineral investments, which typically do not make for mass markets, and which is the case with three of the top four FDI recipients in Africa. Second, services continue to grow as the largest sector of FDI. Services as well as other important areas of FDI presuppose a literate society and a significant techno-scientifically-savvy stratum in the workforce. So while the region has made the necessary adjustments politically and legally in its campaign to attract FDI, governments of the region have yet to make the necessary adjustments in terms of prioritizing science and technology to realized their FDI goals.

Numerous long-term S&T development models exist. Two competing, though certainly not mutually exclusive, models are displayed in Figures 1 and 2. Figure 1 is a prospective model advocated by SADC. The model suggests starting with an umbrella macro-level national S&T plan. This overarching national S&T plan would be run out of some portfolio within an influential ministry. All of the principal stakeholders from industry, universities, professional societies, labor, and research centers would collectively debate in symposia on needs and priorities. From there, legal institutions with appropriate statutory provisions, adequate funding, regulatory bodies, and means for ongoing consultations with stakeholders would constitute the aspirations and machinery of the S&T plan. Once the destination and direction of how the economy and other aspects of society would become more technological, specific sectors would next develop meso-level plans that would enable them to both meet their sectoral objectives and contribute to the macro-level technological objectives of the nation. The third box involves the short-, medium-, and long-term more micro-articulated and highly quantifiable technological development criteria for each sector. Meeting each of the micro objectives would be another step towards the meso and macro objectives.

Figure 2 is a three-tiered *ex-post* model that has taken shape in China since the mid-1980s (Bhagavan 1997). The first tier concentrated on applying the existing S&T capabilities to the production sector. Rather than abruptly abandoning all of the technologies bequeathed from earlier eras, based on cost-benefit analysis, systems analysis, and operations research, the Chinese skillfully allowed some older facilities and equipment to be optimized and continue in operation. This allowed investments to be recouped, avoided needless unemployment, and kept engineers and technicians technically and analytically engaged.

The second tier involved meeting the challenges of bringing on board the new generic and other advanced enabling technologies through technology transfer. Legal instruments here are as important as technical production facilities. Thus, China has sought to protect itself from transfer pricing, dumping, tunkey

Figure 1. Prospective SADC S&T Development Model

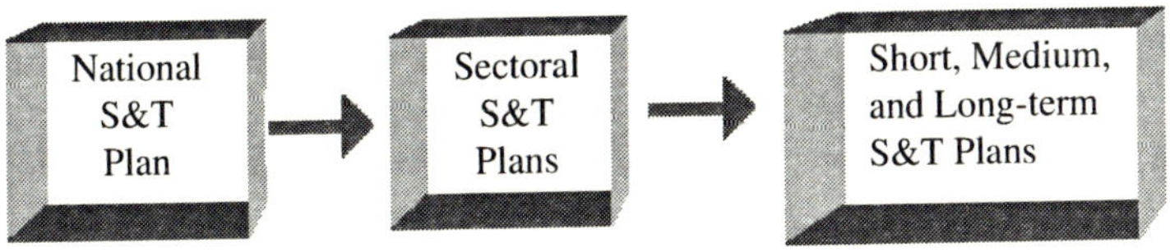

Figure 2: Ex-Post Chinese S&T Development Model

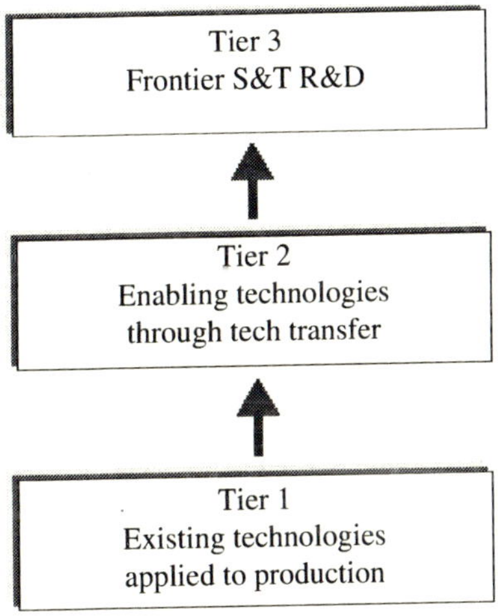

deliveries, and undue dependence of technology supplies, all of which are associated with large-scale international technology transfer. The final tier has been dedicated to fundamental and long-term research and development at the frontiers of science and technology. In other words, to be successful in the global arena, China has articulated a vision for redeploying its exports around increasingly higher valued-added technological production. Because it is all but impossible to be a world-class competitor in all of the crucial emerging technologies, China is attempting to specialize in targeted areas whereby policies can be tailored to turn the vision into reality.

Two assessments are offered to conclude this section on S&T

prospects and activities by the SADC secretariat. First, in addition to the practical short-term objectives that the coordinating body is working towards, the body should also consider a sagacious integration of the elements represented in Figures 1 and 2, and consider adopting the model as the secretariat's official recommended long-term technology development model. Second, the secretariat may also consider the mechanics and the implications of a region-wide reverse engineering agenda. India, the R&D superpower of the Third World (Bhagavan 1997), has made substantial gains technologically through, among other actions, an aggressive campaign of reverse engineering. Other countries, such as Brazil, China, and Taiwan, have also made notable gains through reverse engineering. And this technological sleuthing is not just limited to the Third World. Essentially all of the blue-chip technology firms indulge in this competitive process. Nevertheless, state-sponsored reverse engineering campaigns remain a dicey issue, particularly in the era when global trade law covering intellectual property rights is privileged over knowledge dissemination for development as an aspect of human rights. The secretariat could investigate how some of their Third World counterparts have succeeded with reverse engineering. What were the specific mechanisms for achievement? What is the curriculum by which reverse engineering students are trained? Since this also borders on volatile international political and legal issues, the secretariat could also be a clearinghouse for counsel on such matters as well.

Botswana and Zimbabwe

Botswana

Botswana's economic success over the past three decades has been understandably widely heralded. The nation has avoided the problems of the "curse of natural resources" that have plagued many African nations as well as others in the Third World. The ruling Botswana Democratic Party (BDP), has run a largely frugal and uncorrupt government for more than a generation. Diamond sales in conjunction with effective government have allowed the country to experience the world's fastest economic growth for many years.

After decades of sustained successes, however, Botswana faces two particularly long-term challenges that require tremendous leadership. First, Botswana has the world's highest HIV infection rate. The country could lose a large share of its professional and technocratic workforce, which would simply arrest the economic growth and sociopolitical order witnessed to date. BDP could also be forced politically and morally to deplete its sizeable foreign exchange reserves to care for the growing number of AIDS patients. Second, BDP now appreciates that the future national economy will not experience the growth witnessed in past years through diamond sales. There is now an effort to turn this adversity into a propitious opportunity for constructive change. Leadership in Botswana, in the BDP and in contending parties, agree that the nation's economy must be diversified away from minerals and made competitive through science and technology. A new fixation now exists with the fact that less than one percent of GDP is spent on science and technology in Botswana, while successful developed and emerging industrialized countries allocate between two and three percent. Table 4 is a summary of the activities unleashed to address this second long-term problem, and the anticipated objectives of a national blueprint for S&T presently under construction.

Table 4: Itemized National S&T Activities and Expectations in Botswana

Itemized Chronology	Activities for S&T Enhancement
	Accomplishments to Date
1	President commissioned Botswana Technology Center (BoTech) to develop proposal for national S&T plan
2	BoTech studied growth needs of local firms
3	Held national S&T stakeholders forum
4	Submitted and gained approval from cabinet of proposed national S&T plan
5	Parliament approved proposal in July 1998

6	Defined principal need: implement a national science & technology research and development council (NSTRDC): a secretariat
7	Hired New Zealand consultants to recommend implementation plan
8	Cabinet and Parliament approved implementation plan
9	Expect to have a secretariat in place by 2002
Concrete Expectations From New National S&T Plan	
1	Establish umbrella institution for national S&T organization Develop Indigenous R&D capacity
2	Develop Indigenous R&D capacity
3	Increase expenditure on R&D (e.g., 2.5% of GDP)
4	Investment in S&T infrastructure
5	Commercialize R&D innovations
6	Attract technology transfer

Zimbabwe

Unlike most African countries, Zimbabwe has a plethora of S&T institutions. A principal problem for Zimbabwe, however, is the disarticulation of these scientific and technological support systems. The R&D institutions of higher learning do not impact the policy formulation process, and the results of these institutions are not applied in the productive sectors of the economy. The disarticulation among the network of S&T institutions as well as with the economic planning agencies is a direct product of an absence of a committed state supported national S&T plan. While there are many state-of-the-art planning and R&D research activities occurring, both publicly and privately funded, these isolated activities lack the synergy that would be experienced if there was greater coordination under a national plan. All of the most successful and heralded NICs (i.e., Singapore, South Korea, and Taiwan) formulated national plans, which gave direction and coordination to S&T activities within those respective nations. Late-coming nations to industrialization, particularly at the start of the Information Age and twenty-first century globalization, can ill afford to proceed with S&T activities that do not constitute a "blue-printed" critical mass that produces synergistic effects.

The Research Council of Zimbabwe (RCZ) is the nation's central institution for S&T policy formulation, coordination, R&D activities, popularization of scientific research, and organizing of biannual symposia. While RCZ is ensconced in the office of the president and cabinet, it lacks the funding and regulatory authority to critically affect R&D and support private sector innovation. Moreover, RCZ lacks the gravitas and a "stakeholder-produced" national S&T blueprint sufficient to influence official economic policy. In fact, the Confederation of Zimbabwe Industries (CZI), the trade association and voice of industry, has no linkages to RCZ and, hence, the existing fragmented "S&T policy" from the council oftentimes does not reflect the needs and interest of industry. Instead, CZI choreographs the needs and interests of its members with policies from the Ministry of Commerce and Industry as well as the national economic policy planning agencies. As far as meaningful, real-world policy that involves Zimbabwe's technological activities is concerned, RCZ is marginalized.

To address this lack of national S&T coordination between policy makers and industry, movement got underway in the early 1990s to link public S&T funding with industry needs. First, national consultations occurred with all stakeholders who debated and deliberated extensively on the S&T needs of the country and how best to address them. One broadly embraced conclusion was the need for establishing a research center whose mandate would be to spearhead technological development and technology transfer from abroad and from research centers to industry. That center now exists as the Scientific and Industrial Research Development Centre (SIRDC). SIRDC was founded on the 1988 Ottawa-based International Development Research Centre (IDRC) model of "results-utilization." The results-utilization policy involves the following principals: target beneficiaries, user needs, user participation, utilization promotion, and results dissemination.

Results-utilization has significantly reduced, though certainly not eliminated, the estrangement between CZI members and publicly funded research efforts. One means of accomplishing this was to deliberately limit SIRDC's budget just enough to compel the institution to seek industry-sponsored research. This way, SIRDC focuses on the needs of industry, and industry capitalizes on subsidized research sponsored through the center. SIRDC has seven research institutes: biotechnology; building technology;

informatics and electronics; energy technology; environment and remote sensing; national metrology; and production engineering. Through these institutes, SIRDC is helping local affiliates, locally based firms, and new entrepreneurs benchmark their products successfully against global standards.

Nevertheless, the manufacturing sector, which is dominated by foreign capital, has steadfastly resisted conducing manufacturing R&D in Zimbabwe. Zimbabwe African National Union-Patriotic Front (ZANU-PF), the ruling party, has been reductionist by concluding this is simply the intransigent indulgence of an international division of labor in TNCs. This reductionism is becoming increasingly antiquated in light of the today's growing international production as a replacement for exporting globally. Accompanying TNCs' globalizing production facilities are their global R&D facilities. Zimbabwe, in the framework of a broader SADC S&T agenda, will have to compete more effectively to get a greater share of foreign technology and privately funded R&D. This observation is not asserted out of sanguinity, for the difficulty and time requirement for Zimbabwe to attract greater foreign R&D and technology are appreciated.

Despite these impressive S&T improvements through SIRDC, Zimbabwe has been limited by macro-politicoeconomic events. The principal focus of economic policy has been towards short-term stabilization at the expense of long-term growth. Thus, "science and technology policy as an instrument for structural transformation and long-term dynamism of the economy has been peripheralized" (Zwizwai, et al. 1997: 48). The travails of ZANU-PF with respect to the land redistribution question has also adversely impacted technology transfer into the country. For instance, while the country received only $59 million in FDI in 1999 when the crisis approached a new crescendo, Zimbabwe received $444 million in FDI in 1998, a time when the country was in a more quiescent state. The latest adverse impact of the continuing instability was the announcement in April of 2001 that large mining concerns would suspend production until political stability and economic growth resumed. Industrial production and FDI declines will have both short-term economic effects as well as long-term S&T effects.

Conclusion

Since independence, all African governments have, with varying degrees of success, tried to integrate Western analytical science and production technologies into their respective indigenous cultures and societies. In the 1960s, the impetus for incorporating these Western techniques was the notions of "economic self-reliance" and advancing from primary production to manufacturing. The prevailing wisdom at the time was that such incorporation was the best means for African nations to catch up with the West, and as a direct consequence, improving the quality of life for their peoples. At the start of the twenty-first century, the economic gap between the West and most African nations is even greater than when these nations first embarked on the road of independence. Many facts account for this, among which has been the difficulty of surmounting the challenge of embedding a rich scientific and technological tradition in African societies. Some of the plausible factors that have inhibited S&T development in SADC and other African regions have been discussed in this chapter.

This chapter has also provided as much prospective analysis as it has offered retrospective analysis. In other words, since the challenges and opportunities brought on by S&T are fundamentally different today than during the 1960s, the impetus and the policy prescriptions for adopting Western techniques are equally different. Today, governments think less in terms of being economically self-reliant and more in terms of being successfully integrated into the global economy. Making globalization work for a given Third World nation or a region of such nations requires going upstream in terms of knowledge-intensive production. But, at this juncture, the TNCs have a monopoly on much of the state-of-the-art production technologies, accounting for an estimated two-thirds to 90% of international technology transfer between nations. This is particularly an acute problem today as the global gap in knowledge production is even greater and more consequential than the global gap in income.

The necessity of SADC tapping into international production at a higher level is matched by TNCs' search for "locational assets" and "created assets," both of which SADC nations have a great deal to offer. Due to the increase in knowledge-intensiveness of

FDI, attracting FDI more and more requires a literate society with a sizable techno-scientific stratum of workers. Data in this paper illustrate that SADCs' youthful citizens are receiving considerably more and higher quality education. College students are also studying technical subjects at rates similar to those of advanced East Asian NICs.

Since SADC nations are only (1) in the early stages of providing the requisite educational foundation for success in S&T; (2) recently considering policy shifts that could eventually reconstitute the countries' comparative advantages from extractive production to technological production and exportation; and (3) witnessing continued political instability, the region does not receive much FDI, which is increasingly of a knowledge-intensive variety. To facilitate long-term S&T advancement in SADC, two key changes were identified in this paper. First, the S&T coordinating body within SADC's secretariat should consider integrating sagaciously SADC's existing prospective S&T development model with China's *ex post* S&T development model. The Chinese model offers new insight into navigating the contradictions of technology transfer in the broader context of economic liberalism and active state intervention for S&T advancement. Should the new synergistic model survive critical scrutiny on conceptual, analytical, and empirical grounds, it should be considered for adoption throughout the region. Second, SADC should consider investigating specific curricula and extensive institutional support for state- and quasi-state-sponsored reverse engineering campaigns. Studying the reverse engineering campaigns of India, China, Brazil, and others may provide insight into both the technological aspects as well as the politico-legal aspects.

ENDNOTES

1 Britain in the 1790s, the first to advance through industrialization, had no foreign industrial competition from which the state may have felt similarly compelled to protect domestic industry.

2 It is recognized that the international science indexes are fraught with many well-documented problems. Nevertheless, there is enough convergence of consistent evidence to know that a gulf exists between Sub-Saharan Africa and India with respect to scientific and technological advancement.

CHAPTER 4

THE AFRICAN INFORMATION SOCIETY INITIATIVE: 1995-2000

Nancy Hafkin

Introduction

In May 1996 the Conference of [African] Ministers of Economic and Social Development and Planning adopted the *African Information Society Initiative: An Action Framework to Build Africa's Information and Communication Infrastructure* (ECA 1996). Despite Africa being the least advanced of all the regions of the world in information technology, the African ministers were the first of any of the developing regions to adopt a master plan and a declaration to move their countries into the information age. This chapter examines the origins of the Initiative, its content, what it has done, how it was implemented, what the region did in the five years following its adoption, and what needs to be done next.

Wake Up Call to African Policy Makers

A series of events in 1995 appear to have moved the 1996 meeting of Ministers to the realization that the time had come to put information technology on the list of major development issues that the African region had to deal with. One event that took place in Brussels and two in Addis Ababa resonated across many African capitals.

In February 1995, Ministers from the Group of Seven Highly Industrialized Nations (the G-7) and Members of the European Commission (EC) met in Brussels at the G-7 Ministerial Conference of the Information Society. This Conference emphasized the need for all countries, including developing countries and countries in transition, to be integrated into the global information society (GIS). The keynote speaker at the Conference was South Africa's then-Executive Deputy President Thabo Mbeki. In his keynote address to the Brussels G-7 Information Society Conference, Mbeki underlined Africa's delay in entering the information age, pointing out that there were 'more telephone lines in Manhattan than in all of sub-Saharan Africa' and that 'half of humanity has never made a telephone call.' Mr. Mbeki challenged the G-7 and the EC to work together to help bring the information society to Africa and other developing regions. The statement comparing Africa's teledensity with that of the United States struck an alarm bell all over the region. That phrase has been repeated tens of thousands of times since: clearly it set off considerable thinking that the time had come to deal seriously with this issue.

Addis Ababa 1995

In April 1995, some 450 participants from all over Africa and the rest of the world as well came together at the headquarters of the United Nations Economic Commission for Africa (ECA) in Addis Ababa at a conference entitled 'African Regional Symposium on Telematics for Development.'[1] The gathering was a watershed in the realization that information and communications technology could be a force for development and that the Africa region was ready to consider it (Hall 1995:1).[2] Those who attended were government policy makers, post and telecommunications officials, system operators, equipment suppliers, non-governmental organizations, educational institutions, users, and donors. Some 50 telematics initiatives underway in Africa were discussed and demonstrated.

The key messages from the meeting were:

- There was a surprisingly significant amount of information and communication technology activity underway in Africa.

- African governments needed to become aware of the information revolution in Africa and the rest of the world and adopt their policies to take advantage of it, not thwart it.

A number of participants at the meeting expressed the idea that if African countries did not surf the great wave of the information revolution, they might be crushed by it. The main obstacle was not obsolete technology, but obsolete policies and regulatory frameworks. 'The real challenge,' said one paper presented to the meeting, 'is not technical or financial, but organizational and political.[3]

The energy and enthusiasm of the Symposium were harnessed into an attempt to persuade African authorities to catch the wave of technological change. The participants put up a communiqué and a challenge to the ECA Ministers of planning and development who would be meeting in the same hall a month later. The communiqué said:

- Unless African countries become full actors in the global information revolution, the gap between the haves and have-nots will widen, opening the possibility of increased marginalization of the continent. This gap will increase the likelihood of cultural, religious and ethnic ghettos leading to regional and inter-regional conflicts.
- Telematics offers African scientists and researchers unparalleled participation in the global scientific community through direct access to the Internet, the global network of networks. Developments in the telematics field make it possible for Africans, particularly in rural areas, to have dramatically increased access to communications and information, accelerating and bolstering sustainable development.
- Education, health, trade, and commerce as well as regional and international cooperation will be vastly strengthened by an enhanced telematics infrastructure in the region – with an investment cost less than the price tag of a modern jet fighter.
- The increased flow of information will increase Africa's participation in the global dialogue on issues such as the environment, human rights, and democratization (ECA 1995).

1995 Conference of Ministers

African policy makers heard the call of the Telematics Symposium. In May 1995 the twenty-first meeting of ECA Conference of Ministers, which consists of the fifty-three African ministers of social and economic development and planning, adopted Resolution 795 (XXX) entitled 'Building Africa's Information Highway' that called upon the ECA to appoint a High-Level Working Group on Information and Communications Technologies in Africa to draft an action framework on utilizing information and communications technologies to accelerate the socio-economic development of Africa.[4] It was the first time that the ministers of social and economic development and planning had dealt with the issue of information technology in Africa. Many of them felt that IT was still inappropriate to the Africa region, but enough were convinced that the time had come for Africa to move to pass the resolution. The essential point that the Ministers made was the importance of Africa setting its own directions for the development of an African information infrastructure.

High-Level Working Group and AISI

The High-Level Working Group that the ECA appointed consisted of eleven experts on Information Technology in Africa. The Group met in Cairo, Dakar, and Addis Ababa between November 1995 and April 1996 and communicated further by electronic mail. The result of its work was the document entitled 'Africa's Information Society Initiative (AISI): An Action Framework to Build Africa's Information and Communication Infrastructure' that was submitted to the twenty-second meeting of the ECA Conference of Ministers in May 1996 and adopted by *Resolution 812 (XXXI)* entitled 'Implementation of the African Information Society Initiative' (see Annex I).

Shortly after its adoption by the ECA Ministers, the (International Telecommunications Union) ITU-organized African Regional Telecommunication Development Conference endorsed AISI as being complementary to the ITU African Green Paper.[5] African member states of the ITU requested the Union to work closely with the ECA and other partners in the establishment of national information and communication infrastructure policies. The Executive Secretary of the ECA K.Y. Amoako launched AISI at the Information Society and Development Conference, organized for 44 developing countries by the Government of South Africa one week after the meeting of the ECA Conference of Ministers. In introducing AISI to the Conference, Mr. Amoako said:

> To move into the information society, Africa must be clear on what it wants and make its desires clear to others. If we are clear, if we have a compelling vision for ourselves, then it is likely that our choices will turn into our own reality. If we are not clear, then we will either be perpetual observers of the information highway or find ourselves on a road not fit for our needs (GIIC 1996).

AISI was subsequently endorsed by the Organization of African Unity Summit of African Heads of State held in Yaounde in July 1996 as well as by the G8 at their Summit in Denver in 1997. The OAU member States declared their intention of establishing national awareness programs 'on the move to the Information Society.'

Africa's Information Technology Situation in 1995-1996

African information infrastructure in 1995 was marked by 'barriers of over-regulation and artificially high charges, in a clatter of outdated telexes and $30-a-page faxes' (Hall 1995). Zambia was the only sub-Saharan country outside South Africa with full Internet connectivity.[6]

Only four other countries had full Internet connectivity: Egypt, Algeria, Tunisia, and South Africa. There was no local dial access outside the capital cities. About half the African countries had daily dial-up electronic-mail-only links to the Internet, with a total of 3,000 users on the continent! Few Africans were trained in computer networking.

Telephone density in sub-Saharan Africa in 1996 was at the very low rate of 0.5 lines per 100 inhabitants. The waiting list for telephone connections was 3.6 million, with an average wait of nine years. The estimated 'PC-density' (number of PCs per 100 habitants) ranged from 0.3 in Egypt to 2.1 in South Africa.[7] There were high tariffs and complex customs procedures that constrained the import of computers and related equipment (especially telematics equipment). Many countries maintained heavy controls on the import of fax machines and modems. There were no cellular telephone networks. All African public telephone operators (PTOs) were run as state monopolies (Jensen 1995). However, in Uganda and Sudan, there were some signs of telecommunications liberalization.

No African countries had plans or a policy for the development of national information and communication infrastructure (UNECA 1995b).[8] Many countries, however, had enacted a number of laws and regulations to govern computer technology, which became constraints to the optimal utilization of information technology. Two countries had banned the importation of computers for various periods of time on the grounds that they brought unemployment. In others, high tariffs and complex customs procedures limited imports. Regulations requiring type approvals of equipment and bans or other limitations on communications equipment (faxes, modems, satellites, and satellite dishes) also prevented information technology utilization at anywhere near the levels of developing countries in other regions of the world.

Until 1995 the technical and non-governmental organizations involved in the installation and operation of low cost electronic communications systems, largely FIDO, had supplied the major energy in the area of information technology for development in Africa. The 1995 Telematics Symposium marked the awareness of the technical and NGO community that what they were doing would remain small-scale and marginal unless it gained the attention of African policy makers.

FidoNet

1995 was the high point in Africa of FidoNet; low-cost electronic information services store-and-forward e-mail systems, run largely by non-governmental organizations and universities. However, thousands of enthusiastic users of these systems were already creating the demand for full Internet.[9] The ECA had been in the forefront of this area, helping to develop the first e-mail nodes in some 24 countries, first through its Computer Networking in Africa project (1987-1991) and then through its Capacity Building for Electronic Communication in Africa (CABECA) project (1991-1995). At the time of the Telematics Symposium, the ECA staff said:

> Over the past five years we have proved we can work in many difficult situations, setting up networks with system operators deep in the countryside – and we've come to know that the technical problems are not as difficult as the administrative and policy problems. So much networking has been going on either semi-legally, or with PTTs and ministries of information or communications at best turning a blind eye, at worst, being constraining. We feel this is because they are not familiar with the Internet, with the merging of the computer and telecommunications worlds, so they go on in the old way. And if they go on thinking like that, Africa will be left out of the information age (Hall 1995).

Content of AISI

The AISI was not meant to be a regional plan, but rather a vision and an exhortation to wake African policy makers to the importance of information technology for development and stimulate them to draft national policies, plans, and strategies to take full advantage of the opportunities that the information age offered. The vision was that building an information society would help stimulate African economic growth and provide new opportunities in education, trade, healthcare, job creation, and food security, helping African countries leapfrog stages of development and raise their standards of living.

The action framework called for the elaboration and implementation of national information and communication infrastructure plans involving development of institutional frameworks, human, information and technological resources in all African countries and the pursuit of priority strategies, programs, and projects that could assist in the sustainable build up of an information society in African countries.

Central to the vision was the conviction that information technology was not a luxury for the elite but a necessity for development. AISI foresaw a sustainable information society in Africa by the year 2010, where: 'Every man, woman, child, village, government office and business can access information and knowledge resources through computers and African information resources available reflect the needs of every sector.'

Elements of AISI

Other central elements of AISI included:

- Working with African countries on developing national plans for building information and communication infrastructure
- Eliminating legal and regulatory barriers to the use of information and communication technologies
- Establishing an enabling environment to foster the free flow and development of information and communication in society
- Developing policies and implementing plans for using information and communication technologies in the public sector
- Identifying information and communication applications in areas of highest impact on socio-economic development
- Facilitating the establishment of locally based, low-cost and widely accessible Internet services and information content
- Preparing plans to develop human resources in information and communication technologies
- Adopting policies and strategies to increase access to information and communication facilities with priorities for rural areas, grassroots society, women and youth
- Raising awareness of the potential benefits of information and communication infrastructure.

AISI encouraged the liberalization of national telecommunication and public broadcasting services. It envisaged the central role of government as shaping an enabling environment for the use and development of ICT.

The Ministers directed the ECA to incorporate AISI into its work program and to undertake its implementation in collaboration with a broad range of partners. For the ECA it marked an important change in its development information activities, from a technical focus (promoting and managing the Pan African Development Information System and establishing electronic communication nodes) to the policy arena of promoting information technology as a tool for African development.

Implementation

In working with African States for the implementation of AISI, the ECA joined a large number of partners. The first meeting to coordinate implementation of AISI was held in Addis Ababa in October 1996 and attended by some twenty-nine-partner organizations. A Web site was developed and a listserv set up for communications among the partners (aisi-hitd-cl).[10] In April 1997 at a meeting in Rabat, Morocco, the Partnership for Information and Communication Technologies in Africa (PICTA) was established, of which the ECA assumed the role of secretariat. PICTA brought together the many organizations working to promote the use of information technology for African development. By 1997, the U.S. Government Leland Initiative had become prominent among them. Leland Initiative and the Carnegie Corporation of New York, and the ITU took the leadership in calling for collaboration among the multiple initiatives working for shared aims. Among the members of PICTA were multilateral and bilateral aid and development organizations, foundations, NGOs, and representatives of the private sector. PICTA meets in semiannual meetings and communicates through a listserv (picta-cl) outside of formal meetings. It developed a joint collaborative work program and an AI-AIMS database of information on projects undertaken by partners.

AISI Implementation Areas

Van Audenhove, et al. (2001) make a sober assessment of what the ECA, as a multilateral organization, could do vis-à-vis its member States in implementing AISI.

> [The ECA can] influence policy formulation as well as the actual implementation of ICTs in developing countries through several mechanisms, e.g. programs in policy assistance, institution-building and private sector support; specific ICT related projects; international and regional conferences; and the spread of documents and scientific articles (van Audenhove et al. 2001).

In trying to realize the AISI vision, the ECA and its partners identified seven areas of concentration: Policy awareness, connectivity, national information and communication infrastructure (NICI) policy and planning, training and capacity building, democratizing access to the information society, sector applications, and development information. The following section describes the major thrust of activities in each as well as their impact.

Policy Awareness

In the implementation of AISI, the ECA published the AISI document in English and French and distributed it widely. Both alone and with numerous partners, it held policy awareness workshops in some twenty-six countries of the region. Its staff presented AISI at dozens of meetings and conferences in the Africa region, particularly those that convened African policy makers.

A significant event in creating policy awareness was the special session during the 1997 Conference of Ministers organized by the ECA and the Global Information Infrastructure Commission (GIIC) on 'Meeting the Challenges of National Information and Communications Infrastructure in Africa with Public and Private Sector Cooperation' (Cogburn and Hafkin 1997). As the first time that the Conference of Ministers had devoted a day of their annual meetings to a specific theme, this gave elevated status to the issue of information technology and development.

In the five years following the adoption of AISI a great deal of sensitization and awareness took place on the importance of building information infrastructure and bringing Africa into the information age. One measure of that is the content of the debates on information technology issues at the annual meetings of the ECA Conference of Ministers.

In the 1996 and in 1997 meetings, there was still a fair amount of resistance to the idea of the relevance of the information age to Africa. By 1998, ministers comments tended more towards 'give us suggestions on how to get there.' In 1999 and 2000, more and more countries began to take the floor to detail their accomplishments in adopting information technology in their countries and jockey for the position as one of Africa's leaders in this area. Another indicator of the importance of the issue is the number of ministers and high government officials (including Heads of State) who attended the ECA meetings on this subject. There were some 32 ministers at the 1998 Global Connectivity for Africa Conference, and a larger number at the African Development Forum of 1999 on the theme *The Challenge to Africa of Globalization and the Information Age,* along with two Heads of State and the Vice President of Malawi. At the national level, high-level officials opened and attended the information technology policy workshops. This demonstrates that the awareness of national policy makers was growing stronger and that a majority became convinced that ICTs could have a genuine impact on national development in Africa.

Some African countries are moving forward in the adoption of new technologies as development tools to an extent that would not have been predicted based on their GDPs and income levels.[11] In these cases, the key element is leadership: In particular, where the Head of State becomes a champion of information technology and gives it a high place on the national development agenda. President Konare in Mali, President Kagame in Rwanda and President Mbeki of South Africa were African Heads of State who made a strong commitment to promoting information technology as a development priority. African ICT expert Michael Jensen (1999) concludes that AISI's policy awareness activities have made a significant impact on African government officials:

> AISI has created significant internally generated pressure from the responsible ministries [throughout Africa] to urge their administrations to adopt appropriate regulatory, tariff and service provision policies.

Connectivity

Although the ECA reduced its technical assistance activities in connectivity after 1996, it intensified its actions to promote connectivity in the policy arena and, through PICTA, worked with the Leland Initiative and the United Nations Development Programme African Internet Initiative to put the connectivity infrastructure in place. In 1998, with the World Bank, ITU, and the African Development Bank, the ECA organized the Global Connectivity for Africa Conference that brought together some 250 African ministers and heads of communications authorities and others involved in information infrastructure deployment. The aim was to assist African governments and the domestic private sector in dealing with the new fiber optic, wireless and satellite technologies to increase the possibilities for low-cost rural connectivity. As will be seen in the assessment of the current information infrastructure situation in Africa, this is the area in which the greatest progress was made in the years 1996-2001.

National Information and Communication Infrastructure (NICI) Policy Planning

This area concerns development of demand-driven policy programs that reflect national priorities for action in utilizing information and communication technologies to accelerate socio-economic development. The basic NICI policy elaboration process guidelines are shown in Annex III. NICI policy planning is an area of particular AISI achievement. By 2001, ECA and its partners were working with more than half of African countries on elaboration of NICI plans. Countries that are relatively advanced in their NICI planning process include Gambia, Ghana, Mauritius, Mozambique, Rwanda, Senegal, Tunisia, and Uganda (UNECA 1999f). The ECA documented the progress made in this area in the paper presented to the African Development Forum (ADF) 1999 (UNECA 1999a).

Training and Capacity Building

Training and capacity building are essential remedies to enhance African technical know-how and the spread of information and communication technologies. Activities in this area included the establishment of regional institutions and the development of demonstration courses, introduction of information and communication technologies into public schools, training of information workers (including librarians) in electronic information management, and training in Internet systems operation, administration and maintenance. The ECA and its partners introduced a number of courses in this area from 1996-2001. Of particular note is the course module that the ECA developed and delivered in the master's degree program at the *Institut pour le développement économique et la planification* (IDEP) in Dakar that combined information technology literacy with information technology policy awareness for African planners

Democratizing Access to the Information Society

To the credit of the authors of AISI, democratizing access to the information society was a core area of the initiative from the beginning, three years before the developed world became aware of the issue of the digital divide. From its inception, AISI was concerned with creating universal access to ICTs in Africa, emphasizing public over individual access. Through the work of PICTA, AISI partners (in particular, UNESCO and ITU) established the first telecenters in Africa. The International Development Research Centre's Acacia Initiative, using ICTs to improve the lives of marginalized communities in Africa, defined itself as a modality in support of AISI (IDRC 2001). Providing IT access for women and youth became a major concern of the ECA and its AISI partners. In 1998, the ECA with the International

Development Research Centre (IDRC) organized the sub-theme on gender and the information revolution in Africa for the ECA's fortieth anniversary conference on Women and Economic Development in Africa (Rathgeber and Adera 2000).[12]

Sector Applications

This area concerns software applications to help countries overcome stubborn development problems in various sectors of society. The test of relevance of information and communication technologies is the use to which they are put — from distance education to decision support systems to telemedicine and public information databases accessible to every citizen. Each country determines the most relevant sector applications to its national priorities.

The ECA involved itself particularly in promoting the utilization of health sector applications. Other partners chose to work with African countries on applications in other sectors. There has been considerable advance in the utilization of these sector applications, particularly in support of health and education; however, more needs to be done before measurable impact can be declared.

Development Information

This area concerns information on social and economic aspects of African development. It comprises, among others, collecting and organizing African information sources by African countries, the development and use of software that addresses the variety of African languages and levels of literacy, and the establishment and promotion of electronic development information exchange in Africa. For connectivity to be relevant, Africa must be able to find information from Africa, and Africans must be involved in creation of content for electronic dissemination.

This is an area that the ECA had been working in for nearly two decades prior to AISI through its Pan African Development Information System, established in 1979, on the principles of information management as the basis for development information exchange. The basic elements of content creation and dissemination over the Web are information management and a commitment to information sharing, which were the essential elements of the PADIS approach. The ECA, with UNESCO and with the support of the Carnegie Corporation of New York, worked with dozens of African institutions to promote African content development and train them in its techniques. The ECA's annotated compendium of some 5000 African development information Web sites presented to the 1999 African Development Forum illustrates the progress made in this area.[13]

African Development Forum 1999

The ECA's largest single activity in implementation of AISI was the organization of the ECA's first African Development Forum on the topic *The Challenge to Africa of Globalization and the Information Age.* The first Forum marked the beginning of a process led by the ECA to position an African-driven development agenda reflecting a consensus among major partners and leading to specific programs for implementation at national, sub-regional, and regional levels. The ADF was based on a conviction that Africa cannot achieve sustainable development unless Africans themselves design and own their development policies, strategies, and actions.

The issue of Globalization and the Information Age was chosen for the first ADF because of the importance of defining African-owned and African-led strategies to join the global information economy. It underscored the importance that the ECA accorded to the issue of information technology in the African development panoply.

The major goal of ADF '99 was to develop and support African initiatives shaping the African Information Age. ADF '99 provided the first occasion for a broad range of African policy-makers and practitioners to come together as a group to assess the opportunities, confront the challenges, and develop a plan of action for Africa in the Information Age. It was the culmination of more than three years of activities to build an African Information Society and the occasion to set future directions for the AISI.

The Forum was unique because it demonstrated how much was being done in Africa by Africans in harnessing Information and Communication Technologies (ICTs) towards development.

Much was already underway, and yet it was not well known, even within Africa itself. Some of the most vibrant and innovative applications of Information and Communication Technologies (ICTs) for development occupied center stage during the Forum. Some 1000 participants attended the Forum from government, the private sector, civil society (including the academic sector and media), bilateral and multilateral organizations. Participants ranged from community activists and technical experts to national and international policy makers, private entrepreneurs, and heads of state.

The themes of ADF '99 were information infrastructure building, the information economy, democratizing access to information society, and ICTs and governance. The discussion on the information infrastructure theme focused on rural communication development, the use of new cable and wireless technologies for greater connectivity, the creation of an enabling environment to speed up infrastructure development, the development of human resources, and the development of an appropriate institutional and regulatory framework (Jensen 1999). The presentation on the information economy was unique, as it was the first time that African ministers dealt with this issue on a regional level. The considerations of the issue included:

- the impact of the global information society on national economies
- the transition to a knowledge-based economy
- the creation of the basic elements of an information and knowledge industry
- national, sub-regional and regional requirements for promoting the growth of the information economy in Africa.

Many tools available for democratizing access, including those in telemedicine, community telecenters and distance learning, were demonstrated at the Forum. Particular attention was paid to the needs of women and youth, both of whom constituted vocal and articulate Focus Groups before, during and after the meeting (UNECA 1999g). The potential of improving governance in Africa through the use of ICTs, with emphasis on involving civil society in political debate, was the focus of the governance theme (UNECA 1999e).

Follow up to the Forum has concentrated on the areas of policies and strategies, electronic commerce, ICTs and health, and ICTs and education.

The Situation in 2001

By 2001, Africa made dramatic improvements in African information infrastructure and the use of information technology as compared to the situation at the time of the adoption of the AISI in 1996. Whereas five countries had full time connections to the Internet in 1996, in 2001 the figure was 52 out of 53 member states of the Organization of African Unity (Jensen 2000b).[14] Most of the growth in connectivity, however, came in capital cities, although 16 countries had points of presence in some secondary towns. Of particular note is the policy that 15 countries adopted to make Internet access a local call across the whole country (Jensen 2000a). In January 2001 the number of Internet users in Africa was estimated at 3.11 million, less than one per cent of the 407.1 million users worldwide (NUA 2001).[15] While African Internet usage percentages remained small, the number of users has increased enormously over the 3000 users in 1995.[16] With the liberalization of the Nigerian Internet market, this figure was expected to grow even more rapidly. By 2001 the Nigerian telecommunications regulator had licensed thirty-eight Internet service providers, of which 12 were active (Jensen 2001).[17] However, few countries had a critical mass of users. Half the countries in Africa had fewer than 1000 dial-up subscribers, while only eleven countries had more than 5000 subscribers (Jensen 2000a).[18] The largest numbers of users were found in South Africa (1.82 million), Egypt (400,000) and Morocco (30,000) (Cyberatlas 2001).

Costs for Internet connections remain very high throughout Africa compared to those of developed countries, and most ISPs still price their services on a connect-hour basis. The average for five hours a month (a very low usage rate) is about $50/month. This cost puts the yearly cost of a five-hour-month connection at greater than the annual GDP per capita of all the thirty-three

LDCs in Africa! In Ethiopia, the cost of a twenty-hour/month connection is 8.4 times the GDP per capita.

There were more than 450 Internet Service Providers (ISPs) in Africa by 2000. Some twenty countries had only one ISP, but in Ethiopia and Mauritius there was only one ISP by government policy (Jensen 2000a). The largest of the multinational ISPs was AfricanOnline that was operating in eight countries at the end of 2000.

Most users access the Internet from work, not home. The largest number of users work for non-governmental organizations, private companies, universities and international organizations. The majority of users are well educated, and most are men (Menou and Hafkin 1997). Undergraduate university students rarely get access. At many universities, access is limited to senior staff. Only twenty universities have full Internet connectivity; the rest rely on dial-up connections.

While African content development is increasing, the Web is growing so fast worldwide that percentages of content of African origin continue to fall. Interesting areas of content development are African Web sites that reach out to their compatriots living in developed countries with news, music, and e-commerce. However, few African institutions have made the commitment to fully disseminate information on the Internet and fewer use it interactively for their activities. In most cases, African institutional Web sites are limited to basic description and contact information. Perhaps surprisingly, Francophone African countries have a higher profile on the Web and also tend to have greater institutional connectivity than non-French speaking countries. This is largely due to the substantial assistance provided by various Francophone support agencies and the Canadian and French governments (Jensen 1999).

Much of the development described results from the liberalized telecommunications environment that has emerged in the last five years (Wilson II and Wong 1999). There is a clear movement from state-controlled centralized domestic telecommunication monopolies to market-driven, decentralized enterprises with foreign participation. The movement began with cellular telephony, but it is expanding to other areas as well, including broadcasting and even voice telephony in some countries. Liberalization in the area of value-added services has led to the emergence of IT-enabled businesses such as phone shops and *teleboutiques*, many of which are starting to add other information technology, including faxes and Internet access.[19] All but eleven countries have started some liberalization of the telecommunications regime in their countries. A number of countries have lowered import duties on computers and related information technology. Separate regulatory authorities that have been set up in some twenty countries are struggling to increase their professional competence and to maintain their independence and competence (Jensen 1999).

Computer dissemination remains very low in Africa compared to other regions, largely because of the high cost of computer equipment relative to the low levels of economic development and availability of computer skills. Despite the lowered tariffs in several countries, the majority of countries still treat ICTs as luxury items subject to high taxes. ITU statistics put the number of PCs in Africa at about 8.4 per 1000, as compared to 511 in the U.S. (ITU 2000b). Countries with user numbers above the mean include Mauritius (95.7), South Africa (54.7), Namibia (29.5), Togo (17.7), Senegal (15.1), Tunisia (15.3), Egypt (12.0), and Morocco (10.8). However, the number of users is much higher than the number of computers because several users are likely to share a single computer. Computers are found most commonly in private companies and least frequently in the public sector and in homes. Where they are found, they tend to be underutilized, with applications concentrating on word processing and accounting. The use of databases is less widespread as are Geographical Information Systems. However, GIS is beginning to be used more frequently by universities, planning departments at ministries and large municipalities. A major reason for their underutilization is that most PCs tend to be standalone, with little use of Local Area Networks (LANs).

Deepest of the Digital Divide

While much progress had been made within Africa by 2000 in comparison to its position five years earlier, there is no question that Africa remained the deepest part of the global digital divide. Table 1 shows African countries in which Internet users constitute less than one per cent of the population. Slightly more than half (twenty-seven countries) are in this category.[20] Not surprisingly, the Least Developed Countries remain furthest outside the information loop. No African LDC has more than one per cent of its population connected to the Internet, and most of the LDCs cluster near the lowest end of countries ranked by percentage of Internet users. Other negative aspects of Africa's current information technology situation are the lack of growth in teledensity and, with the exception of southern Africa, the absence of regional backbones. Most African international communication goes between African and the developed world and not intra-Africa.

Overall, the period since the adoption of the African Information Society Initiative Africa has seen spectacular growth in the spread of information technology in African terms. However, in global perspective Africa is falling further behind in relation to the rest of the world, particularly with respect to low investment levels in telecommunications that provides the infrastructure necessary for information technology development. Throughout the region, too many disenabling environments remain.

Table 1: African countries with less than one per cent of population accessing the Internet

Country	Internet Users per 10,000 Population	Internet Users as % of Population
D. R. of Congo	0.04	0.00
Somalia	0.10	0.00
Burundi	0.23	0.00
Liberia	0.38	0.00
Guinea	0.65	0.01
Algeria	0.66	0.01
Eritrea	0.84	0.01
Tanzania	0.93	0.01
Mali	0.94	0.01

Ethiopia	1.01	0.01
Rwanda	1.21	0.01
Mozambique	1.85	0.02
Malawi	1.90	0.02
Madagascar	1.94	0.02
Angola	2.07	0.02
Nigeria	2.82	0.03
Ghana	3.13	0.03
Zambia	3.42	0.03
Benin	3.45	0.03
Kenya	5.17	0.05
Uganda	7.30	0.07
Senegal	8.33	0.08
Zimbabwe	8.80	0.09
Tunisia	10.71	0.11
Morocco	14.44	0.14
Egypt	15.16	0.15
Namibia	30.12	0.30

Source: (ITU 2000a, UNCTAD 2000)[21]

Conclusion: What Needs to be Done

Does this discouraging lack of progress mean that the AISI and other efforts have been in vain? The author thinks that this is not the case. Without the AISI and other related efforts, Africa would be further behind than it is now. This is not a game that Africa can choose not to play; there is no longer a choice. The forces of globalization have made this one world and one economy. The costs of being left out are a permanent sentence to poverty and marginalization. The new technologies are a lifeline that can enable African countries to join the battle for economic, social and political empowerment. While there are risks in joining the global information society, the greatest risk of all is in not joining.

The obvious prerequisite for entry into the information age is infrastructure development. In order to finance infrastructure development, there has to be foreign direct investment. An enabling environment is necessary to give investors confidence.[22] National information and communication infrastructure plans, policies and strategies provide the basic framework for the creation of an enabling environment that will facilitate infrastructure development and the release of national and diaspora entrepreneurial energy. The task that AISI was created to tackle remains a major challenge worth confronting.

Annex I

Resolution 812 (XXXI):

Implementation of the African Information Society Initiative (AISI)

The Conference of Ministers,

Recalling resolution 795 (XXX) on building Africa's information highway which requested the Executive Secretary of the Economic Commission for Africa to constitute a high level working group to develop an action plan on information and communication technologies to accelerate socioeconomic development in African countries;

Appreciative of the proposal contained in document E/ECA/CM22.6 related to the building of Africa's Information Society;

Convinced that building Africa's Information Society will help Africa to accelerate its development plans, stimulate growth and provide new opportunities in education, trade, healthcare, job creation and food security, helping African countries to leapfrog stages of development and raise their standards of living;

Sharing the vision of an African Information Society which calls for the building of information and decision support systems to enhance policy formulation, fostering a new generation of Africans capable of using information and communication technologies, building African information resources and linking Africa regionally and globally to the information age,

- **Expresses** its thanks to the high-level working group for developing the AISI action framework;
- **Adopts** the principle of the African Information Society Initiative as an action framework for building Africa's Information Society and calls upon member states to incorporate the African Information Society Initiative in their national development plans according to national needs and priorities;

- **Calls upon** the ECA, in consultation with member States, along with the International Telecommunication Union, UNESCO, UNCTAD, the World Bank, and international and bilateral partners in the harnessing information for development priority area of the United Nations System-Wide Special Initiative on Africa to use the African Information Society Initiative as a guiding framework;
- **Urges** member states to utilize the capacities and experiences of national, sub-regional and regional African centers of excellence in the enrichment and implementation of the African Information Society Initiative;
- **Requests** the Executive Secretary of the Commission to:
  - Take action, in particular mobilizing funding and technical assistance, in cooperation with member States, regional organizations such as the Pan-African Telecommunications Union and the Regional African Satellite Communication System, non-governmental organizations, the private sector, bilateral and international partners, to realize the goals of the African Information Society Initiative;
  - Incorporate the action framework into the secretariat's work program under the new strategic directions of the Commission program area on harnessing information for development;
  - Set up a regional mechanism to assist in development, follow-up and coordination of the implementation of African Information Society Initiative programmers and projects;
  - Establish a gender-balanced African technical committee to advise on programs and projects and to evaluate results.
  - the Executive Secretary of the Commission to report to the next Conference of Ministers on the progress in implementing this resolution.

Adopted at Addis Ababa , 7 May 1996

Annex II

Informatics Policy Instruments in Africa Prior to 1996[23]

The study was undertaken in a total of ten African countries, five primarily anglophone and five francophone. From the five anglophone countries, the findings included:

- None of the countries had formulated and implemented an integrated policy on informatics.
- Economic reform programs of the preceding five years had resulted in a new socio-economic environment more conducive to the growth of informatics, with specifically easing of the restrictions and difficulties in importing informatics products.

While there was awareness of the importance of informatics for development, there were no examples of integrated approaches that brought together national perspectives with national resources and institutional capabilities.

All of the countries suffered from a shortage of trained manpower in informatics; this was particularly marked in the area of computer networking. This impacted the introduction of IT in the curriculum and had obvious implications for the efficient use of IT potential and equipment maintenance.

All had policies and instruments to promote the development of human resources in information technology, but implementation was difficult due to lack of trained manpower and other policy and sectoral constraints on the import of hardware and software.

At the policy level, none of the countries saw the policy importance of linking national information systems to external networks; nowhere was there awareness of the importance of regional networking. Only one country had started a database for policy makers.

There was a total absence of policies to promote the development of software industries (despite low salary levels and unemployed technical graduates everywhere) and virtually no research programs in informatics (except as applied to national languages). Nowhere were there policies to promote the use of IT as a private sector development activity.

Nowhere was there any attention to software, either in standards or in intellectual property rights. None of the countries surveyed had telecommunications systems that could support a modern IT sector.

There had been strong growth in this sector over the last few years, but public and private sector utilization and exploitation of this medium fell far below world standards and far below the standards of developing countries in other regions of the world as well.

In the five francophone countries, most of the countries had elaborated informatics policies. However, almost all of these concerned only the public sector and thus there were no policies to influence national global optimization of the potential of information technology. In some of the countries the elaboration of policy had pre-dated the arrival of microcomputers and had remained relatively unchanged despite the changes in information technology made possible by this phenomenon.

The general pattern found in the five francophone countries was the creation of a National Commission to formulate policy, a national office to execute it (although in one country its execution was entrusted to one person!) and the elaboration of a national Informatics Plan. In one country (Senegal), the Plan has been implemented continuously. In another country, (Cote d'Ivoire), two different plans sporadically; in a third one plan implemented for a time, Congo. In the fifth, there was no plan.

The studies found that whenever responsibility for implementing national informatics activities had been entrusted entirely to one office, resulting in centralization of activities, that had retarded the growth of the sector overall in comparison to national growth rates. In the countries where the policy was in nature promotional-indicative and initiative, rather than controlling, appreciable results had been obtained (the cases of Senegal and Cote d'Ivoire). The general characteristics of policy that emerged were monopoly (Congo) where a single agency (until very recently) had employed all informatics specialists in the country, Although this country had been one of the earliest users of informatics in the Africa region, it had now become marginalized by comparison to others and has a very low overall usage of informatics.

Two other countries had policies of severe control of information technology (Madagascar and Cameroon), and their usage suffered as well. Madagascar, for example, in order to keep costs down, limited imports, insisted on employing only nationals in the area and demanded that all equipment imported be compatible, leading to domination by certain brands. The regime has liberalized in the last several years, but there is still no plan for promotion of IT, no regulation at all (in the important areas of confidentiality, intellectual property rights), no coordination with telecommunications and no quality control on the private sector.

In another country, the lack of national human resource development in IT allowed the country to be at the mercy of foreign contracts and the acquisitions of IT was supply and not needs or demand driven.

Senegal was found to be the example of the best practices in policy, reflecting a situation of policy where it was needed and openness where control was not necessary. The national structure he saw as a necessary organ of reflection, co-ordination, establishment of priorities and policy implementation that has given the country the necessary push towards the realization of the potential for informatics in all areas of government.

Annex III

Guidelines for Developing National Information and Communication Infrastructure (NICI) Policies and Plans

1. Define vision, mission, strategic objectives, scope.
2. Define the institutional framework.
3. Define the regulatory framework.
4. Define the information technology business sector.
5. Define the development priorities (five-year plan, cabinet directions, etc.).
6. Define the economic and business sectors as well as the market trends.
7. Define education and science and technology infrastructure.
8. Define needs and priorities for information, decision support, networking, information services, etc.
9. Define the data/information and decision support agencies, actors, etc., at the different levels: national, local.
10. Define the technology infrastructure.
11. Identify, formulate and develop programs and projects.
12. Formulate strategies for development of NICIs.
13. Develop a detailed action plan, time schedule, priorities and budget.
14. Determine the implementation agencies in charge.
15. Solicit the commitment of policy makers, industry leaders, etc.
16. Formulate the public awareness campaign.
17. Implementation.

ENDNOTES

1 In 1995, telematics was the word commonly used to connote the convergence of computers and telecommunications. Since that time, telematics has gone out of use and been replaced by either information technology or information

and communication technologies. The term information technology (IT) tends to be used interchangeably with information and communications technologies (ICTs). The latter recognizes the multiple technologies involved as well as the ubiquitous convergence of communications with information technology. However, the term information technology has come to include the idea of many technologies as well as the communications element. IT tends to be more widely used in the United States, while multilateral development organizations prefer ICT, which is more widely used in developing countries.

2 ECA co-organized the symposium with the UN's International Telecommunications Union (ITU), UNESCO, Canada's International Development Research Centre (IDRC), and Bellanet International.

3 Key papers, in terms of their impact at the Symposium, were those of the World Bank and Mike Jensen: Eugene Boostrom, et al., 'Increasing Internet Connectivity in Sub-Saharan Africa: Issues, Options, and the World Bank Group Role', http://www.un.org/Depts/eca/divis/disd/padis/telemat/africa03.htm; Michael R. Jensen, 'Telematics for Development: Discussion Paper,' http://www.sas.upenn.edu/African_Studies/PADIS/telmatics_Jensen.htm. Nearly all the papers from the Symposium are archived at the University of Pennsylvania African Studies Web site. http://www.sas.upenn.edu/African_Studies/Padis/menu_telematics.html.

4 The language of the 1995 resolution is also a study in the history of terminology. In 1995, few people were talking about the Internet or information infrastructure. The conventional usage was 'information superhighway.'

5 The African Regional Telecommunication Development Conference met in Abidjan in May 1996, one week after ECA Ministers. 'Telecommunication policies for Africa, The African Green Paper,' 'http://www7.itu.int/treg/reform/Policy_Papers/Green/green_eng.htm. focuses more on policies concerning network and physical infrastructure

development, i.e., the development of the physical infrastructure while the AISI concentrates on economic and social aspects of information society.

6 The sole ISP in Zambia in 1996 was a commercial company Zamnet, which had been started by academics at the University of Zambia who leased a line with a World Bank loan.

7 The ITU report *African Telecommunication Indicators 1996* provided information for only five African countries. The countries for which the information was available were amongst the best performing countries.

8 A summary of the findings of this study is presented in Annex II.

9 The history of the pioneers of electronic communication in Africa is documented in Wendy White (1996).

10 A listserv® is an electronic mailing list that allows people to send e-mail to one address; their message is then copied and sent to all the other subscribers to the list.

11 In general on the level of countries information technology usage correlates with per capita income.

12 IDRC has published the papers presented at the conference on the theme of gender and information technology in the Rathgeber and Adera (2000) volume.

13 UNECA (1999) 'Africa on the Internet: an Annotated guide to African Web sites, 'http://www.bellanet.org/partners/aisi/nici/index.htm.

14 Liberia was also connected in 2000, but the private ISP was not financially viable. The Ministry of Posts and Telecommunications has been planning to establish a link with MCI in the U.S. but has not done so to date.

15 Obviously, these can only be educated guesses, but the NUA Internet surveys are respected as the best of the educated guesses. African users were 0.76 per cent of world users, while African accounts for about 13 per cent of the world's population.

16 In Africa, the number of users is much bigger than the number of Internet subscribers because so many people use public facilities to gain access to the Internet. An ECA study showed an average of three users per subscription, but this figure seems too low in view of the explosion of public access points (usually cyber cafés) in capital as well as secondary cities all over the region. Credible observers put the number of users in Africa at up to 50 per subscription. Jeffrey Zuckerman, Director, GeekCorps, presentation at Information Technology Group, Center for International Development Harvard University, 19 April 2001.

17 At the beginning of 2001 there were an estimated 10,000 Internet subscribers in Nigeria.

18 The countries with more than 5000 subscribers each are Cote d'Ivoire, Egypt, Morocco, Kenya, Ghana, Mozambique, Nigeria, South Africa, Tunisia, Uganda and Zimbabwe.

19 This phenomenon is most notable in Ghana, Morocco and Senegal.

20 By comparison, in the United States about 60 per cent of the population are Internet users.

21 Bold denotes a Least Developed Country.

22 Growing investment from the African digital diaspora is a part of foreign direct investment. Like its Indian counterpart, the African diaspora (both current generation and historical) wants to promote information technology as a development tool in the motherland.

23 Hafkin, 'Final Report: Study of the Effectiveness of Informatics Policy Instruments in Africa,' presented to the First meeting of the High Level Working Group on Information and Communication Technologies in Africa, Cairo, November 1995, http://www.bellanet.org/partners/aisi/sumnancy.htm. The ten countries covered by the study were Ethiopia, Kenya, Nigeria, Tanzania, Zimbabwe, Cameroon, Congo, Cote d'Ivoire, Madagascar, and Senegal. The ten component studies and the summaries by language area are available on the web. The studies are:

1. Study of the Effectiveness of Informatics Policy Instruments in Africa, by Elliott Zwangobani.
2. Informatics Policy Study in Ethiopia, by Teferi Kebede, Addis Ababa, Ethiopia.
3. Effectiveness of Informatics Policy Instruments in Africa, Country Report for Kenya, by George Okado, Nairobi, Kenya.
4. Informatics Policy Instruments in Nigeria, by Dr. G. A. Alabi, Ibadan, Nigeria.
5. Informatics Policy Studies in Africa, by Howard M. Shila, Arusha, Tanzania.
6. Study of Effectiveness of Informatics Policy Instruments in Zimbabwe, by Hazel Moyo, Harare, Zimbabwe.
7. Etude des Politiques Informatiques en Afrique, sous la coordination de M. Marius T. Francisco.
8. L'efficacité de la politique informatique au Cameroun, par Boyom Sop Flaubert.
9. Etude des instruments de politique informatique en Afrique, Cas de la Côte d'Ivoire, par Safoura Fadiga.
10. Etude prospective sur les instruments de la politique informatique au Congo, Parts I et II, par Ange Nambila.
11. Inventaire des outils politiques informatiques à Madagascar de 1953 à ce jour, par Roger Andrianasolo.
12. Etudes des Politiques Informatiques en Afrique, le cas du Sénégal, par Moustapha N'Diaye.

CHAPTER 5

INFORMATION TECHNOLOGY, EMPLOYMENT, AND EQUITY IN SOUTH AFRICA: THE ROLE OF NATIONAL ICT POLICY

Chris Benner

Introduction

The rapid development and diffusion of information and communication technology (ICT) is dramatically transforming opportunities for economic development in Africa, as in the rest of the world. Access to information technology has become critical for generating wealth, expanding employment opportunities and promoting development. Information technology, however, is clearly not a panacea for the problems of poverty, unemployment and underdevelopment and indeed represents many dangers as well. The rapid global pace of change in information technologies can contribute to the deepening of technological dependency and economic underdevelopment. In addition, the diffusion of ICT can also disrupt existing employment opportunities, potentially entrenching inequality and social exclusion, and leading to heightened social conflict. To maximize the employment benefits of the information revolution, therefore, public policy cannot be limited to promoting growth in information technology industries and expanding application of information technology in other economic sectors. In addition, it must also simultaneously address inequality and the dislocation associated with diffusion of information technology through the economy.

This ambitious agenda has been explicitly articulated in most information technology related initiatives in South Africa in recent years. With a well-developed telecommunication infrastructure and deep integration into global economic networks, South Africa is better positioned than any other African nation to take advantage of growth opportunities in information technology industries. Meanwhile, the country's recent history of mass-based social struggle and newly formed democracy creates greater pressure on the government to prioritize equity. Nonetheless, it is still not clear to what extent it is possible to balance growth and equity in ICT policy in this context.

What strategies has the South African government developed to promote growth in information technology industries while simultaneously addressing the social problems associated with ICT? What is the likelihood of these policies having a significant impact? This chapter addresses these questions through examining two critical policies related to ICT in South Africa: first, the South African Information Technology Industry Strategy (SAITIS) Project, a broad consultative policy development process that produced an integrated strategic framework for the ICT sector in November 2000; and second, the South Africa's National Skills Development Strategy and its implications for skills development in ICT occupations.[1]

In examining these two policies, this chapter argues that despite a stated concern with socio-economic development and addressing inequality, there are significant problems with these government policies in this regard. The ICT sector development framework seems likely to contribute to continued economic growth in ICT industries. It pays little attention, however, to the potentially negative impacts of information technology on other industry sectors and provides little guidance on how to help mitigate the negative impacts of these technologies. Similarly, the Skills Development Strategy should provide significant new resources for expanded training opportunities in some sectors and companies. Unfortunately, it is poorly suited to the ICT sector, where rapidly changing skill requirements, networked production relations and critical tacit knowledge are unlikely to articulate well with the complex and more formalistic company-centered training systems being prioritized under this strategy.

The chapter is organized in the following way. Section two provides a context for the rest of the chapter, briefly discussing some of the social implications of information technology and the urgent need for simultaneous attention to both growth and equity in ICT policy in the South African context. Section three then summarizes the current character of information technology industries and workforce in South Africa, highlighting some of the structural imbalances that exist within the industry that hinder its growth. Section four provides an analysis of the SAITIS project, while Section five discusses the skills development process. Section six provides concluding comments.

Information Technology, Economic Restructuring and Equity

Information technology industries have been the most rapidly growing industries in the global economy in the 1990s. The United States has benefited the most from this technology-based boom, experiencing its longest unbroken period of economic expansion in the twentieth century, largely as a result of this rapid development and diffusion of ICT (Margherio et al. 1998; Mayer, Price et al. 1999).[2] The U.S. is not alone, however, in benefiting from the boom in global information technology industries. A range of countries have become critical players in the global division of labor in ICT industries, and benefited greatly from this role. This includes many countries that twenty years ago would not have been predicted to play a significant role in global ICT industries, such as Taiwan,[3] Singapore,[4] and Ireland.[5] Even poorer developing nations, such as India[6], and Malaysia,[7] have seen significant job growth associated with information technology industries, and not simply low-wage labor, but jobs that show signs of innovation and long term developmental potential.

Information technology, however, has far more impact on the economy than simply being a rapidly growing industry. Information is critical to all forms of economic activity and thus information technology has potential applications in an almost unlimited number of economic contexts. Efforts that focus simply on promoting growth in ICT industries, without addressing the broader transformations of work and employment, risk contributing to significant economic distress.[8] There are two central issues of concern for the analysis here. The first issue is the question of the relationship between information technology and job creation and job destruction. The debate often gets framed in terms of whether technology is creating jobs by improving productivity and stimulating overall demand, or whether it is eliminating jobs by replacing workers with machines and automated production processes.[9] Framing the debate in this way, however, neglects the fact that the rapid diffusion of information technology has speeded up *both* the creation and destruction of jobs. The increasingly important role of information and knowledge in economic activity of all kinds increases the importance of innovation in the competitive success of firms and industries. As a result, the 'churn-rate'[10] of jobs and firms has increased, resulting in a higher-risk economy (Mandel 1996). Policy related to information technology must recognize that there will be simultaneous job destruction throughout the economy at the same time as job growth is occurring.

The second issue has to do with skills development. It has now become widely accepted that "life-long learning" is essential in the new, information-based economy, and nowhere is this more true than in information technology industries (Paye 1995). This obviously creates substantial challenges for skills development systems which are typically rooted in an older industrial paradigm in which the formal education system was designed to provide people with the basic education they needed for entry-level position with most of the skills needed for advancement learned in the workplace. In addition to the challenges for the training system, the need for life-long learning also creates serious challenges for workers in the form of skills obsolescence. Rapidly changing technology, along with rapidly changing product markets and intense competition, means that workers face a more highly risky labor market, even if they have advanced skills (Benner, Brownstein et al. 1999). Adjusting to these rapid changes, for both training systems and for workers, is difficult. It requires a recognition that while formal, certifiable skills are important, social networks and the circulation of tacit knowledge are even more important in ensuring both successful industrial development and successful career outcomes (Nonaka and Takeuchi 1995; Ernst and Lundvall 1997; Lawson and Lorenz 1999).

The key concept in both these areas—jobs and skills—is creative destruction. Jobs and skills both grow and die at a more rapid pace than in the past. While this creates tremendous opportunities for economic growth, innovation and change, it also makes workers highly vulnerable to these rapid changes in this new, 'high-risk' economy (Mandel 1996). This is especially important in a context like South Africa, with its glaring economic inequality, now estimated as the third most unequal society in the world.

South Africa: Economic Restructuring, Employment and Equity

South African has gone through a significant period of economic restructuring in the post-apartheid era, as it has rapidly integrated into the global market. One of the most striking aspects of this economic restructuring is the significant decline in employment levels, which declined by nearly 9% between 1994 and 1999 (see Table 1). Unemployment in 1999 stood at 36.2 percent.

There is a popular perception that the cause of this employment decline is simply the result of increased trade—that foreign competition is destroying jobs in South Africa's domestic industries. In fact, however, evidence seems to point to technology, and particularly information technology, as having a much greater impact on employment than trade itself. Edwards (2000), for instance, compares four demand side factors in explaining changes in occupational and sectoral employment: domestic demand expansion; export expansion; import substitution; and technological change. In the 1993-1997 period, he finds that of the total employment decline of 3.63%, technological change resulted in a –14.82%, which overwhelmed the employment gain of 10.23% attributed to growth in domestic demand, and the net employment gain of 0.96% attributable to trade. More importantly, this technological change primarily impacts lower-skilled occupations, as evidenced in Table 2. Thus, there have been significant employment declines in labor-intensive industries, and a related shift from low-skilled to higher skilled occupations (Bhorat and Hodge 1999; Edwards 2000).

Table 1: Employment in the Formal Non-agricultural Business Sector (1000s)

Industry	1994	1997	1998	1999	Change	% Change
Mining	601	524	443	416	-185	-30.8%
Manufacturing	1,432	1,374	1,321	1,291	-141	-9.8%
Electricity	40	39	37	34	-6	-15.0%
Construction	355	301	255	225	-130	-36.6%
Trade	740	786	831	884	144	18.9%
Transport	285	268	249	238	-47	-16.5%
Finance	196	221	214	207	11	5.6%
Community/ social services	1,607	1,577	1,563	1,497	-110	-6.8%
Total	5,256	5,091	4,914	4,792	-463	-8.8%

Source: Statistics South Africa, October Household Survey, 1999, reproduced from (NALEDI 2000)

Table 2: Occupational Impact of Economic Growth, 1993-1997 (% of 1993 Employment)

Occupational Category	Final Demand	Exports	Import Penetration	Net Trade	Technology	Total Employment Change
Skilled	1.39%	0.75%	-0.74%	0.01%	1.70%	3.09%
Semi-skilled	1.91%	1.20%	-1.32%	-0.12%	-3.63%	-1.84%
Unskilled	1.77%	1.23%	-1.13%	0.10%	-5.18%	-3.31%
Elementary	4.97%	2.01%	-1.03%	0.98%	-9.66%	-3.72%
Total	10.23%	5.28%	-4.32%	0.96%	-14.82%	-3.63%

Source: Adapted from (Edwards 2000)

Thus, while technological change contributed to 1.7% growth in skilled occupations, it resulted in a decline in employment in all other categories, particularly impacting unskilled and elementary occupations. Thus, while information technology does provide important growth opportunities for the South African economy, including long-term important employment opportunities, it is also leading to serious restructuring with decline of employment in many sectors of the economy.

ICT Sector in South Africa[11]

In contrast to the overall economy, the South African ICT sector has grown rapidly over the past decade. Between 1997 and 1999, for instance, it is estimated that growth rates were over 17% per year (Table 3). The combined market for ICT hardware, software and services (excluding Telecommunications Services) is estimated to be R34.5 billion in 2001. South Africa's R&D expenditure was estimated at about 0,8% of GDP (R 5,72 billion / US$942 million) in 1997, a slight recovery from the stagnant economy of the 1980s. Though below most OECD countries (2,5 3% in Japan, Germany and the USA), this compares fairly well with many newly emerging countries and developing countries (Taiwan 1,1%, Mexico 0,6%), and is well in advance of other African countries.

Table 3: Annual Growth in IT Market, South Africa, 1997 2001

Sector	1997	1998	1999	2000*	2001*	Projected value2001 (Million)
Computer Hardware	19%	13%	13%	12%	13%	R 15,200
Software	22%	18%	21%	19%	20%	R 6,700
IT services	22%	20%	21%	20%	21%	R 12,600
Total ICT	20%	17%	17%	16%	17%	R 34,500
Total GDP	2.5%	0.7%	1.9%	3.5%		

Source: SAITIS (2000) * Projections

While South Africa has world class technology development and infrastructure in certain sectors of the ICT industry, it is also highly dependent and distorted in other sectors and in other ways. Within the computer hardware industry, for instance, imports are high and local production is primarily limited to final assembly.

There is essentially no local hardware component manufacture of any significance. South Africa does have some greater capacity in the manufacture of telecommunications equipment. This is largely due to heavy state support, which between 1958 and the early 1990s, took the form of a series of long-term contracts between the South African Posts and Telecommunications (SAPT) and a small number of firms. SAPT offered exclusivity of supply and guaranteed a minimum number of sales per year, resulting in relatively high local content levels. This lasted at least up until the late 1970s when electro-mechanical technology started to be replaced with digital, which began to place South Africa at more of a disadvantage (Kaplan 1990). South Africa has not been very successful in exporting telecommunications technology, though it is now facing some expanded opportunities within the region.

The telecommunications industry is in the midst of significant restructuring, which is likely to transform opportunities in the coming years. Telkom is a government-controlled parastatal that currently holds a monopoly over fixed line telephony, but one which is in the process of being phased out as a result of the 1996 Telecommunications Act. This is occurring in a staged process, but on the condition that Telkom meet certain targets for providing telecommunications infrastructure to disadvantaged areas that may not be serviced in a competitive environment. One component of the deregulation of the sector is the privatization of Telkom itself. In May 1997, a 30% share of Telkom was sold to a combined consortium of SBC Communication of the U.S. and Telekom Malaysia. An additional 10% share will be sold before the end of 2001 in a public offering. In addition, the Department of Communications has announced plans for licensing a second national operator for land lines, with plans to assess the feasibility of additional operators by 2005.[12] The cellular telephone industry in South Africa has exploded in recent years, growing to over 2 million subscribers in 2000, from a base of nothing in July 1994. South Africa has two mobile cellular operators Vodacom and MTN, in 1994. There were about 5m cellular customers by mid-2001. Vodacom is owned by Telkom (50 percent), Vodafone Airtouch (31.5 percent), Rembrandt (13.5 percent), and black empowerment group Hosken Consolidated (5 percent). IT has

60 percent of the market. MTN is owned by a combination of media interests (primarily M-Cell), the state-owned rail company, and black empowerment groups. A third cellular license was awarded in April 2000, but following widespread disquiet about the decision, including accusations of corruption and preferential treatment, the decision was under review by mid-2001.

In the software industry, again most of South Africa's technology is imported. The sale of imported packaged software applications alone accounts for more than 50% of total revenue of software sold and distributed in South Africa. Operating system software and application development tools are also largely imported, though with most custom development and package adaptation performed locally. The services component of the ICT industry is dominated by locally-based firms, though many of whom are also owned by or associated with foreign firms.

South Africa has a well-developed Internet infrastructure in business and academia, and its degree of connectivity places it in the top twenty-five in the world. After rapid expansion in the mid-1990s, however, the pace of expansion has slowed substantially. The growth in the number of internet host domain names registered in South Africa (.za) has slowed to less than 20 percent a year (see Table 4), while a recent survey estimating the number of internet users in South Africa still places it at less than 5 percent of the total population. Though many firms express plans for expanding their use of the Internet and particularly use of e-commerce, firms in many sectors of the economy have been slow to take advantage of the new technology (Moodley 2001a; Moodley 2001b).

Table 4: South African Internet Statistics

	1997	1998	1999	2000	2001
.za Domain Names	99,284	122,025 (23% growth)	144,445 (18% growth)	167,635 (16% growth)	187,649 (12% growth)
Estimated Internet Users	700,000 (1.6% of pop.)	800,000 (1.7% of pop.)	1,622,000 (3.7% of pop.)	1,820,000 (4.2% of pop.)	N/A

Sources: Domain Names from Internet Software Consortium (http://www.isc.org/). Internet Users from NUA Internet Surveys (http://www.nua.ie/surveys/how_many_online/africa.html)

South Africa does have a highly prominent global success story in the Internet industry. Thawte Consulting is a small firm focusing on security certification on the Internet. It was established in 1996 in Cape Town by Mark Shuttleworth, who essentially dropped out of an Honors Information Systems degree at the University of Cape Town in order to start this company. The need for secure Internet certification obviously expanded rapidly with the expansion of the Internet in the 1990s. The particular technology developed by Thawte Consulting rapidly became accepted in global markets, and was compatible with all the major browsers and software packages. By 1999, Thawte had captured 40% of the global market for secure Internet certification, with sales growing at nearly 300% per year. At that point, they were bought out by the market leader, VeriSign of California, in a stock swap valued at $575 million shares when it was completed. This made Mark Shuttleworth the first and still the only South African Internet billionaire (in Rand). VeriSign has continued to maintain the South African development office and recently announced an expansion in employment in the Cape Town area (Wood 1999).

This success of a highly profitable product in global markets, demonstrates the potential of South Africa's ICT sector, but in general South Africa occupies a much more dependent role in the global ICT division of labor. The largest ICT companies on the Johannesburg Stock Exchange[13] account for approximately 80% of the market capitalization of all ICT companies in South Africa, and they all generate a significant amount of their revenues from the sales of services and products overseas. The majority of these companies' activities, however, are not in product development, but are in marketing and providing value-added integration services for products developed overseas, thus limiting their potential for innovative growth.

In addition to the challenges of promoting a truly innovative, globally competitive ICT industry, South Africa also faces the daunting challenge of addressing the significant racial disparities that exist in the industry.

Racial Composition

The IT industry in South Africa is still predominantly characterized by white-owned companies and by a white professional workforce. There are some signs that black-owned companies are growing rapidly, partly facilitated by government affirmative action policy providing access to government ICT contracts for black-owned companies. In their 1999 Black ICT Profile Monitor Report, for instance, research firm Forge Ahead BMI-T profiled 120 IT companies they identified as having at least 51% black ownership. Total turnover amongst these companies in the black ICT sector rose from R1.6 billion in 1998 to R4.5 billion in 1999, accounting for an estimated 15% of the total turnover in the ICT sector.[14]

This figure is somewhat misleading, however, since the largest company, which accounted for R2.3 billion, was Kunene Technology Limited an investment holding company, which had investment interests as a 'black empowerment partner' in a number of other predominantly white-controlled technology firms, including Grintek, Siltek and PQ Holdings (later renamed Comparex holdings).[15] The next largest firm, which can genuinely be categorized as a black-owned and operated firm, is Kwetliso,[16] which provides IT training, web site design, and IT reselling services and had turnover of over R100 million in 1999.[17] Another larger black-owned and operated firm is Lechabile,[18] with turnover of R80 million, gains most of its revenue from value-added hardware reselling services, providing installation, system integration and training services.

The majority of black-owned firms, however, are quite small, with 54% of companies having less than 20 employees, and only 55% having a web site (in 1999). Despite the small size and number of black-owned IT firms, it still represents a rapid growth from the apartheid days. For example, of the 180 black-owned firms in the Forge Ahead-BMI-T database, only 9% were established prior to 1994. Over half (59%) were established in the 1997-2000 period.

The workforce in IT occupations is similarly dominated by whites, with only minimal representation of Africans. As Table 5 shows, only 13% of total workers in skilled IT occupations in a recent survey were African, despite accounting for 78% of the total population of the country. Interestingly, Asians are significantly over-represented in IT occupations, accounting for 8% of workers while accounting for only 2.5% of the total population.

Table 5: IT Skill Domains by Race

IT Domain	African	Asian	Coloured	White
Computer Operations	17.3%	9.9%	10.3%	62.5%
Data Communications and Networking	17.5%	5.8%	11.1%	65.6%
End User Computing	15.4%	9.5%	8.5%	66.7%
Hardware and Computer Architecture	16.2%	7.3%	8.6%	67.9%
IS and IT Management	11.5%	7.4%	6.2%	74.9%
IT Education, Training and Development	7.7%	4.0%	2.5%	85.8%
Information Technology Sales and Marketing	9.4%	13.6%	7.4%	69.6%
Systems Development	10.2%	8.2%	7.0%	74.5%
Other	17.5%	8.9%	11.3%	62.3%
Total IT Domains	13.4%	8.3%	8.1%	70.2%
Total Population	*77.6%*	*2.5%*	*8.7%*	*10.4%*

(SAITIS 2000; Statistics South Africa 2001)

Location

Another challenge facing the country in regards to promoting growth in ICT is the heavy concentrated of these sectors in Gauteng, the central province that includes Johannesburg, Pretoria and the surrounding primarily urban regions. Rogerson (2001), for instance, confirming previous investigations (Hodge and Miller 1997; Hodge 1998; Rogerson 1998; Hodge and Driver 1999) found that the Gauteng region contains approximately 107,000 employment opportunities in high technology manufacturing or 73 percent of the national total. By contrast, the Western Cape, the second most high-tech region, contains only approximately 10,000 high technology manufacturing jobs or 6.8 percent of the national total. In IT services, of the national total of 64,109 jobs, 78 percent were clustered in Gauteng, with 70 percent of national IT employment (nearly 45,000 jobs) concentrated in just two magisterial districts: Johannesburg (23,684) and Randburg

(21,004) magisterial districts. The third ranked magisterial district in terms of employment opportunities is Cape Town (5,000 jobs) which has a 7.8 percent share of national IT employment. In all the major segments of the IT service economy, the dominance of Gauteng is in evidence. In terms of national employment opportunities the share of Gauteng was as follows: systems integrators (82 percent), distributors (95 percent), netware specialists (91 percent), software development (93 percent) and Internet service providers (68 percent). Outside Gauteng, local specializations appear to exist only in the Western Cape for Internet development and Internet service provision. At a fine-grained level of analysis, Johannesburg and Randburg are by far the leading centers for most aspects of the IT service economy; the only notable exceptions are the relative importance of Cape Town in Internet services (second to Johannesburg) and the leadership of the highly specialized area of Internet development by enterprises based in Pretoria and Cape Town-Stellenbosch.

Promoting the decentralization of information technology industries has not been very successful anywhere in the world, but in the absence of efforts to extend the industry, it is likely to reinforce the geographic inequalities in South Africa between urban and rural areas, and between Gauteng and the rest of the country.

Government Initiatives to Promote ICT Growth

There are a wide range of initiatives and policies by various national ministries in South Africa aimed, in one way or another, towards promoting growth and opportunities in information and communication technologies. These include: the Department of Communications efforts to develop policy to facilitate e-commerce, with the publication of an E-commerce Green Paper in November 2000;[19] a wide range of initiatives falling under the title "Info.com 2025", which include public Internet terminals in a wide array of public spaces and the development of a telecommunications software training institute in the Western Cape at a former Denel Aerospace Division facility in Houwteq; the establishment of the Universal Services Agency, an independent government supported agency;[20] the creation of an Innovation Fund within the Department of Arts, Culture, Science and Technology, with the purpose of generating new products, processes and services, particularly those addressed at problems "serious enough to impede socio-economic development or affect our ability to compete";[21] and the creation of the State Information Technology Agency by the Department of Public Service and Administration, to improve the management and use of ICT tools in the public sector.[22]

Obviously with this wide array of initiatives in a range of different ministries, coordination and communication between the various initiatives are a critical challenge, as well as the need for this all to fit within a broad national strategy towards the IT sector. Thus, the most significant overall initiative to promoting growth and opportunity in the IT sector is the South African Information Technology Industry Strategy (SAITIS) Project. This project is a three-year, $3.5 million policy-development initiative, originally conceived in 1995 by Jay Naidoo, then Minister of Posts and Telecommunications and formerly General-Secretary of the Congress of South African Trade Unions (COSATU). The project was sponsored by current President Thabo Mbeki, then Deputy President, and thus has support at the highest levels of government. The idea for this project emerged out of discussions surrounding the Information Society and Development conference, a joint conference of the OECD, G-7 countries and which South Africa hosted in Johannesburg in 1996. As part of the preparations for this project, the need for an overarching government strategy towards the information society was identified.[23]

The formal project itself was launched in 1999, though by this time its goal had narrowed from promoting an information society to focusing on the ICT sector: "To develop a strong South African ICT sector to contribute to sustainable economic growth, social upliftment and empowerment. It was launched as a bilateral development assistance project between the Canadian International Development Agency (CIDA) and the South African Department of Trade and Industry (DTI), with consulting firm Price, Waterhouse, Coopers providing management, research and technical assistance to the project.

From the beginning, the goal of the project was not to develop a policy in isolation, but to instead engage in a broad participatory policy development process that would engage key stakeholders around developing inputs to the policy. This included both a top-down and bottom strategy development process as outlined in Figure 1. Government, industry, unions, academia, and a range of other sectors were represented on a 37 member stakeholder Advisory Committee meeting, and a Project Steering Committee. Working groups were established on the ICT Industry, ICT Usage, Human Resources, and Innovation. In addition, a series of workshops were conducted around capital mobilization, infrastructure, international markets, women in ICT, small, medium and micro-enterprises, and management skills, along with provincial workshops designed to gain a greater appreciation for the status of ICT companies in different parts of the country. The report produced a broad "Baseline Study", which conducted research on the trends in Global ICT industries, analyzed the current status of the industry in South Africa, as well as identified broad opportunities for development of the industry.

It is important to note that in the process of focusing on developing an industry strategy, the explicit goal was to directly support and contribute to the government's macro-economic framework known as GEAR (Growth, Employment and Redistribution). This GEAR framework, despite a stated goal that includes social upliftment and empowerment, is a highly controversial policy that prioritized public sector fiscal austerity, deregulation in the process of international integration, over domestic social programs. This narrowing of social goals is also reflective in the process of developing the SAITIS industry strategy.

In November, 2000, the SAITIS project released their ICT Sector Development Framework, laying out a broad framework for development of the ICT sector. The overall framework is an ambitious agenda covering four broad goals, with a number of sub-objectives:

SAITIS South African ICT Sector Development Framework, November 2000

Goals and Objectives

Goal 1. Robust, growing and sustainable South African ICT Sector with equity

- Build local ICT Sector capacity
- Establish the Enabling Policy/Support Environment that makes South Africa an Attractive Place for ICT Development
- Build a World-Class ICT Sector Support Infrastructure
- Stimulate Export Penetration to Capture a Growing Portion of the Global ICT Market

Goal 2. Increase the Use of ICT as an Enabler for Socio-Economic Development, with Equity

- Stimulate ICT Adoption through Partnerships
- Make a Modern Information Infrastructure Available to all South Africans
- Stimulate ICT Adoption Through Government as Model User
- Bring Disadvantaged Communities into the Main Stream of ICT Development and Use

Goal 3. Knowledgeable and Growing ICT-Skilled Workforce

- Comprehensive Understanding of SA ICT Labor Market
- Establish Environment to Retain Skilled Workers
- Establish HR Development Infrastructure to Support ICT Development
- Make HR Development Integral to ICT Development in SA

Goal 4. World Class Culture of ICT Innovation

- Establish ICT Innovation, Entrepreneurs and Risk Management Culture

- Stimulate ICT R&D for Local and Global Markets
- Increase ICT Technology Transfer for R&D to Commercialization
- Ensure Intellectual Property Protection.

Source: (SAITIS 2000)

More detailed strategies and specific action items are laid out in more detail in the framework document. The overall framework is obviously a broad, strategic vision, and includes strategies that will take many years to address, as well as those that can be addressed in the more immediate term. In an effort to prioritize these issues, to ground these quite broad goals and strategies into more concrete actions, the Framework also specifies a number of projects to be pursued immediately. They include the need to, first, establish an effective governance and organizational approach: The SAITIS project itself will end when funding stops at the end of 2001. Discussions are underway to develop an appropriate institutional framework for pushing these initiatives through, to guide implementation of the Framework, including day-to-day strategy implementation activities. The exact governance structure is not yet clear, but clearly the Department of Trade and Industry will play a leading role, and the goal is to integrate it into on-going activities of the Department in as organic a way as possible, along with ensuring multi-stakeholder involvement.

Second, conduct an impact analysis of current ICT initiatives. This involves assessing the impact of other ICT initiatives currently underway in South Africa, in an effort to document the impact of SAITIS related initiatives and identify potential synergies with other initiatives underway. Building the ability of the various initiatives to network with each other, through greater knowledge of each other's activities, and building opportunities to interact, is clearly part of the goal. Third, accelerate ICT cluster development: The SAITIS project is in the process of gaining additional information about ICT clusters in Gauteng and the Western Cape, with the goal of identifying a specific project in each area that can bring together key players in business, industries, university and government R&D to promote new developments in the cluster.

Fourth, develop an ICT portal. The concept here is to provide Internet based information, services and applications for ICT companies to assist the growth of the ICT sector. The exact content of the site is still in the process of being developed, and the SAITIS project is in the process of developing a survey to ask potential users to elaborate their needs for a portal. One initial need identified is a job linking service, with companies listing job openings and individuals listing their CVs. Other likely services include news on the ICT sector, information about financing opportunities, linkages to training providers and tertiary education programs, and information about networking organizations and professional associations. The portal was expected to be launched in July 2001, with the IISA (Information Industry of South Africa) taking responsibility for on-going development.

Fifth, improve the collection of ICT labor market statistics. There is an urgent need to develop better statistics on trends in the ICT labor market, in order to help make decisions on human resource development and planning. This will involve working closely with the Human Science Research Council (HSRC), Statistics SA and the Information Technology Sectoral Education and Authority (ISETT) to improve data collection and analysis processes. Sixth, establish an ICT Youth Internship Program: The purpose of this initiative is to provide opportunities for recent qualified graduates to gain work experience in the industry and thus improve their skill sets. Finally, promote Black ICT SMME skills development. This is focused on helping Black small and medium firms compete more effectively for government ICT work by providing developmental assistance to form consortiums, prepare bids, and manage projects.

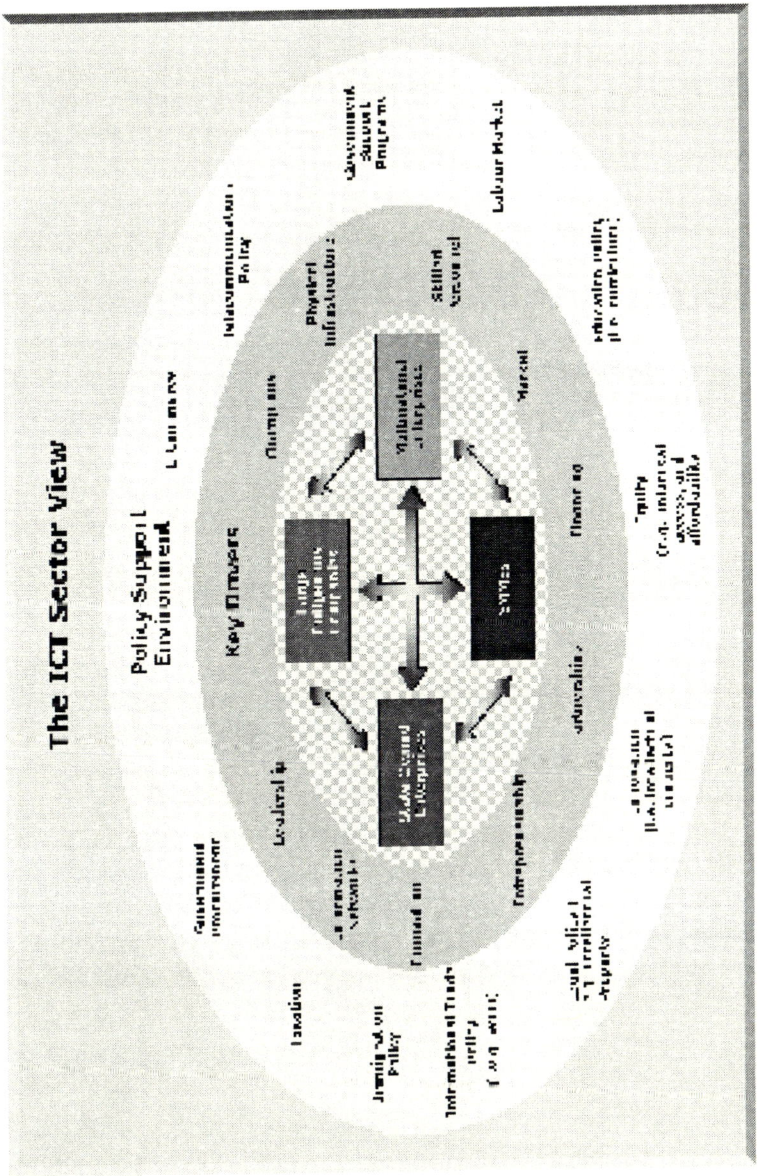

Figure 2:

One thing that is clear, however, is that very little attention is being paid to any of the potentially disruptive impacts of information technology. The portions of the strategy focused on stimulating usage of ICT in non-ICT sectors of the economy and amongst community organizations make no reference to any disruptive elements this strategy may have. The work-plan for the final year of the project does include brief reference to a project for retraining retrenched workers, but work on this project had not gotten underway, and it was clearly a low priority. In discussions about promoting the adoption of ICT in other sectors, no mention is made of trying to do this in ways that expand and promote employment, rather than replace employees.

Skills Development and Levy System

Human capital is obviously a central component of any strategy to promote growth in the high-tech sector. The National Skills Development Plan is an ambitious effort to improve South Africa's skills development and training system. Its primary mission, "to equip South Africa with the skills to succeed in the global market and to offer opportunities to individuals and communities for self-advancement to enable them to play a productive role in society" clearly reflects the dramatic need South Africa has for upgrading the skills of the workforce. The Skills Development Act of 1998, along with the Skills Development Levies Acts of 1999, created a system for promoting training in the country as a whole. The central purpose of the legislation was to promote increased training and skills development in the economy, by simultaneously creating an incentive for employers to invest in training, and to provide a pool of money for the public sector to develop training in strategic sectors and for particular targeted populations. A National Skills Authority was established in 1999 and is made up of representatives from organized business, labor, government and organizations that reflect community and training provider interests. Its purpose is to advise the Minister of Labor on broad skills development strategies, liaise with Sectoral Education and Training Authorities and to monitor and evaluate implementation of the skills development strategy.

At the core of the Skills Development Strategy legislation is a "skills levy" that employers must pay every month (see Figure 3). When the levy was first being debated, COSATU originally proposed that the amount of the levy be 4% of the total wages paid by firms, but in the final passage of the act this was reduced to 0.5% in the first year, and 1% in subsequent years. After being collected by the South African Revenue Service, these funds are transferred to the Department of Labor which directly administers 20% through a National Skills Fund (NSF). The purpose of the NSF is to develop training programs that address youth, the unemployed, retrenched workers and other targeted communities. The remaining 80% of funds collected under the levy is distributed to 25 Sector Education and Training Authorities (SETAs) who are responsible for promoting training within their particular sector of the economy. In the original design of the legislation, it was anticipated that a majority of the funds transferred to the SETAs would eventually be returned to employers directly, to support specific training initiatives. There are several steps employers must go through in order to receive these funds. They must first appoint a Skills Development Facilitator at the worksite whose responsibility it is to oversee skills development programs. The next step is to develop a Workplace Skills Plan (WSP), laying out plans for providing skills to employees in the workplace. With these two steps, employers are able to claim back 15% of the funds they have paid for their levy. Once the WSP has been implemented, employers are then able to get reimbursed another 50% of their skills levy. Thus, in principle, if employers engage in this system fully, 65% out of the 80% funds transferred to the SETA's would be returned directly to the employers who paid them, with the remainder going to support the activities of the SETAs themselves. This activity includes administrative expenses (including administering grants), developing a Sector Skills Plans to identify training needs in the sector, and certifying eligible training providers and courses within the sector, thus ensuring the quality of training providers throughout the country. Employers apply to their SETA for grants—the SETAs administer the grants. The SETAs also have the ability to provide discretionary grants, either directly to employers or to other agencies, for learnerships, skills programs or sector priorities. These grants are not limited by the amount an employer has paid in the skills levy.

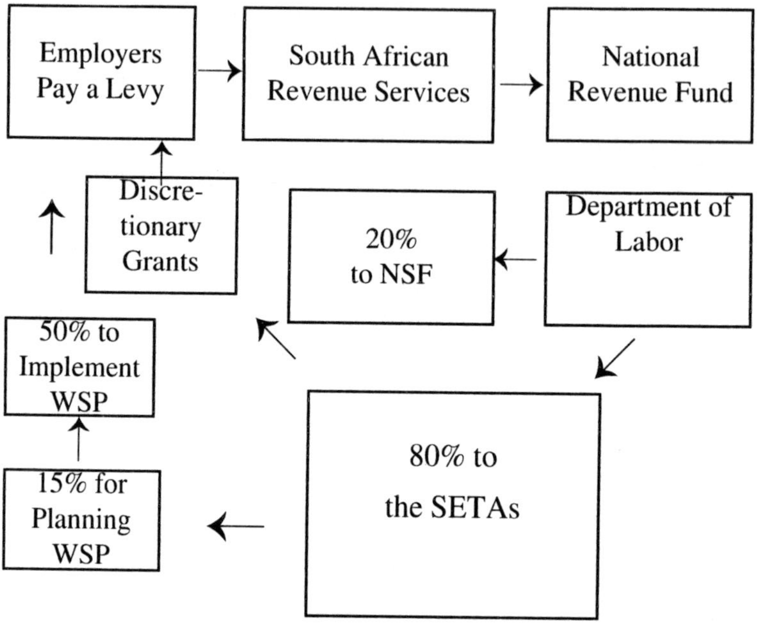

Figure 3: Skills Levy System

The overall system is driven by the fundamental goal of providing employers with an incentive to provide training. Even for those employers who choose not to provide training themselves (and thus get partially reimbursed for their skills levy), they are still providing resources for expanded skills development, particularly within their sector but also within the economy as a whole, through the NSF. The approach also recognizes that many skills sets are specific to specific economic sectors, and building expertise within sectors helps expand the knowledge of how to effectively develop training programs.

The legislation was only implemented in March 2000, and as such it is too early to fully evaluate the effect of the legislation. Nonetheless, it provides a significant pool of resources specifically devoted to skills development, and seems likely to have some positive impact in that area. There are clearly a number of problems with the program as well, some related to the difficulties of implementing a complex system, but some having to do with the very structure of the program itself.

Implementation

A clear problem with the entire system is its complexity. One prominent Department of Labor official described it in this way:

> It is like we have the brand-new, latest model BMW, and, by the way, there is no petrol in the tank, we don't have any qualified drivers, and the road is full of potholes. We just don't have the capacity to implement this system. (Interview, March 2001)

The act requires employers to choose membership within a single SETA. Prior to the setting up of the SETAs, there were extensive negotiations between employer bodies and trade unions about the appropriate industries to include in each SETA. Employers are clearly finding it difficult to figure out which SETA they belong to, or to choose between which SETA they should belong to. In many cases, companies might legitimately be located in multiple SETAs. South African Airways, for instance, would most logically located in the transportation SETA. But the training needed in their telephone service centers would be more closely linked with the ISETT (Information Sciences, Electronics and Telecommunications). Many employers, especially the smaller ones, have clearly not spent the administrative time needed to determine which SETA they want to belong to, and thus are simply treating the levy as a tax with no intention of requesting reimbursement for training. An indication of the problem of specifying SETAs is the fact that a large portion of the funds collected from the Skills Levy are not specified for any SETA. This so-called "SETA zero" actually has more revenue than any other SETA, amounting to more than R100 million in March 2001.

The system is also complex in terms of the certification of particular skills. These are designed to fit within the National Qualifications Framework, based on eight levels of qualifications, designed to ensure that qualifications are recognized throughout the country. SETAs in many cases are having problems in the administrative task of not only determining skills needs within their sectors, but developing the appropriate qualifications and certifications to recognize those skills. This is made even more complex in industries, like information technology, in which skill requirements change so rapidly.

Some problems of capacity and implementation are likely to be addressed, at least partially, as the policy is developed over time, with the SETAs, employers and unions gaining more expertise in administering the systems. Nonetheless, there are a number of more serious problems within the entire structure that have yet to be addressed and that potentially undermine the system's ability to effectively improve skills development in the country.

Employers' Role and Worker Representation

The Skills Development System is problematic in the way it represents both employers and workers. In relation to employers, the system reinforces training in large employers in industries with high wages, typically those industries in which there is already a high level of existing skills development. Small employers in all industries are disadvantaged in the system. They frequently lack the administrative resources to develop the appropriate proposal for workplace skills plan, and to get reimbursed. While larger employers will clearly invest the amount of resources needed to reclaim resources through levy grants, many of the smaller employers will simply treat it as a tax, since it would require more time and cost than they would be able to reclaim. In information technology industries, where there are many small firms, training is unlikely to be developed by firms themselves, and yet the entire system is geared towards employers taking the initiative in this area.

In terms of representing workers, the structure of the SETAs also minimizes the opportunities for workers to be formally represented in the process. In principle, unions have to be included in the governance structure of all the SETAs, and prior to being approved, the SETAs had to demonstrate that trade unions had been involved in the process of approving the SETAs business plans. The SETA process, however, was setup entirely separate from the bargaining councils that already exist in many industries. Thus large union like NUMSA (Metalworkers), is spread across multiple SETAs. They have representation in the electronics sector, a portion of which is included in the ISETT

SETA with information technology industries, and thus NUMSA has representation on ISETT SETA. In practice, however, they entirely lack the capacity to simultaneously engage in training issues in multiple SETAs that is emerging entirely outside of their bargaining issues. Furthermore, the SETA process does not build on existing unionization at all, and in fact allows multiple forms of employee representations. Thus, an organization like the Black IT Forum, whose membership includes owners and middle-managers, as well as IT professionals can be in the governance structure representing employees. In addition, despite being a system designed to improve individual's skill levels, it provide remarkable little resources for individuals to be able to get training they may need to improve their employment opportunities, if such advancement involves moving away from their current employer.

Lack of Attention to Multi-firm Careers

With increasing levels of out-sourcing and the use of contract and temporary employees, workers are moving much more frequently from company to company, rather than spending a lifetime with the same firm (Klerck 1994; Bezuidenhout 1996; Kenny and Webster 1998; Kenny and Bezuidenhout 1999). Internal career ladders are decreasing, yet this trend is not recognized within the SETA process. Employers will only invest in training for their workforce if they are confident they can gain the returns from that investment in the form of a more trained workforce. Frequently, however, with employees moving to other firms, such investment is lost. Thus, there is greater need, especially within the information technology sector, to promote the strategic intervention of training organizations that can take a multi-employer career path perspective.

Furthermore, the SETA system provides few resources for individuals to access training resources. The system is designed in the first case for employers to be reimbursed for training. The system does provide some opportunities for training providers to develop strategic interventions to promote non-firm specific training, but this is limited and so far has not been widely developed. Opportunities for individual workers to apply for bursaries to improve their own training opportunities are not prohibited in the training system, but are clearly not a high priority and have not been significantly developed.

Neglect of Tacit Knowledge

Numerous studies of skills development and learning processes, as well as of new forms of competition and innovation in industry dynamics, recognize the critical importance of tacit knowledge, both in promoting the advancement of individuals, and in improving the competitiveness of firms (Nonaka and Takeuchi 1995; Howells 1996; Ernst and Lundvall 1997; Lawson and Lorenz 1999). This tacit knowledge is built through collective interactions of individuals, through social networks and through the creation of 'communities of practice'. The South African skills development strategy, however, is primarily designed around providing formal certification of individuals, rather than promoting this collective development of social capital. One reason for this is clearly to provide workers with greater leverage in the labor market, by certifying their skills and thus giving them greater ability to negotiate higher wages for themselves. Ultimately, however, the success of workers in the labor market is not limited to formal certification, and is actually more directly related to their ability to act collectively, and have that social capital.

Conclusions

South Africa's information technology sector is well positioned to continue its rapid growth of recent years. Government policies to promote growth in this sector, particularly through the SAITIS project, seem likely to contribute in important ways to the acceleration of this growth. With skills development such a critical component of growth in high-tech sectors, it is also encouraging to see the ambitious skills development initiatives being put in place in the country. Nonetheless, as this chapter has tried to argue, attention to growth alone is not adequate in a country in which inequality is so high, and in which such a large portion of the population is being negatively impacted by the rapid dissemination of ICT. In this environment, policies must also specifically address the negative consequences of ICT, and attempt to mitigate their negative impact. In this area, as this chapter has argued, policy in South Africa is more limited. The SAITIS project almost entirely neglects any possible negative implications of the diffusion of ICT, and minimizes opportunities for implementing ICT in ways that can maximize economic competitiveness without job loss.

Skills development in the ICT sector, especially for disadvantaged sectors of the labor market, is of critical importance. Yet the Skills Development Strategy developed in the country seems highly problematic in the development of these training programs. Neglecting the small business sector, and neglecting the development of tacit knowledge and the creation of social networks, this initiative is missing valuable opportunities to improve skill development in the country. In the U.S., policy initiatives are increasingly recognizing the importance of supporting the organizational development of associations of workers—either through unions or through professional associations—as a critical component of development training programs (Parker and Rogers 1995; Benner 2000). Such initiatives would likely prove beneficial in the South African context as well.

ENDNOTES

1 This chapter is based on a detailed review of appropriate government and industry documents, policy statements and industry reports, along with a series of 40 interviews with key stakeholders conducted in July 2000 and March 2001.

2 Silicon Valley, the core of innovation in global information technology industries, has particularly benefited from this economic boom, becoming perhaps the greatest wealth generator in history, Saxenian (1994); JV:SVN (1999); Kenney (2000).

3 Callon and Asia/Pacific Research Center 1995; Hsu and Saxenian (2000).

4 Callon and Asia/Pacific Research Center (1995).

5 O'Riain (1997); Sweeney (1999); O'Riain (2000).

6 Heeks (1996); Kenniston and Kumar (2000); Parthasarathy (2000).

7 Gourevitch, Bohn et al. (2000); Rasiah and Best (2000).

8 But see ILO (2001); Benner (2002).

9 Kaplinsky (1987); Bessant (1989).

10 This refers to the rate of both creation and destruction of jobs and firms, see Atkinson (1998).

11 Much of the analysis in this section follows Miller (1999) and Day and Miller (2000).

12 Department of Communications, Telecommunications Policy Directions, 20 March 2001 http://docweb.pwv.gov.za/docs/policy/telpoldir.html.

13 M-Cell, Dimension Data, Comparex, Datatec, Altron, Softline, Ixchange and Computer configuration Holdings.

14 Figures were provided in an interview with Forge-Ahead BMI Research Analyst Teboho Mokoena on July 13, 2000. Full research report is available from http://www.bmi-t.co.za, and is reported on at http://196.30.226.221/Sections/monitor/2000/0009140945.asp?CiRestriction=kwetliso

15 http://www.grintek.co.za, www.siltek.com, www.comparex.co.za

16 Kwetliso means empowerment in Southern Sotho. http://www.btimes.co.za/99/0228/comp/comp04.htm

17 www.kwetliso.co.za, www.mandla.co.za

18 www.lechabile.co.za

19 www.ecomm-debate.co.za, Despite having a very broad definition of e-commerce, the policy paper deals primarily with legal and contractual aspects of e-commerce, including contractual and trade laws, taxation, intellectual property rights, consumer protection and security. Significant social issues, such as those around implications for job creation/destruction, use of information technology in the workplace, and representation of workers and disadvantaged sectors of society, are almost entirely neglected in the department's approach. They have been heavily criticized for this, especially by COSATU in a formal response to the green paper. http://www.cosatu.org.za/docs/2001/ecommgp.htm.

20 www.usa.org.za. At its inauguration in November 1997, the USA promised to deliver 150 new centers by March 1998, and a further 100-150 centers every year. The project, however, has largely failed to deliver on intended delivery targets, and by December 1999, only 28 centers were in operation, of which 12 were actually set up by the International Development and Research Center of Canada, see Kaplan and Kaplinsky (2000). Services in existing centers are also limited. Reasons for failure in the program include poor management and disputes within USA, turf battles with the Department of Communications, and lack of co-operation from Telkom, which has not been co-operative in connecting call centers, see Stavrou, et al. (2000); Benjamin and Dahms (2001); and Zongo and Benjamin (2001).

21 Key objectives of the fund are to reallocate funds from historical patterns of government sciences towards key issues of competitiveness, quality of life, environmental sustainability and harnessing IT. The first round of funding in 1998/99 chose 19 projects out of 177 applications for R30million per year, with many projects gaining multi-year funding. A total of R11 million was allocated for five projects promoting the "information society", including: development of a multi-media culture preservation software product for tourism, education and publication awareness; a project examining use of information technology in the education system; development of a high speed microwave radio modem; creation of a national health information network; and an integrated GIS project including agricultural, marine, and biodiversity monitoring, assessment and decision support systems, see DACST 1999.

22 www.sita.co.za

23 Interviews, Charley Lewis, and Peter Benjamin.

CHAPTER 6

THE EMERGING ICT INFRASTRUCTURE AND APPLICATIONS IN UGANDA

Nora Kasirye Mulira

Introduction

Information and Communication Technology (ICT) has emerged to offer unprecedented opportunities for development. It is generally recognized globally that ICT is the tool with a potential to promote economic and social development in Africa, a region associated with underdevelopment. To its champions, in such a vast and sparsely populated continent ICT can assist to reduce isolation by providing an immediate link for inter-regional communication as well as global access. Furthermore, ICT allows Africa to bypass many of the traditional stages and old technologies of development, which can accelerate its development.

Uganda presents challenges and opportunities both in terms of developing an ICT infrastructure and using ICT to drive development because of two critical factors. First, we note that the overwhelming majority (about 80 per-cent) of Uganda's populations of about 21 million people live in rural areas, many of whom derive their income from small-scale agricultural production. This raises questions about the potential impact of ICT in terms of rural connectivity, rural health, rural education, diffusion of improved farming methods, provision of government services, etc. Second, there is the question of the role that a significantly liberalized telecommunications policy environment (which is one of the most liberalized in Africa), combined with the government's strong "pro-development" stance has played in the implementation of information infrastructure initiatives. This chapter explores the emerging ICT infrastructure and applications in Uganda, which recognizes the potential of ICT in overcoming the country's problems of poverty, isolation, illiteracy, and poor service delivery especially in the rural areas.

Infrastructure Limitations

Needless to say, there are significant complementarities between electricity and communications particularly in their commercial applications. At present, only about 2 percent of rural households have access to electricity. Under the Energy for Rural Transformation (ERT) project launched in 1999, rural electricity coverage is expected to rise to 10 percent within a period of ten years. Rural telecommunication is also very limited. By 2001, teledensity in the country, including mobile phones, was 1 line per 100 people, much of it concentrated in the urban centers. The 1997 Communication Act, which established the Uganda Communication Commission (UCC) to act as the regulatory body for the sector, also provided the basis for national duopoly in the provision of basic telephony services. The two main operators in the sector are the recently privatized incumbent, Uganda Telecom Limited (UTL), and the Mobile Telecommunication Network (MTN), which was licensed in 1998. In return for a five-year exclusive period stretching from 2000 to 2005, the two operators are expected to undertake a significant "rollout" of services to the rural areas. The operators were given until 2001 to identify those areas of the country not deemed to be commercially viable, in which they would automatically lose their exclusivity, and whose servicing would then become the responsibility of the UCC's Rural Telecommunications Fund financed by a 1 per-cent levy on all operators in the sector.

As for Internet Service Providers (ISPs), by 2001 there were about half a dozen, all of them based in the capital, Kampala. As the telephone operators rollout their operations across the country, dial-up access to ISPs in Kampala is gradually becoming feasible. Nonetheless, the high cost of cellular dial-up and the very slow speed of the resulting connections severely limit the use of the Internet outside Kampala. Until points of presence are created outside of the metropolitan area, VSATs remain the only reliable mode of Internet connection.

Emerging ICT Infrastructure and Applications

In spite of the infrastructural challenges and the lack of central coordination, a number of ICT initiatives have been undertaken in all sectors. Various government ministries and agencies, including the Ministry of Information, the Ministry of Works, Housing and Communication, the Uganda National Council for Science and Technology, the Uganda National Commission for UNESCO, just to mention a few, have addressed the issue of developing a national ICT policy. In August 2001 a draft policy was submitted by consultants from the Uganda Communication Commission (UCC) and other stakeholder organizations for discussions by the Uganda National Council of Science and Technology (UNCST) Policy Committee. The policy was later submitted to a parliamentary committee for approval. Besides these efforts at the national and official policy level, there are numerous sectoral ICT initiatives in the fields of education, health, business, agriculture, and government.

In education the notable initiatives include SchoolNet Uganda, the National Curriculum Development Centre, the Educational Management Information System, and the Makerere University ICT project. SchoolNet Uganda is a national body which seeks to support educators and learners by providing leadership, expertise, resources, and coordination, and developing effective partnerships in the areas of Internet connectivity and appropriate technology; development of online electronic content, teaching and learning resources; human resource development and capacity building; and advocacy for ICT in schools. However, this initiative continues to be bedeviled by several constraints, such as the lack of electric power, especially in the remote rural areas; high telephone costs or inadequate telephone services in some parts of the country; lack of skilled manpower to handle content development; lack of an Internet Service Provider (ISP) near some of the schools resulting in long distance dial-up connections; and lack of Ministry of Education (MOE) policy on ICT in schools. Moreover, IT teachers are overworked because they are still expected to carry their normal teaching load.

The National Curriculum Development Centre (NCDC) is the statutory body charged with the development and implementation of the primary, secondary and tertiary school curricular in Uganda. The NCDC has produced a syllabus for O-Level Computer Studies (four years). The syllabus is awaiting approval by the Academic Steering Board of the Ministry of Education. The ICT curriculum was launched in January 2001 for those schools that have computer access. The Educational Management Information System (EMIS), created in 2000 is charged with providing statistical and administrative support at the head office of the Ministry of Education Headquarters and at the district level throughout the country's forty-five districts. Specifically, the system is meant to provide education managers in the districts improved access to information, familiarize them with elements of a personnel system critical for their operations, and enable them to design the necessary financial systems. The system is currently operational with limited access from upcountry districts.

The Makerere University ICT project was undertaken in 2002 in order to bring Uganda's oldest and main university in line with global institutions of higher learning. With funding from donor agencies and development partners, such as the Swedish International Development Agency Sida/Sarec and the Norwegian Agency for Development (Norad), the university embarked on installing a university wide backbone network, which connected faculties on the main campus as well as other remote campuses. The project was driven by three main objectives. First, to facilitate both university wide internal communication (Intranets) and external communication (Internets). Second, to use ICT in pedagogical and administration functions. Third, to avail faculty and students online research and information resources.

Currently an 8km fiber optic backbone has been installed around the main campus. A wireless network provides internal back up and connectivity to the remote campuses, which are not yet linked. Email/ Internet services are prevalent in all faculty and administrative units. The main setback is inadequate computers for access by all. Makerere University has completed a systems analysis for the central systems (Academic Records ARIS, Library Information System LIBIS, Financial Management FINIS, Human resource Management HURIS). These systems will be fully operational by the end of 2003. The challenge for the university will then be to extend technology into improving the central functions of academic staff and research within and outside the borders of the university. An E-learning project to train academic staff in developing online content for instruction and assessment has been undertaken at the Technical University of Delft, funded by the Netherlands fellowship program.

In the field of health, the Ministry of Health put in place the Health Management Information System (HMIS) to coordinate and disseminate information to the districts. Its main objective is to improve access to quality health care and continuous medical education in the remote rural areas. The Ministry's website was particularly informative and was used to solicit international assistance during the Ebola outbreak in Northern Uganda in September 2000. It is envisaged that the HMIS will be linked up with Telecenters to support improved medical services through telemedicine to communities in the sub-districts. In December 2000 a Telemedicine pilot project was launched to provide connectivity from three selected sites namely Mulago Hospital (the main referral government hospital in Kampala), Mengo Hospital, and Nsambya (private hospitals). A performance evaluation exercise will be completed in May 2002 to improve service.

In the business sector, e-commerce has been slow to develop because of several constraints to do with the fact that electronic signature is not yet legalized in the country. There is also lack of credible and consistent e-payment mode. Moreover, traditional business attitudes and practices serve as inhibitors against appreciating the benefits of e-transaction. Nevertheless, e-commerce is slowly picking up, especially in the banking sector, where ATMs are now available in some private commercial banks. SWIFT as a mode of international funds transfer has been adopted within the last few years by some commercial banks including the country's leading bank, the Uganda Commercial Bank (UCB), all of it made possible by the country's steadily improved communications infrastructure.

As for the agricultural sector, as the Ministry of Agriculture adopted the policy of decentralizing its dissemination of information to the rural farmer by liberalizing the agricultural extension scheme, the potential role of ICT to disseminate agriculture information from research centers through community radio and to local telecenters was increasingly appreciated and emphasized. The Ministry aims to expand and strengthen this dissemination structure by expanding rural access to ICTs.

Finally, some government ministries and agencies have taken the lead in constructing an ICT infrastructure. Among them are the Uganda Computer Services (UCS) and the Uganda Bureau of Statistics (UBOS). The Ministry of Public Service (MPS) is involved in the process of developing both user and system specifications with a view to acquiring an integrated personnel and payroll system (IPPS). In parliament a Department of Library, Research and Information Services has been set up to conduct, among other things, research on all matters related to the operations of parliament. A website http://www.parliament.go.ug has been set up for the purpose of disseminating information on parliamentary activities and a VSAT connection has been established for electronic mail and Internet services. The judiciary has embarked on a computerization program of its registries with a view to enabling the High Court and Court of Appeal to have access to a reliable and accurate history of cases referred to them by lower courts distributed throughout the country. The expected linkages with district courts will be through telecommunications infrastructure or by courier where telephones do not work. The major problem facing the computerization program is lack of skilled manpower to use the system. Although the judges have been given some introductory courses in the use of ICTs, more resources are required to get everybody fluent in the use of the system. The President's Office, the executive branch of government, has also joined the ICT bandwagon. It has set up a government network (GOUEXECNET) linking the office of the president's internal departments and the public.

Assessing the Successes and Challenges

Previously, radio was the main channel for disseminating news and information, including government policy. In recent years there has been an enormous expansion of both the old and new information and telecommunication technologies. For example, in addition to Radio Uganda which provides 100 percent coverage of the whole country and Uganda television which covers 60 percent of the country, the number of private FM radio stations increased from 14 in 1996 to over 50 in 2001, while television stations increased from 4 in 1996 to 13 in 1999. It is estimated that there are between 300-500,000 TV sets in the country. The importance of FM radio coverage cannot be overemphasized because, in the short term, it remains the most effective way of disseminating information to the rural areas. Coverage through FM stations, especially through the phone-in programs, has raised public debate on national issues to unprecedented levels. The concept of the openness and accountability of government has empowered even remote communities, which have learnt that citizens can call public servants to account.

Clearly, Uganda's information and telecommunication infrastructure has grown considerably since the mid-1990s, especially following the introduction of competition in national service provision. Table 1 below gives some data summarizing sector expansion since 1998. There has been remarkable growth in the national telecommunications sector. The Uganda Telecom Limited (UTL), formerly state-owned and the company with the widest national coverage, was privatized in 1998. The second national operator, the Mobile Telephone Network (MTN), covers most of the country's districts. While it is clear that there has been a lot of growth in the sector, the problems of unequal access remain severe. The number of lines, for example, has grown three fold from about 70,000 in 1998 to about 250,000 in 2000 (190,000 mobile phones, and 60,000 fixed lines, see figure 1 below). The cellular networks have been a boon to rural communications.

Table 1: Telecommunications Sector: Comparative Figures For October 1998 and February 2001.

	Service Provided	DEC 1996	OCT 1998	DEC 1999	FEB 2001
1	Wired telephone lines (UTL)	45,145	55,749	57,913	58,880
2	Fixed wireless lines (MTN)	Not Operational	447	148	932
3	Fixed wireless pay phones (MTN)	Not Operational	0	*200*	1650
4	Mobile cellular subscribers (MTN, CelTel, and UTL)	3,000 - Celtel	12,000 *Celtel: 8100 MTN: 3900*	72,602 *19,074 – Celtel 53,528 – MTN*	188,568 *MTN: 146,634 Celtel: 32,934: UTL: 10,000*
5	Internet/ email subscribers (dial-up)	500	1300	4150	4469
6	Internet/ email subscribers (wireless)	4	8	98	1229
7	National Telecommunications Operators	1	2	2	2
8	Mobile Cellular Operators (CelTel, MTN, UTL)	1	2	2	3
9	VSAT International Data Gateways (Includes UTL and MTN)	2	3	7	8
10	Internet Service Providers	2	7	9	11
11	Public Internet Access Providers (Cafes)	0	3	18	24
12	Public Payphone Licenses	1	7	12	18
13	Private FM Radio Stations	14	28	37	100
14	Private Television Stations	4	8	11	19

15	Private Radio Communication Licenses	453	530	688	770
16	National Postal Operator	1	1	1	1
17	Courier Service Providers	2	7	11	10

As they attempt to cover the main road corridors and towns, they have achieved signal availability to many rural areas so that the only hindrance to immediate communication is the availability of a hand set.

Table 2 shows the growth in Internet access based on the available international data capacity. Gateway operators increase this according to demand for Internet access so it is a fairly correct indicator. The table does not include UTL's planned capacity which is (1mbps) supposed to come on line anytime.

TABLE 2
TOTAL DATA UPLINK AND DOWNLINK CAPACITY

	UPLINK	DOWNLINK
1998	256kbps	384kbps
1999	640kbps	768kbps
Feb 2000	1.2mbps	1.7mbps
Feb 2001	4.088mbps	5.216mbps

Despite this impressive growth in ICT, the development of ICT in Uganda continues to face several challenges. Four critical ones can be identified. First, there is the question of unequal access. ICT is largely confined to the elite and the urban areas in terms of class and location, respectively. Ownership is probably limited to 1 percent of the country's total population. Shared access in public and private enterprises as well as government offices does increase the actual number of people with access to telecommunications probably to one million but the truth is that the country is still a long way from universal access. Altogether, no more than 10 percent of the population has access to telephones and related telecommunication services. As for location, 70 percent of the communication services are concentrated in the urban centers. ICT in the rural areas is limited by the absence of and/or unreliable power supply. Yet for any ICT infrastructure to be established and succeed, there is need for adequate and reliable power supply. Not surprisingly, all the stakeholders call for an improvement in the supply of power to rural communities. Given the high costs and challenges of establishing fixed telephone lines, it seems certain that rural communication development can best be achieved via wireless technology. Related to this, second, are the connectivity issues. While some effort is being made to extend telephone services through the privatization of telecoms, it is generally agreed that these services remain out of reach of the rural communities because of costs. It should be recognized, therefore, that there is a need to provide an infrastructure that supports both telephony and data services at affordable cost to rural residents. Another important aspect of connectivity involves the lack of an Internet Exchange Point (IXP) in Uganda. All traffic not on the same ISP has to be routed outside Uganda and back again. Recently, ISPs and the national operators have come together in a voluntary association to set up an internal IXP. UCC has encouraged this initiative and will ensure that there are no regulatory or financial barriers to its implementation. The only requirement is that it should not in any way restrict entry by new ISPs. Where the information infrastructure exists, the bandwidth offered is too low to meet the various user requirements. Both the telecommunications and Internet service providers need to consider bandwidth issues so that they can attract as many users and as wide an applications base as possible. In this connection, the lack of a high capacity data network also needs to be dealt with. This is combined with the fact that local access costs are time-based, making the cost of local access a limiting factor in Internet usage. It is evident that a flat local access charge has been a major factor in the rapid growth of Internet usage in the USA.

The third set of issues concern the question of sustainability. This involves the challenges of maintaining and sustaining the various initiatives that are currently being implemented or that are planned for implementation through external donor support. The government is being called upon to waive taxes on ICT equipment to make it initially affordable and consider lowering charges for telecom services. In this context the lack of an official information policy presents constraints to the orderly growth of the ICT sector. For quite some time, there have been debates and proposals on the need for a national ICT policy and who should guide its development. Although we have seen a new era of co-operation and information exchange among the major public and private agencies involved in ICT development after a long period of disjointed efforts, a coordinated national policy is still missing. It is important to note that a policy is developed to focus and create synergy among initiatives. Critical to any effective policy is the question of ownership, which *is an essential ingredient in the success of any policy*. Policies must be developed by government in partnership with the private sector and with all stakeholders. But the articulation of policy, by itself, will not remove inertia. Policy is most effective when used to coordinate initiatives that benefit from a controlled pool of resources. Currently the biggest problem is not lack of a policy, but inertia.

This inertia is the result of the fourth set of factors that constrain the development of ICT. These include limited public awareness about the importance of ICT to the country's potential development. Much of the population and many of the policy makers are still ignorant of the potential uses of ICTs. There is also what can be termed an "analysis paralysis," the tendency to discuss issues endlessly without acting on them. The need to change old ways of doing things is imperative. Too many of the current public and private leaders are accustomed to the old ways of doing things. They do not want to see any change, and will indeed resist or frustrate change in many subtle and unsubtle ways. This is connected to fear of venturing into the unknown: Shall we be able to cope? Does using ICT mean unemployment and retrenching? All the sectors identified need a systematic and deliberate program of sensitizing the communities and in particular the policy makers on the usefulness and importance of ICTs. It is generally appreciated by all the experts in the field that the education sector has the highest multiplier effect and that the Ministry of Education should expedite the provision of ICT skills to teachers at all levels of education. These teachers would then play a significant role in imparting appropriate ICT skills to their pupils and students via the approach identified earlier as Technology Enhanced Instruction and Learning (TEIL). It is also recognized that the involvement of the private sector is crucial since government alone cannot provide the required infrastructure. It is recommended that the government provide an enabling environment and facilitate partnerships between government and the private sector.

Solutions and New Strategies

The solutions proposed here, are a result of several consultation processes and focus group discussions with various ICT stakeholders by the Developing an Information Infrastructure Agenda in Uganda Project (DIIAUP). Among the proposed solutions are:

- The need to set up of an information coordination agency. The importance of this agency cannot be overemphasized. There is an opportunity to establish, build capacity, and define the roles of this agency based on international best practices. A one-stop information center (web based) will coordinate all ICT initiatives. This will reduce the duplication of efforts and enhance learning from precedents.
- Formulation and development of an ICT-policy and institutional framework. ICTs are a potent factor in all aspects of the economy and social life and a critical component of effective public sector governance and control. There is need to coordinate the various efforts aimed at developing this policy and also to help formulate the policy in accordance with international best practices.
- Basic infrastructure development. There is need to set up appropriate infrastructure to connect the sub-counties to the districts and the districts to the center. The infrastructure to support governance in Uganda's decentralized government structure has been noted as critical to development. Multi-sectoral telecenters have been recommended as the community information dissemination and sector integration points. There is also an urgent need to provide connectivity

to rural communities to access both local and international information at affordable costs.

- Increase public awareness by developing suitable programs aimed at sensitizing policy makers, individuals and communities about the use of ICTs in improving service delivery to them by the government, as well as government improving on its efficiency in service delivery and transparency.
- Empowering disadvantaged groups through computer mediated communication. There is need to teach disadvantaged groups like women, youth and the disabled how to access information on their empowerment using ICTs.
- Manpower development. There are opportunities in identifying ICT training needs in various sectors and conduct or organize relevant training for them.

Besides the ongoing formulation of ICT policy by the Uganda National Council of Science and Technology (UNCST), a consultative strategy in supporting and coordinating Uganda's emerging ICT infrastructure and applications is being undertaken through the Developing an Information Infrastructure Project (DIIAUP), a project funded by InfoDev (World Bank) and implemented by the Institute of Computer Science at Makerere University. The project has researched the information needs of Uganda and aims to develop a National Information Infrastructure (NII) agenda, which includes the following:

- A list of priority project proposals to showcase possibilities for ICT enabled change in bridging the local "digital divide"
- Sector policy guidelines
- Strategic information systems for Uganda.

The DIIAUP project specifically addresses rural communications and human well being by focusing on three specific InfoDev objectives, namely, reducing poverty and exclusion of low-income social groups, improving education and health, and increasing efficiency, accountability and transparency of government. A concurrent expectation is that the project will provide a framework for coordinating existing information infrastructure initiatives. The underlying objective–a truly daunting challenge–is that the rural population must be developed to a level of information literacy that permits them to gain unmediated access to government information.

The overall approach of the project has involved undertaking comprehensive research in Uganda's information needs for six sectors with a high development impact: agriculture, business, education, health, government and ICT. By 2001, working papers were produced which served as a basis for a series of consultations and focus group discussions with stakeholders. Also, priority projects were identified in each sector and sector policy guidelines were developed for the government to follow through. The identified projects are meant to demonstrate ways of using ICT to deliver services to rural communities in particular with the objective of alleviating rural poverty and improving the quality and access of social services. Specific priority project proposals have been developed by the DIIAUP for the selected sectors.

In the agricultural sector, there is the Agriculture Management Information System (AMIS), which addresses the specific issues of nationwide dissemination and utilization of information by rural farmers. The broad objective of the project is to coordinate different stakeholders in the agricultural information packaging and delivery system using information and communication technologies. Specific objectives under this include the need to provide information on weather so as to enable farmers make decisions on planting, clearing, maintenance, harvesting, processing, etc. (production of information); receive and disseminate information on prices of agro-inputs (price information); provide information on recommended farm practices (management information); credit information (resource availability and accessibility); and receive and disseminate information related to prices of agro-products (market information).

In the health services sector, the Sustainability Model for Telehealth/Telemedicine has been developed. To the typical resident in rural areas, health services are non-existent. A lot needs to be done to address the inequity in access to medical services. This project proposal focuses on increasing the potential of Telehealth in improving rural access to primary health care in particular, and strengthening and complimenting the already existing telemedicine project in Uganda's main referral hospital and its linkage to rural hospital sites. Indeed, telemedicine was one of the major issues in the "African Connection," the name for the continental initiative launched by the African ministers in charge of communications in Johannesburg in 1998. The private and public sectors can take up initiatives here. As for the education sector, there is the Teacher Internship Project designed to spread ICT capacity from high ICT schools to low non-ICT schools. The exchange of teachers enabled schools to provide training.

The priority for the government sector has been on developing district data banks, which are seen as pillars for planning to support e-governance in the decentralized structure at the district level. The project proposes a model to harmonize the existing data collection efforts so that only one reliable stop center that contains reliable data and statistics is maintained in all Uganda's 69 higher local governments. For the business sector the Kikuubo E-Business Focal Point (KEBFP) project was launched. The project, situated in the heartland of Uganda's business area known as Kikuubo, seeks to create a pool of ICT literate business people; network rural business with a central nucleus; boost local e-commerce efforts; reduce security risks by enabling remote ordering of goods; and accelerate formal definition of e-payment in Uganda 's context.

There are a number of cross-cutting issues, particularly those involving the rural areas. It has been widely agreed by stakeholders that multi-sectoral telecenters will be the centers for sectoral integration, dissemination and development of content for rural communities. The Rural Communication Centers and Coordination Agency project reviews the role of existing telecenters and proposes models that will increase community ownership.

Conclusion

It stands to reason that while a lot has been achieved in laying the ICT policy and physical infrastructure in Uganda, much remains to be done to further facilitate ICT enabled development in the country. A winning strategy has been identified: There should continue to be *partnership between the public and private sector* in recognition of the complimentary needs and roles in ICT enabled development. Public/private sector support and commitment are required for the implementation of the National Information Infrastructure Agenda (NIIA). This will entail *global sensitization* starting with the policy makers. The NIIA, which is being formulated by the InfoDev Project, will guide, coordinate, and demonstrate the emerging ICT infrastructure and applications in the identified critical sectors for national development. Specifically with reference to infrastructure development, the Uganda Communication Commission's (UCC) Rural Development Communication Fund (RDCF) needs to accelerate the construction of the infrastructure to facilitate universal access to information. This fund should urgently support identified projects that alleviate poverty in the rural areas. It is further recommended that alliances with experienced operators from outside Uganda who will also have access to financing at reasonable interest rates should be encouraged.

The success of the ERT project is crucial in providing minimal infrastructural support for ICT in the rural areas. There is need to effectively exploit existing infrastructure for wider access. For example, the rural schools that have been connected have outreach programs, which the surrounding communities can utilize. Some have set up school telecenters to exploit the existing connectivity. It is clear that dialogue rather than policing is needed to foster the development of ICT in the country. As a result of the coordination and sharing of information by all stakeholders, the Uganda Communication Commission (UCC) is better informed and in a position to do more to help remove or minimize the regulatory barriers that impede their success. It is quite evident that the training of human capital in all aspects of information society management is essential, so is developing content appropriate to society's needs. Underlying it all must be commitment to equitable nationwide access to ICT for all Ugandans. It must be recognized that some conditions, like the duopoly in telephone service provision, were put in place to compensate the service providers for their obligatory requirement to extend services to areas that are not commercially viable in the medium term. This policy must be appreciated if we are concerned about closing the internal digital divide. However, the two national operators, MTN and UTL, are challenged to ensure that their dominant position is not abused. They should ensure that Ugandans are positioned to take full advantage of new and emerging technologies at fair prices.

PART II:

Information Technology, Development and Democracy

CHAPTER 7

HARNESSING INFORMATION TECHNOLOGIES FOR TRANSFORMING AFRICAN PUBLIC AGENCIES FOR DEVELOPMENT

Muta A. Tiamiyu

Introduction

Some basic facts can hardly be disputed. Firstly, the opportunities for using information technologies (IT) as lever for transforming organizational processes are greatest in information-intensive activities. Secondly, government and governance processes are information-intensive. Consider for instance the following government and governance processes: public policy formulation, implementation and evaluation; law making and enforcement; registration, regulation and control of citizens and businesses; statistical and data banking services, etc. Much of the information processing is also often voluminous and repetitive, thereby providing great scope for IT to facilitate effectiveness and efficiency. Thirdly, decades of structural adjustment programs and privatization of public enterprises notwithstanding, the public sectors of most African economies continue to be a major source and determinant of socio-economic resources and initiatives. Indeed, the poor growth and development records of African countries can be attributed as much to the low effectiveness and efficiency of the dominant public sectors as to the absence of large and vibrant private sectors. The implication is that, until the balance of socio-economic resources and initiatives shifts to the private sector, African public sectors and agencies must improve significantly their capacity to initiate, stimulate and coordinate national developmental processes. And, as noted above, there is great potential for transforming and improving their information-intensive processes through the development and exploitation of information and IT.

Accordingly, this chapter reviews relevant frameworks, issues, constraints and prospects of IT for transforming African public agencies into vibrant participants in African governance systems. The chapter addresses two main research questions: What technical, economic, socio-political and other factors (might) facilitate or constrain IT development in African public agencies? How can the visions, culture and processes of the agencies be exploited or modified to promote rapid and sustainable IT development in the agencies? The chapter attempts to answer these questions by analyzing the socio-technical dimensions of IT transfer and development, as well as interpreting illustrative data from recent and on-going surveys of Nigerian public agencies. The chapter is concluded with recommendations on IT development strategies for African governments and agencies.

Decision-making for Organizational IT Development

Organizations are usually described as goal-seeking entities. Organizations are expected to make "rational" decisions on relevant issues (including IT development) toward achieving their goals. Among the usually mentioned goals are effectiveness, efficiency, stability, and growth. These goals are often assessed from the perspective of profit-seeking private firms, and in terms of technical and economic variables, such as input-output ratios, sales volume, market share, profitability, return on investment, fixed and current assets, etc. However, most of these indices of performance are difficult to assess for non-profit organizations because unexpected socio-political and parochial stakeholder considerations can overshadow the bottom-line technical and economic imperatives that the agencies might need to satisfy their effectiveness, efficiency and growth goals. The danger is also greater in African countries with underdeveloped democratic institutions and weak socio-political checks and balances.

IT Development Frameworks for African Public Agencies

A growing number of frameworks exist in the literature for analyzing the variables pertaining to concepts of organizational change through IT transfer and development. We will here build a framework for analyzing the potential facilitators or inhibitors of IT development in African public agencies by integrating three complementary frameworks proposed by Nolan (1973, 1979), More and Irwin (1995), and Baark and Heeks (1999).

Nolan (1973) was the first to suggest an evolutionary process for IT development in organizations. Variants of the model have been proposed and evaluated by Galliers and Sunderland (1991) and others. The models suggest that organizations usually pass through different IT development stages, sometimes described as initiation (of computerization), contagion, control, integration, repositioning, and maturity (Kanter 1984). Nolan-type models highlight the changes that usually or might occur in IT development variables such as: IT philosophies, leadership and users; IT planning framework and focus; and types of computing applications and resources. Proponents of the models also suggest that knowledge of these stages in terms of the status of key variables could be used to identify each organization's level of maturity with IT implementation, and to guide further development (Galliers and Sunderland 1991).

Baark and Heeks (1999) categorized potential IT development variables in the context of the different stages of the technology project life cycle: design/choice of technology, purchase/installation, assimilation and use, adaptation, diffusion and innovation, monitoring and evaluation, and long-term sustainability. They used the model to highlight the various variables that must be considered to facilitate the implementation, sustainability, use and growth of donor-funded IT projects.

More and Irwin (1995) applied a framework originally developed by Madu and Jacob (1991) to the processes of global technology transfer to Australian organizations. They categorized the relevant variables in terms of key entities into: (a) *technical* (pertaining to the technologies to be transferred

—input requirements, output quality, reliability, safety, cost, etc); *organizational* (pertaining to the target organizational structure, culture and process); and *personal* (pertaining to the goals, attitudes and behaviors of the various individuals who would implement decisions about the development and use of the technology). They did not separately identify *environmental* variables, such as local culture and social infrastructure, physical environmental conditions, and socio-political imperatives. These variables are crucial in African countries where favorable environmental conditions often cannot be assumed.

Table 1 combines the three frameworks towards highlighting the diversity of potential perspectives and socio-technical variables pertaining to IT development in African agencies. Nolan-type frameworks provide the long-term planning and evolutionary perspective to IT development, while the project cycle framework draws attention to specific IT projects that must be implemented to move an organization through each of the Nolan stages. Accordingly, the relevant variables (technical, economic, political, etc.) within each project cycle (bottom row) in each Nolan stage (last column) could then be analyzed in terms of the key entities – organization, stakeholders, environment and technology, and/or in terms of the need to balance the socio-technical equations of African agencies.

Basically, applying the model requires identifying and analyzing the different variables associated with each relevant Nolan-type stage or project cycle phase in specific African agencies to determine how critical they are, and how they interrelate with other variables in other phases or stages. The interrelationships among variables might then be used to simulate, predict, and guide both IT development projects and programs in each agency.

Socio-Economic Dimensions of IT Development

We will now highlight the socio-technical issues pertaining to the visions, culture and processes of African public agencies, their stakeholders and environments, as well as IT development in the agencies.

Table 1: Framework for analyzing the socio-technical determinants of IT development in African public agencies. [After Nolan (1979), Baark and Heeks (1999) and More and Irwin (1995)]

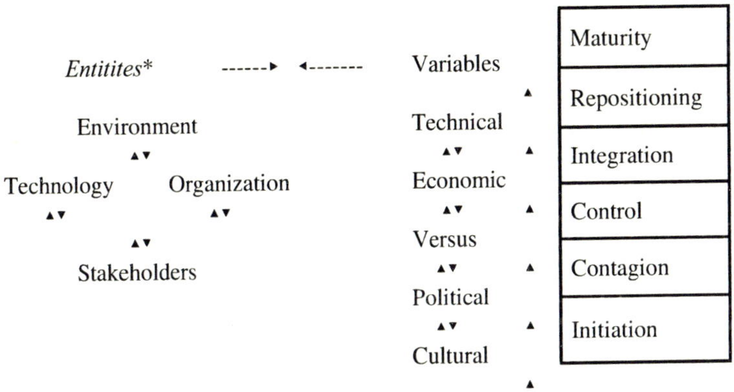

Organizational Visions, Change and Learning

The development of IT capabilities in African public agencies would require, first of all, significant changes in their visions, processes and technology learning capabilities. It would be difficult to get support for investments in costly IT systems in an agency, if agency personnel believe that they are performing as best as is necessary or possible in terms of types, volume and quality of services. Similarly, if an agency's stakeholders do not take seriously information accuracy, timeliness, analysis and sharing in decision-making, they are unlikely to accord much importance to IT systems and business processes that seek to assure these usually valued attributes of information.

African agencies must see IT as providing opportunities not only to automate and informate their operations, but also to transform the quality of agency business processes and services to society. A business process is described as how an enterprise provides product(s) or service(s) to its publics; how it acquires what inputs from which suppliers; how it transforms the inputs into outputs; how it delivers the outputs to which customers/ clients; how it acquires, processes, stores, transfers and uses data and information in its activities. Each business process utilizes various organizational and environmental resources – people, organizational structures and culture, equipment, information technology, data/information, etc. Changes in business processes would require changes in the nature and uses of these resources. In other words, we need to remind ourselves of the mutually complementary components of IT capability – hardware (computers and telecommunications), software, peopleware, organizational structures (institutions and policies), and data/information (as raw material/output of IT systems). Not only must African public agencies be infused with modern and computer and communications equipment, they must invariably also undergo commensurate changes in organizational philosophies, goals, structures, culture and dynamism to facilitate the adoption, integration and sustainability of new equipment, software and associated technologies and skills into improved business processes.

African agencies must also improve their capacity to acquire, learn and innovate ideas, technologies and business processes. The concept of the organizational learning curve is really about how eager and quickly organizations and their personnel are able to imbibe, master and innovate new ideas. Emergent environmental imperatives, technologies and skills of the information age imply that African agencies must view organizational change as inevitable; else, they will become more and more irrelevant to their environment and publics. Moreover, only agencies that seek and manage change proactively would reap strategic leading-edge benefits from change, not those that only react to environmental imperatives such as new competition or public discontent.

Technical issues and variables

Major technical issues associated with IT development pertain to such variables as the suitability of specific IT components or systems for required data processing tasks, input-output and other performance measures of the components or systems (i.e., outputs, speed, accuracy and reliability, etc.) relative to existing or alternative technologies, operating procedures and requirements, safety designs, scalability and networkability of components, and sustainability of coupled components and systems. (Madu and Jacob 1991; More and Irwin 1995). Technical variables are connected with the nature and relationships between the various inputs and outputs in the IT system design, installation, operation and upgrading sequence. Inputs include equipment, software, human resources, data resources, social infrastructure (e.g. public electricity and telephone), physical environment (humidity, temperature, security), organizational structure and culture (institutions, work ethics, etc.). IT equipment and generic software are usually imported to most developing countries, and might not be perfectly suited to the physical, social and infrastructural conditions of these countries. Additionally, some software must be imported, but all other inputs are expected to be sourced mostly locally—personnel, custom software, social infrastructure, environmental conditions, and organizational practices.

The fundamental technical problem then is how to select, synchronize and integrate foreign and local IT components into the business processes and information systems of African agencies for improved effectiveness, efficiency, and growth. The available options are: (i) consider local environmental conditions and input quality as given (within limits) and select/configure only foreign IT components that can integrate with local inputs and conditions; (ii) consider foreign components as given (within limits) and develop local conditions and inputs to the standards that would assure reasonable performance of designed IT systems; and (iii) modify both foreign and local components, standards and conditions as necessary and feasible. Technical choices entail appropriate decisions about the types, scale, phasing and synchronization of foreign and local IT components and resources. Such decisions must be facilitated by adequate knowledge of organizational information processing requirements and status, technical features of available IT components, local physical, infrastructural and social conditions, as well as the decision-making flexibility to choose, modify and integrate foreign and local IT components as appropriate.

Technical decisions cannot however be completely isolated from the economics and socio-politics of a country or public agency. Technical choices have implications for the cost, affordability and benefits of IT systems; the more complex and sophisticated the chosen technology, the higher would be the initial and maintenance costs. Socio-political factors often also serve the useful purpose of providing vital political or donor funds and support to IT development projects. But they can also lead to the displacement of critical technical and economic imperatives required for project sustainability. Socio-political factors are often unpredictable, and might not assure the sustainability of the IT projects that they initiate. It is imperative that IT technical decisions are influenced by long-term economic realities such as efficiency, cost, affordability, and fund flow.

Economic Variables and Issues

The key economic variables are cost-benefit, affordability, and fund flow. Cost-benefit analyses of IT investments in public agencies are constrained by various factors. It is usually difficult in non-profit, service-oriented agencies to quantify the value of services being or to be provided toward determining whether IT investments are or would be cost-beneficial. In large agencies, the impact of IT on organizational effectiveness and efficiency might also be imperceptible until a critical mass of networked technologies is established. This often would require high and sustained funding which might be unavailable on a consistent basis due to the influence and unpredictability of socio-political imperatives on public/agency budgets. Interdependent IT projects might have to be implemented and coordinated until the critical mass is achieved, thereby often making it difficult and probably inappropriate to assess cost-benefit on individual project basis.

Cost-benefit analysis of IT projects in African public agencies is also complicated by the issue of appropriate micro- or macro-economic framework for such analysis (Tiamiyu 1993). Private firms usually consider profits as the bottom line criterion for IT investments. They can ignore such macro-economic objectives as unemployment reduction, improved balance of trade, and development of the local IT industry. Public policy is often not unmindful of the danger of expensive and foreign IT components displacing cheap local human resources and worsening trade balances, and might require that public agencies emphasize local and labor-using than foreign and labor-saving technologies in their business processes. Hence, although government and top management might be aware of the potential productivity impact of labor-saving IT investments, they might be constrained to sustain or emphasize less efficient but labor-intensive processes. African agencies are usually expected and politically pressured to provide employment for citizens, sometimes much beyond optimal levels. In other words, IT development in African agencies might not be justifiable on microeconomic considerations alone, but rather on macroeconomic and system-wide imperatives.

Two other critical economic issues are affordability and fund flow. Affordability relates to the capability and readiness of agencies to devote resources consistently over time for initial, recurrent and improvement IT expenditures. Affordability is intertwined with the income and cost profiles of public agencies, and as well as their expectations of the impact of IT systems. Related to affordability are the costs of foreign IT components relative to local components and wage levels, as well as between the start-up and maintenance costs of IT systems and other costs of the agencies. The costs of IT components and systems often appear to be too high compared to wage bills and other costs. Unfortunately, there appears to be a computer hardware pricing paradigm in the industrialized countries that ensures that the cost of basic microcomputer workstations hover around 1,500 USA dollars, although with significantly improving computing capacities with each new generation of microcomputers prices should go down. This raises the question of whether, for African and other developing countries, this paradigm should not be replaced with that of 'basic computing capability at ever-declining prices' so that computing components and systems would be more affordable. Local IT component manufacturers and system integrators have key roles to play in this respect: They should consider realistic tradeoffs between sophisticated and expensive IT systems on the one hand, and basic and cheap systems which might indeed satisfy the entry-level data processing requirements of some agencies. For instance, there would be no point recommending a highly sophisticated and expensive database architecture for an agency that is only about to initiate computerization.

Fund flow bottlenecks are a bane of African public sector projects. The problems arise from either deliberate under-budgeting, inadequate projections of fund flow, or unnecessary delays created by socio-political or personality conflicts. IT projects are exceedingly vulnerable to such problems because of the high obsolescence potential of IT components, skills, and systems.

Stakeholder and Socio-political Interests

The power-politics models of organizational decision-making processes mentioned earlier emphasize that large organizations usually contain diverse social groupings each with its own agenda and objectives, competing over resources, influence and autonomy, and using every means available (Peled 2000: 47). For many organizations, and particularly those directed more by socio-political and group than technical and economic imperatives, the IT decision process becomes a platform as much for socio-political bargaining among stakeholders as for the technical re-engineering of processes towards delivering revolutionary new and higher quality products and services (Kling 1980:63-65). Consider the potential conflict among stakeholder goals in public agencies in developing countries. Competing goals would include genuine desire to improve society through agency activities and services; desire by politicians to manipulate agency activities for political gains; desire by particular units or employees to monopolize organizational resources including information; desire by suppliers and consultants for undeserved contract terms; desire by development assistance agencies to use public agencies to pursue their own agenda; and desire by some stakeholders to create or protect opportunities for corrupt practices. And, depending on the dynamics of the organizational culture and processes, specific stakeholder goals might dominate organizational action. In particular, if key agency personnel are motivated more by personal than collective organizational goals, they are more likely to support only IT projects that promote, or at least do not hurt, their interests.

Let us consider the behavioral inclinations of some of

these stakeholders, particularly, politicians/top management, bureaucrats and lower-level personnel, IT systems suppliers, and donor agencies. Top management preferences are usually significant facilitators or inhibitors of IT development in organizations. Hence, IT decisions would often crucially depend on the personality and political variables associated with politicians, chief executives and top management. Private firms often cannot afford to allow personal and socio-political considerations of their leaders to continuously override the technical and economic imperatives of IT development for profitability and survival in competitive environments. By contrast, most African agencies are monopolies providing services to politically ambivalent publics in a patronage-based socio-cultural environment. Hence, personal and socio-political variables associated with the politicians, chief executives and top management team might consistently over-balance the technical and economic imperatives for cost-beneficial IT investments. The result might then be unnecessarily large, inadequately planned, over-priced, unsynchronized and unsustainable IT projects or systems that satisfy the rationality of dominant stakeholders, but fail to meet organization-level rationality criteria.

Technocrats and other lower-level public servants constitute another key group of stakeholders of public agencies whose goals, behavior and motivation cannot be assumed as always attuned to IT development in African agencies. Well known is the legend of the poorly paid African public servant. Poorly paid staff usually develop various survival strategies, including moonlighting, corrupt enrichment through deliberate bureaucratic red tape and deliberate low productivity, and lack of commitment to organizational improvement programs. The agencies have themselves contributed to the entrenchment of these strategies through overstaffing and inadequate attention to staff discipline. Consequently, the strategies have become endemic features of the bureaucratic norms of the public agencies to the extent that they tend to persist for some time even after public service pay packages are improved. These are the aspects of the bureaucratic "culture" mentioned by Galliers et al. (1998) as constituting constraints to IT development in Pakistani public agencies.

Two other key external stakeholders of African public agencies – donor agencies and local system integrators/vendors – often also exert significant influences on IT development choices of the agencies, often with negative consequences. Donor agencies often provide large funds to start off IT projects, assuming that the agencies would be able to source funds to sustain and improve the projects later. They sometimes also impose foreign experts who might lack adequate knowledge of local inputs and conditions. In order to maximize their fees and profits, local IT system integrators/vendors often recommend highly sophisticated and costly systems that are much beyond the technical, managerial, and long-term financial capabilities of the agencies.

Bureaucratic Culture

Several authors (e.g., Braa et al. 1995; Odedra 1992) have highlighted the necessity to match and integrate foreign and local technological components, standards and conditions toward ensuring sustainable technology transfer and organizational learning and innovation. In this respect, the following question usually comes up: What aspects of the foreign and local IT components, standards and conditions can and must be modified? Foreign and local IT system designers and integrators usually have some choice in selecting or recommending foreign IT components and designs that meet client requirements, although they might be biased towards specific technologies and designs. However, even assuming well-intentioned designs, there is always the problem of how to implant and sustain the designs within modified or unmodified African bureaucratic cultures that often exhibit resilient traits.

Let us consider some of these traits such as the stakeholder interests and behaviors highlighted in the previous section. Most African countries gained political independence about 40 years ago. However, the bureaucratic culture of their public agencies still exhibits traces of the colonial era when African public servants never fully identified with the colonialists and their agencies. Hence, in the absence of a politically sophisticated civil society and accountability checks, loyalty by stakeholders to the statutory goals of the agencies is at best lukewarm. An agency might then, at various times, degenerate into a preserve of dominant stakeholders whose interests override all agency decisions.

Another feature of African bureaucratic culture are practices that foster lack of effective accountability and transparency. Firstly, there is commitment to classify all decisions, documents and information as secret, and to create unnecessary barriers to information access, thereby hamstringing effective decision-making even by insiders. Secondly, long delays in the statutory auditing of the accounts and activities of agencies are fostered or condoned, with negative implications for good governance, as noted for Ghana by Akotia (1998). Thirdly, government publications that would have provided the citizenry with some information about various agencies are issued many years late, if at all. These practices make it possible for the agencies and their stakeholders to conceal poor performance, bad decisions, and corrupt practices. Such practices also run contrary to the information age desiderata of information timeliness, dissemination, access and sharing as the basis for organizational dynamism and accountability, and social development.

Environmental Variables

In the developed countries, the widespread deployment of information and communication technologies at work, school and home came after basic telephone and broadcast infrastructure and services had become widespread. Moreover, the level of literacy was high, which meant that people were also already able to communicate among themselves through books, newspapers, and other print documents. High levels of literacy and communication also enabled science, as well as the effective participation by citizens' groups in the democratic governance of those countries. The information-rich environment subsequently enabled and was itself further enabled by the computing revolution as from the late 1970s. Nevertheless, even in such IT-enabling environments, public agencies are often slower than most other organizations in embracing new technology. Hence, in a recent paper, Peled (2000) described the dualism in the IT development experience and status of Israeli private and public sectors respectively as "First Class Technology, Third Rate Bureaucracy." Much earlier, and writing from the Canadian perspective, MacDonald (1986) had noted that government agencies are usually slower than private firms in exploiting IT because they are usually monopolies, non-profit, and often subject to intense political scrutiny. The usually overriding political context of public agencies means that they often cannot individually formulate and implement operational strategies dramatically different from those of other agencies or of the political policy-makers.

Apart from the above generic features and constraints of public agencies, African agencies are additionally constrained by a socio-economic environment characterized by a largely illiterate citizenry, which assures the persistence of primordial cultures; low incomes per capita, which informs societal emphasis on meeting basic needs immediately with existing crude technologies than investing in, and waiting for, the benefits of better information and improved technology; politically ambivalent civil society, resulting in inadequate scrutiny of public institutions; and underdeveloped communication and transportation facilities and services, which constrain socio-political interactions.

Consider for instance, the combined influence of primordial cultures and low literacy and educational levels. These often translates into a socio-cultural belief system comprising a mix of scientific and pre-scientific attitudes to data and technology, as well as such dominant attitudes as "what would be would be, despite human effort". Such attitudes, which have been described as fatalistic (Mabawonku et al. 2001), or as lack of ambition toward exploiting science for development (Salam 1987), tend to limit vigorous individual and societal effort toward revolutionary improvements in decision-making and productivity through scientific information and technology. Almost twenty years ago, Aiyepeku (1982) studied the level of appreciation by Nigerian policy-makers of 'hard, scientific' versus 'soft, cultural' information as the basis for effective policy-making. Although African public servants have improved in education, experience and stature since political independence, attitudes to IT systems and to the key attributes of information produced by such systems

– accuracy, timeliness, objective analysis, sharing, etc. – are still mostly only lukewarm.

Jeppe (1994) went further to characterize the "afro-centric" (in contrast to "euro-centric") orientation to risk taking with such words as "resignation," "contentment," "destiny," "security in groups," etc. Jeppe (1994) and Mabawonku et al. (2001) also described interpersonal relations in the African cultural context as primarily hierarchical, such that orientation to leadership tend to be primarily "submissive," "dependent," "conformist," "permanent," and "ascriptive." These traits beget public bureaucracies wherein employees are acutely averse to, and might be severely sanctioned for, innovation and risk-taking outside the framework of the entrenched leadership and organizational hierarchies.

In summary, the infrastructural and socio-cultural environments of African public agencies are not the best for stimulating IT development and use in the agencies. Hence, Adam and Wood (1999) concluded that, despite the great potentials for IT to improve African growth and development, actual impact had been low and would probably improve only gradually because of a complex web of environmental constraints within African societies. Strategies are clearly needed to systematically improve the environmental conditions of African agencies to stimulate them to vigorously exploit IT for enhancing their effectiveness and efficiency.

IT Development Constraints and Strategies: Some Nigerian Data

We will now examine some data from previous and on-going studies by this author in Nigeria in order to highlight the status of some of the technical, economic, socio-political and attitudinal variables in African public agencies.

Table 2 summarizes data from a national postal survey of Nigerian federal, state and local government agencies undertaken by this author during the last quarter of 2000. Respondents to the survey were mostly top and very senior public servants. We asked them to react to some 30-odd statements pertaining to actual or potential constraints to IT development in their agencies. The statements were phrased to reflect the different socio-technical dimensions of IT development. The responses were analyzed using Friedman non-parametric test thereby classifying the constraints into most frequent/severe (1-10), moderately frequent/severe (11-20), and least frequent/severe (21-31).

Table 2: Relative importance of organizational and socio-economic constraints to IT development and use in Nigerian public agencies.

S/N	*Socio-economic constraint*	*Score*
01	Delayed/inadequate staff training for computers	20.72
02	Inadequate staff knowledge of computers	19.61
03	Poor maintenance culture	19.48
04	Negative attitude to data collection	18.41
05	Traditional staff evaluation	18.40
06	Over-emphasis on political factors	18.20
07	Unbalanced distribution of computers	18.12
08	Over-emphasis on/neglect of IT equipment/other	18.11
09	components	17.98
10	Improper handling/care of equipment	
	Lack of funds	17.77
11	Negative attitude to information sharing	17.75
12	Only very senior staff have access to telephones	17.71
13	Vested interests in old ways of doing things	17.52
14	Entrenched paper-based information processing	17.30
15	Inflated cost of computer equipment and systems	16.93
16	Low organizational recognition of computing skills	16.84
17	Lack of appropriate IT policies/guidelines	16.84
18	Absence of challenging work	16.52
19	Inadequate computer-based work for trained staff	16.04
20	Misuse of computer/telephones for private work	15.52
21	High turnover of computer-trained staff	15.39
22	Over-ambitious computer systems projects	14.40
23	Only very senior staff have access to computers	13.58
24	General Orders not supportive of computerized procedures	13.49
25	Staff fear of being replaced by computers	13.48
26	Fear of information breech in computer systems	13.41
27	Low real benefits from existing computer systems	13.36
28	Financial regulations not supportive of computerized	12.56
29	accounting	11.75
30	Fear of loss of control by senior/older staff	9.41
31	Most work in my agency cannot be improved by computers	9.34
	Bosses frown at time spent working with computers	
	N = 81; Friedman Chi-Square = 181.1656; D.F. = 30 Sig. = .000	

Focusing on the top 10 constraints, it would be noted that skills-related constraints ranked highest: "*01. Delayed/inadequate staff training for computers*," and "*02. Inadequate staff knowledge of computers*." At the time of the study, the availability of computers in Nigerian agencies was very low, with most agencies still wedded to completely manual information systems (Tiamiyu 2000a). At the same time, evidence reported in Tiamiyu (2000b), suggests that agency personnel who go for computer training often complain about being unable to apply and develop their new skills due to lack of computing infrastructure in their agencies. Hence, the top ranking of the two constraints above suggests that top management of Nigerian public agencies were worried by the lack of synchronization of IT resources and skills even in the agencies where computerization had begun. This deduction is confirmed by the high rating also given to "*08. Overemphasis on IT equipment to the neglect of other components*."

Poor maintenance culture is pervasive in African countries as had been elucidated by Woherem (1995). Nigerian agencies are not immune to this problem as confirmed by the high rating given to "*03 - Poor maintenance culture*." Poor maintenance culture is facilitated by a social and managerial culture that emphasizes large initial capital investment usually on tangible assets, but gives little attention to recurrent and down-stream expenditure required to maintain, sustain and improve the initial investments. This cultural trait is often fostered by top management, IT suppliers, and sometimes even donor agencies. Top management and suppliers are often inclined to start and recommend new projects than maintain and build upon existing projects. Donor agencies often abandon old for new donor-assisted projects in line with their own changing foci and strategies.

The constraints "*04. Negative attitude to data collection*," and "*05. Traditional staff evaluation*" is indicative of inertia to desired organizational change for IT in Nigerian public agencies. In addition, "*11. Negative attitude to information sharing*," "*13. Vested interests in old ways of doing things*," and "*14. Entrenched paper based information processing*" also ranked in the top half of the table, indicating an overdue need to transform the bureaucratic culture of Nigerian agencies to enable IT-based business processes.

The magnitude of the inertia problem is also demonstrated by the implementation difficulties of the 1988 Public Service Reforms, which were launched with fanfare in 1988 (Tiamiyu 1993, 2000b). IT projects stand the risk of being quickly stifled by such hostile bureaucratic environment.

Table 3: IT for public agencies: strategies emphasized by top Nigerian public servants in 2000.

Suggested Strategies	*Respondents: Number*	%
Hardware/software		
01 Maintenance culture	23	6.5
02 Procurement & installation of IT equipment	22	6.3
03 Development of appropriate software	1	0.3
04 Efficient electronic equipment	1	0.3
05 Periodic upgrade of computers in ministry/local govts.	1	0.3
06 Honest & sincere repairing	1	0.3
Human resource development		
07 Quarterly staff training on use of computer	47	13.4
08 Staff should be encouraged to be computer literate	9	2.6
09 General enlightenment about IT equipment	7	2.0
10 Motivation & social awareness	6	1.7
11 Attendance of workshops on importance of IT	5	1.4
12 Excellent remuneration and working environment	5	1.4
13 Refresher programmes to up-date skills	3	0.9
14 More staff to be employed	3	0.9
15 Staff should be encouraged to go on computer course	3	0.9
16 Computer training made compulsory for employment	1	0.3
17 Accounting/secretarial staff should have computer knowledge	1	0.3
Organizational development		
18 All ministries should be supplied with computers	8	2.3
19 Computers to reach all sections	7	2.0
20 Create zonal offices for IT training	4	1.1
21 Capacity building for information management systems	3	0.9
22 Computer training centers by federal/state/local	3	0.9
governments	3	0.9
23 Survey of data/information needs of each ministry	2	0.6
24 Creating a department for IT	2	0.6
25 Using computers to process information	2	0.6
26 All heads should have IT	1	0.3
27 Seeing computers as work facilitators than status symbols	1	0.3
28 Offices should have telephone, fax, e-mail	1	0.3
29 Monitoring and evaluation of IT projects		

Finance		
30 Financial support for IT	9	2.6
31 Fund should be voted in the annual budget	5	1.4
32 Adequate funding of IT training programmes & activities	2	0.6
33 Ministries/departments should be given back-up money	2	0.6
Socio-economic and infrastructural policies		
34 Making computer prices affordable to customers	12	2.3
35 Enacting government policies for IT development	8	3.4
36 Adequate power supply	5	1.4
37 Government subsidy for IT equipment	2	0.6
38 NITEL phones should be made available to all	2	0.6
39 Make available radio message, fax, e-mail	2	0.6
40 Creation of data banks	2	0.6
41 NITEL to charge very low for use of internet	1	0.3
42 Positive attitude/policies to IT	1	0.3
43 Emphasize use of computer at local government level	1	0.3
44 Computer knowledge should be taught in secondary schools	1	0.3

Finally, note the high rating of "*06. Overemphasis on political factors*," and the fact that "*10. Lack of funds*" barely made the top 10 list. Although finance is an important constraint, it is overshadowed by problems of organizational inertia toward requisite organizational changes for the sustainability of IT development programs.

Table 3 summarizes the strategies recommended by the respondents for promoting IT development in Nigerian public agencies. The recommended strategies have been categorized under various main headings, but there is no doubt that strategies for staff training in computing, and for improved maintenance culture, got the most frequent mention. Also mentioned frequently was the need for procurement and installation of IT equipment across agencies (Strategies 02, 18, 19), a reflection of the currently underdeveloped IT infrastructure in most Nigerian agencies. Also mentioned, although less frequently, are socio-economic and infrastructural development strategies such as appropriate IT policies and subsidies, improved access to telecommunications, and reliable public electricity supply (Strategies 34-44).

The data summarized in Tables 2 and 3 show that the respondents (comprising mostly leaders of the sampled Nigerian agencies) are keenly aware of the infrastructural and IT skills deficit in their agencies, and would support strategies to solve these problems. However, although they had seen and rated highly some of the culture-related constraints (Constraints 3-6, 11, 13 and 14) earlier on in the questionnaire, they did not themselves recommend strategies for dramatically transforming the culture and business processes of their agencies to facilitate the establishment and sustainability of knowledge and IT-based systems and processes. Either they were satisfied with the status quo, or at best, were unaware or ambivalent about the need to dramatically transform the structure and culture of Nigerian agencies for modern information technologies.

Conclusions and Recommendations

We began this chapter by observing that the effectiveness, efficiency and sustainability of organizations depends on their socio-technical balance. Socio-technical balance is related to concepts of organizational rationality, which determines the relative values that an organization and its stakeholders would place on various technical-economic, political, socio-cultural and other variables in IT decisions. Hence, our analyses and discussion were geared to highlighting the determinants of the organizational rationality and socio-technical balance of African public agencies.

The discussion has highlighted the great potential for bureaucratic traditions and socio-political and parochial stakeholder interests to dominate the decision-making processes of African agencies. A related problem is the possibility of mismatch between foreign IT culture and the existing bureaucratic cultures of developing countries, as had been highlighted by Galliers and Sunderland (1991), Madon (1992) and Ojo (1992). This suggests that either incoming IT culture and/or the traditional bureaucratic culture of African agencies would have to be modified to facilitate IT development in the agencies. Parochial stakeholder interests must also be effectively counter-balanced by organizational-level and social development goals. However, there is the question of whether we now know enough about actual socio-technical imbalances and cultural mismatches to enable recommendations on specific IT development strategies. In this regard, Sahay and Walsham (1995) have called for more rigorous and stakeholder-focused research toward theories of IT development in developing countries, whereas Montealegre (1999) has counseled qualitative case studies of IT implementation experiences in different contexts as the basis of practical IT development strategies.

A number of pertinent questions were asked at various times in the discussion. These can be summarized as follows: Do African public agencies and their various stakeholders agree that rapid and significantly improved effectiveness and efficiency are desirable and achievable organizational goals? Do they also believe that IT can serve as a catalyst for achieving these goals? Are they eager to exploit IT for transforming the vision and operations of their agencies? Assuming 'yes' responses to all the preceding questions, are they eager to implement appropriate strategies for requisite IT development and organizational change? Affirmative responses to each of the above questions is imperative if African agencies are to be able to exploit IT to improve their roles in African socio-economic development processes. In any event, strategies are needed not only to ensure consistently affirmative responses to the questions, but to ensure that progress is actually made toward IT development for enhanced efficiency and growth of the agencies.

In conclusion, we subscribe to the view that there are fundamental socio-cultural, institutional and infrastructural constraints to the rapid and sustainable IT development in African agencies. The constraints subsist both within the agencies themselves, as well as in their external environments, and complementary strategies are needed to eliminate or reduce these constraints. Hence, we suggest here a number of organizational (within the agencies) and environmental development strategies that African governments and agencies should pursue to stimulate IT development in their agencies.

- *Human Resource Development for Organizational Cultural Transformation*: African public agencies need to implement an optimal mix of in-house, off-site, formal and informal IT training program for different categories of their personnel. IT training program must also be synchronized with IT infrastructural development program as much as possible.

In-house IT training would require, for instance, in-house computer training laboratories and trainers. However, such computing-related program should be combined with program for dramatic improvement in the capacities of African agencies to learn, adapt and innovate new philosophies, technologies and business processes toward promoting the following IT-supportive orientations in the agencies: information access; information and decision-making accuracy and timeliness; information dissemination, pooling and sharing; customer-oriented approach to public services; and IT as a lever for business process re-engineering.

- *Sustained Development of IT Infrastructure*: African governments and agencies must implement program for the systematic development and use of IT in public agencies, and particularly in those agencies and information processing tasks and services where effectiveness and productivity payoffs are high. To that end, African governments and agencies should: (i) select and prioritize governance improvement goals in terms of improving quantities and qualities of existing and new public business processes and services; (ii) develop and evaluate alternative IT projects for achieving the goals; (iii) develop viable and sustainable project implementation plans, and (iv) continually monitor, evaluate and update the operational reliability and impact of each project (UNECA 2000:25). African governments and agencies must also articulate and implement short, medium and long-term plans for the development of IT capabilities.

- *Integration of IT Development Programs into Public Service Reform Programs*: IT development program cannot be implemented in a vacuum; they must be properly integrated with public service reform program. Among the usual objectives of public service reforms (Kaul, 1995; Rasheed, 1995) are: down-sizing and improvement in pay packages (IT can be used by the reduced but better remunerated personnel to achieve higher-level output); professionalization of personnel and sound policies for recruitment, training and management of personnel (training of existing personnel in professional IT applications, recruitment of professional cum IT-knowledgeable personnel, etc.); customer orientation and improvement in interfaces with the public (through IT-enabled computerized business and citizen registration services, and databases, websites, publications, information dissemination, etc.); greater accountability and ethics (computerized and networked records can facilitate the oversight functions of statutory bodies such as the legislature, regulatory and auditing bodies, non-governmental and civil society organizations, etc.).

- *Creation of Universal Access to Communication Technologies*: Telecommunications would provide the channels through which civil society, public agencies, mass media, private businesses, NGOs and other stakeholders can be linked together. In particular, affordable telecommunications services will stimulate the communication and sharing of ideas as well as participation in governance among the citizenry through telephone, radio and TV. An information-sophisticated society would in turn exert significant pressure on public agencies toward improving their visions, processes, and services.

- *Promotion of Mass Participation in Governance*: Mass literacy and education, mass communication, mass mobilization, and socio-economic empowerment for participation in governance are mutually reinforcing. Mass educational program, such as Universal Basic Education Program, should be effectively implemented towards raising significantly the literacy and educational level of Africans. Higher literacy will empower the citizenry to appreciate and demand more intellectually nourishing information, as well as their right of access to public information. Governments should also stimulate active participation in governance by the citizenry through mass mobilization program, open government, and improved public access to public information.

- *Promotion and Patronage of local IT industries by African governments/agencies*: African governments can also use their agencies to promote the development of national IT industries by implementing policies to patronize the products of such industries, such as the local assembly and manufacture of IT components, and software and content development. Apart from helping the agencies to develop their IT infrastructure, such a strategy can also be used to promote other developmental objectives, including reduction in unemployment levels arising from the growth of IT industries, and improvement in trade balances initially through IT import substitution and ultimately export promotion. In other words, African public sectors should strike IT development partnerships with their private sector for their mutual benefit and socio-economic development.

CHAPTER 8

INFORMATION TECHNOLOGY AND TELECOMMUNICATIONS: LEGAL IMPLICATIONS

Deborah Akoth Osiro

Introduction

It is rather disheartening to hear remarks from certain quarters discouraging the transfer of information technology (IT) aid to Africa. However, this should not be regarded as a bad omen for the rapid development of IT in Africa. African countries should ensure that other industry donors have no cause to discontinue extending IT aid. Facilitating the dissemination of IT application and benefits to the populace can achieve this. Providing the necessary infrastructure to generate investments in IT and to increase IT usage can do this. Unlike other goods and services, information technology value increases with every new user.

A shortcoming that may hamper the development of IT in Africa is the regulatory framework of the telecommunication sector, a very important infrastructure for facilitating the use and spread of information technology.[1] Governments need to establish appropriate telecommunications policies and regulatory structures to create an enabling environment for investing in IT. The credibility and effectiveness of a regulatory framework and hence its ability to facilitate investments varies with a country's political and social institutions.

The discussion begins with an examination of the interaction between IT and telecommunications and the relevance of the international legal regime in this highly globalized sector. The second part provides a close scrutiny of the basic legal requirements for the telecommunications sector that facilitate the transfer and application of IT.

Significance of IT on Telecommunications

Developments in IT

The significance of IT evolution on the telecommunications industry has partly contributed to the need to restructure the formerly traditional and monopolistic industry. To begin with, the evolution of IT has created an explosion in the number of functions to which the traditional telecommunication networks can be applied. Essentially, it has increased the value of telecommunications as a major infrastructure determinant in the desirability of a country as a potential investment haven. The increase in diversity and amount of IT applications required the upgrading of telecommunication network capabilities to enable them to cope with the traffic. Changes were required to facilitate the efficient allocation of resources on the traditionally monopolistic market. Deregulation and privatization of the sector created an opportunity for more players. This in turn has completely changed the regulatory framework.[2]

Moreover, the IT innovations and the speed at which these innovations are rendered obsolete have created a need to redefine regulatory boundaries. Even the countries which liberalized the sector as early as the 1980s admit that the pace of technological change has been overwhelming. Countries are being forced to come up with a framework that can keep up with the pace and not be rendered quickly obsolescent. IT clearly highlights the global nature of telecommunication systems and this makes it vital for African countries to keep up with the technological advances to decrease the digital divide between them and the rest of the world. Furthermore, for a country not to embrace IT means that a large section of the population will not benefit from IT which increases the digital divide within a country.

IT has affected telecommunications through the convergence

of the traditional modes of communication i.e. telecommunications, computing and broadcasting. This has created a nomenclature problem that is causing regulatory hurdles. What could earlier be considered as an area or a part of telecommunications clearly defined by certain physical characteristics can no longer be so defined. Convergence is rendering the former definition of services and markets within the industry a nullity. Hence, retaining old forms of classification and definitions adversely affect consumer benefits and prevent service providers from utilizing the most appropriate configuration for offering services.[3]

The main issues that have emerged requiring regulatory restructuring as a result of these developments include inadequacy of the general competition policy, interconnection (competition in the local loop, pricing, and rights of way), ownership structure, licensing for third generation (3G) mobile services, and convergence and infrastructure requirements for e-commerce.[4]

Basic Framework

The growth of the telecommunications sector facilitates the exploitation of IT. An ideal market for the industry should include a strong pro-competitive foundation that fosters growth among the existing operators as well as providing incentives for additional investment into the sector. The basic features of this regulatory framework include application of sector specific rules as opposed to generic rules for certain aspects of the industry, autonomous regulation, efficient and equitable authorization procedures, and well established interconnection principles and rules of operations.

Another essential characteristic of the telecommunications regime is flexibility to anticipate the fast changing innovations in IT that affect the industry. The framework should not be difficult to implement and enforce by the relevant legal, social and political institutions. The convergence of the three traditionally distinct communication sectors requires the regime to have room for co-ordination among the different regulators as well as the global nature of the sector. While telecommunication is mainly concerned with mode of delivery of services, broadcasting and the Internet sectors constitute mainly contents being delivered.[5]

An examination of these requirements will draw on the communication legislation of Kenya. This will serve as an illustration of the strengths and weaknesses of telecommunications legislation in Africa. The analysis will also refer to the international practice, namely the European Union (EU) framework and the World Trade Organization's (WTO) basic telecommunication Annex (BTA).[6] The discussion is premised on the fact that the telecommunication industry in question is liberalized or is in the process of being liberalized.

The Kenyan government's objective for the telecommunications sector is to optimize contribution to the development of the Kenyan economy as a whole by ensuring the availability of efficient, reliable and affordable communication services throughout the country.[7] The main legal framework governing telecommunications is the Kenya Communications Act (KCA). This is the primary instrument employed to liberalize the sector. With the restructuring of the former monopoly, Kenya Post and Telecommunication Corporation (KP&TC), three institutions were created: Communications Commission of Kenya (CCK), the industry regulator, Telkom Kenya Limited, and the Postal Corporation of Kenya. In addition, Kenya made commitments to the WTO's framework for the provision of basic telecommunications services. The influence of the organization in hastening the liberalization process for most African member countries cannot be ignored. Also, due to the trans-border and market-based nature of telecommunications, the need to have an effective arbitrative forum for future disputes is increasing the role of this international trading regime.

The World Trade Organization Framework

The WTO framework incorporated within the general provisions of the General Agreement on Trade in Services (GATS) and the BTA plays a central role in the liberalization of telecommunications services.[8] Under the general GATS obligations, member countries are required to accord most favored nation treatment to all other members in the area of trade in services.[9] Service *per se* is not defined but the modes of its supply are described to include factors pertaining to service provision, consumption, production elements and labor.[10] This may sound very liberal, except that most countries in signing the Agreement requested for a number of exemptions from this provision, telecommunications services included.[11]

Another relevant GATS principle is the requirement to accord national treatment (NT) to all services regardless of the nationality of the provider.[12] However, this provision is specific to a sector and only applicable on a reciprocal basis. Mainly developed countries and a few developing countries made commitments to extend NT to the telecommunications sector.[13] The duty to accord market access is a major obligation within the GATS framework as its commitment may surpass NT.[14] Similarly, it operates on a reciprocal basis and most countries have made very few commitments regarding this obligation. Generally though, most developed countries liberalized telecommunications investments and cross-border movements of telecommunications services.[15]

With the ratification of the BTA, the members drew up a Reference Paper on basic telecommunications, which sets out the following regulatory principles for the sector:

- *Competitive safeguards* such as measures against the anti-competitive practice of cross-subsidization and use of information obtained from competitors; and surveillance against exclusionary practices by dominant carriers.
- *Interconnections guarantees* which comprise for instance, imposition of obligations on the dominant carrier to ensure access at any technically feasible point in the network for other operators. The interconnection, sufficiently unbundled, should be provided on non-discriminatory terms, in a timely fashion, and on terms and at cost oriented rates that are transparent and reasonable with regard to economic feasibility. Procedures should be publicly available and a dispute resolution mechanism provided for by an independent entity.
- *Universal service policies* that harmonize competition with regard being paid to members' right to define the kind of universal service obligation they wish to maintain.

- *Transparent licensing* which calls for public availability of licensing criteria as well as the terms or conditions of individual licenses.
- *Need for an independent regulator* who should be separate from and not accountable to any operator. Moreover, its decisions and the procedures it applies should be impartial.
- *Provides for fair allocation* of scarce resources with transparent, objective, timely and non-discriminatory procedures.

The WTO provides a forum for making global telecommunications decisions. However, although its members are governments, it is profit motivated and competitiveness is its catchphrase. On the other hand, the International Telecommunication Union (ITU), the long-established international telecommunications body, is concerned with the efficient use of telecommunications resources. Nevertheless, it has also been affected by the developments in telecommunications, which are forcing the organization to re-examine its traditional role. Membership has traditionally been restricted to states only but the ITU is now under increasing pressure from the existence of many private industry players. Furthermore, it has no formal executive powers, and no powers of enforcement. There is need to redefine the ITU if it is to continue being beneficial to African countries. Otherwise, the WTO may be the sole forum for setting telecommunications standards.

Legal Requirements

Specific Competition Rules

The telecommunications field, unlike the other economic sectors, has several features that make it peculiar. These include the complexity of the market structure, obligations and privileges attached to different operators, the technical nature of the industry as well as the scarcity of some of the resources. These peculiarities tend to render the general competition framework of a country inadequate for ensuring sufficient public access to the telecommunications services and for facilitating the growth of the sector subsequent to a privatization exercise.

Presence of a Dominant Operator

Most telecommunications markets have been made competitive recently leading to the presence of a dominant operator in the market. In principle, competition laws apply to anti-competitive market behavior, hence inapplicable to the telecommunications market, which started as a monopoly structure and is undergoing a transition to a competitive market. The proactive application of rules calls for asymmetric regulation that includes imposing obligations on the dominant operator with a view to assisting the new entrants to become competitive.[16]

CCK, the Kenyan industry regulator, granted a five-year monopoly to Telkom, the incumbent, in the provision of public telephony services for the Nairobi area.[17] This is because a significant proviso concerning competition in the industry is not in effect (i.e. neither CCK nor the Minister can grant a license to create a monopoly in any industry sector).[18] The monopoly granted to Telkom is viewed as a factor that will stunt the growth of this sector.[19] The application of Kenyan competition law in this regard would prove inadequate as it grants broad discretion to the Minister responsible when determining the legality or otherwise of an incumbent's actions. Thus, if applied to the sector this would have adverse effects instead of ameliorating the situation.[20]

This situation makes it rather complicated to determine the abuse of dominant position by the incumbent, or the existence of certain restrictive trade practices by the same in the relevant market merely by applying generic competition law.[21]

Ownership of multiple telecommunications services by the dominant carrier poses another problem. This brings about a conflict between the operator's obligation to provide fair and equitable interconnection services to other telecommunications operators and a desire to see its subsidiary succeed through granting it privileges denied to others. Flexibility to deal with the dominant operator is required to balance the incumbent's obligations to other operators and concessions granted because of, for instance, the universal service commitments.

Universal Service Obligations

Basic telecommunications service is considered an essential facility. The policy maker has to create a regulatory framework that guarantees provision of the service while maintaining a pro-competitive environment. Universal service obligation is generally a condition for the grant of a license particularly for fixed telephony. The concept is concerned with social justice aims to provide universal access to the less well off in society so that they benefit from telecommunications services.[22]

This entails the imposition of certain development obligations. Each country defines universal service to suit its circumstances. Whatever the definition, this would impose certain obligations on telecommunications service providers causing them to lose a competitive edge.[23] To enable them to be at par with other operators they may be exempted from certain actions which otherwise would be considered anti-competitive if the general competition policy was applicable.

Under universal service considerations, public interests arguments tend to be utilized by regulatory bodies to impose a burdensome condition on potential market entrants or to prohibit foreign ownership or control of a corporation.[24] The Kenyan government proposes to enact appropriate regulations and procedures that require all industry operators to contribute towards provision of universal service and access.[25] This arrangement is prone to be abused in Kenya where there are no established regulatory and institutional checks.

Convergence and IT

The fast pace of transformation in the sector, the introduction of new products into the market and the inevitable convergence with the other communication sectors also create a need for sector specific rules for the transition to a competitive market. Mobile service is an area that faces major regulatory bottlenecks presently particularly with the allocation of the third generation (3G) licenses. This license will allow for the development of advanced multimedia services, which implies that there may be a rule for content providers in service provision on the 3G infrastructures, hence an examination of services access to mobile infrastructure.

Problems already exist with these 3G licenses related to the high debts incurred by the operators in Europe that are inhibiting further development for the sector. There are pleas by the 3G licensees to be exempted from certain license's obligations to enable them utilize the licenses.[26]

Moreover, anti-trust principles being generic are not suitable for the telecommunications industry because the sector is highly technical embracing new transactions that had not been anticipated by most existing competition laws in Africa, including technology acquisition and transfer, hybrid services and intellectual property licensing. A review of their telecommunications rules should come up with a better structure that accommodates all the challenges and limitations of the general legal framework.

The EU Framework

The EU has come up with a new framework for the industry, which generally addresses all these issues.[27] The legislation applies only to the distributive networks (contents issues are addressed in another framework). Nevertheless, it is applied across the board. The proposed Directive aims to establish a harmonized regulatory framework for electronic communications networks and services across the EU. It seeks to respond to the convergence phenomenon by covering all electronic communications networks and services within its scope.[28]

Some of the general features of the proposal for the new electronic regulatory framework include:

- Updating the existing legislation to be technologically neutral and to cover new communication services
- Harmonizing the rules for authorizing provision of such services with the aim of creating a single market
- Establishing a framework for access and interconnection agreements across the EU
- Setting out users' rights in particular in respect of universal service, and
- Requiring local loop unbundling.

In coming up with the proposal the European Commission acknowledged the inadequacy of the regulatory framework in place, which was primarily for facilitating the transition from a monopolistic market to a competitive market. The new framework thus applies to a competitive market. The competition rules apply in the context of the framework as its scope generally addresses the common agreements such as those relating to significant market power, market analysis of this, standardization and harmonization procedures, and the powers and functions of the national regulatory authorities.

Independent Industry Regulator

Formerly, there was no division of responsibilities as is apparent today because telecommunications companies were state-owned. With the liberalization of the sector, three distinct institutions have emerged – regulatory, policy-making, and telecommunications operations. In Kenya, the former monopoly, KP&TC was the operator as well as the market regulator. CCK, as the new regulatory body was created with the objective and purpose of licensing and regulating the telecommunications industry through promoting private investment, acting as a link between the market operators to ensure a level playing field, assigning frequencies, and protecting consumer interests through standards maintenance.[29]

The establishment of an independent regulator essentially requires the entity to be separated (need not be physical) from the telecommunications operators and the ministry responsible for the sector. Some countries opt to have the ministry as the regulator but with a distinct division of responsibilities. In rare cases, the competition authority is responsible for the regulatory functions of the sector.[30] The trend began in the 1980s in the UK, U.S. and Canada where the regulatory reform process began with the separation of telecommunications operational functions from policy functions with the creation of a telecommunications operating entity.

The interaction between the regulator and the competition authority may be illustrated through their respective areas of competence. In general, the telecommunications regulator is responsible for technical regulation (type approval, spectrum allocation, number allocation and standard setting), which is difficult to incorporate in general competition rules and requires the regulator to have professional knowledge of the industry that is technology oriented. The regulator also oversees specific economics and social regulations (licensing, price regulation, rights of way, interconnection, universal service, and the interconnection regime).[31] These require a regulator who takes a proactive position (*ex ante)* to develop a competitive environment as opposed to the reactive position of the competition authority. The latter is responsible for the elements that are not exempted from the general competition application or are within the telecommunications regulator's jurisdiction. Still, the presence of the independent regulator in no way diminishes the role of the competition authority in this respect.[32]

The WTO Reference Paper describes an independent regulator as one who is separate from, and not accountable to, any supplier of basic telecommunication services. The interpretation of the degree of independence varies from country to country, but it should not hamper the liberalization of the sector. This can be determined by the power of decision-making, the relative functions of the sector assigned to the body, budgetary independence, and composition of the body. The composition of the regulatory body is an important determinant on whether it will be independent and effective in undertaking its functions.[33] The Kenyan tendering process for the second cellular phone provider was widely believed to have been full of irregularities, lacked transparency and there was a strong claim of regulatory capture from the government.[34]

As telecommunications is gradually converging with the broadcasting regime, it is quite necessary for the framework to delineate the extent of the regulator's power to determine the scope of their jurisdiction. For instance, the CCK mandate in the allocation of radio and television broadcasting frequencies is unclear. The KCA does not give any guidelines with reference to broadcasting frequencies. The CCK is in charge of the content of the telecommunication network, but it is not clear whether, for instance, a cable provider would fall within its jurisdiction.

The regulatory instrument should explicitly spell out the regulator's functions as well as give them a certain amount of discretion. Too much discretion may prove detrimental to the sector particularly in Africa's present environment where there are insufficient political and social institutional checks and balances. Unless a country has such or other similar restraining mechanisms, the discretion should be granted with caution.[35] The US experience provides illustrations of potential pitfalls for a regulatory agency.[36] The CCK has a wide range of regulatory discretion without any apparent restraining mechanisms save for the judiciary.

Market Entry

The industry regulator is normally charged with authorization of telecommunications operators. The basic tool of authorization is licensing which serves to prohibit unapproved operating of systems.[37] This regulatory tool is the primary method which African countries have used to restrict private investors especially foreign ones. For the authorization process to constitute a low entry barrier and guarantee operational continuity, it should be flexible, cost affordable, have minimum and simple procedures, and be enforceable. The process should be transparent with the terms and conditions explicitly outlined.

Licensing

Some jurisdictions have no formal entry requirement while others only require registration. In many countries, licensing is essential for one to provide telecommunications services. A license may be exclusive or non-exclusive, the former tending to be personal and granted for the scarce resources thus requiring restrictions on entry. The CCK regulates market entry for the communications sector through licensing.[38] This license is specific to a certain class or person(s). The terms of the license reveal the conditions to be fulfilled by the licensee and outline the licensee's obligations and rights as granted by the license.

While the KCA outlines the procedure to be followed by an applicant prior to application, the CCK is empowered to determine which conditions are to be fulfilled and these are unclear. As a result, many potential market entrants develop the impression that the operating environment may be subject to vague regulations.

The licensing requirements should be expressly outlined to create an equitable system of authorization. The regulator should adopt a licensing policy that meets the needs of the market for operators and consumers as well as achieving broader national policy goals.

The mode and fees charged in granting the license will depend on whether it is exclusive or non-exclusive. For exclusive licenses, the common methods used are tendering and auctions. The latter is more transparent than the former yet it is proving to be discriminatory in favor of dominant operators.[39] For non-exclusive licenses determinants would include an operator's commitment to provisions of network infrastructure, population of the region, and the type of license already held by the operator. For both types of licenses, the nationality of the applicant is considered.

Foreign Ownership

Most African countries insist on a certain amount of local control. This applies to personnel, equity ownership as well as portfolio investment. As mentioned before, the WTO regime prohibits such measures.[40] Regulating foreign investment is usually justified by the need to protect local industries and to discourage negative externalities that may arise when foreigners obtain control of critical economic sectors.

In Kenya, 60 per cent local ownership is necessary in order for one to be licensed as a telecommunications operator.[41] This provision has proved to be a stumbling block despite the fact that it is subject to review from time to time in line with the country's investment and socio-economic needs.[42] Limiting the amount a foreign investor may invest restricts their degree of control but also discourages investment. Further, the provision only affects initial market entry; it does not impose similar ownership restrictions once the company is operating. Thus it does not make sense to limit the initial grant if the purpose of using this regulatory tool is to encourage local entrepreneurship, as this would fail. Portfolio investment by foreigners is also affected by the local control provision. Attempts to attract foreign financial resources through

African capital markets often fail due to this kind of restrictive regulation that may lead to unpredictable risks for the investors.

Authorization is by no means a guarantee of actual market penetration. Entry is virtually impossible without access to the telecommunications networks of the major incumbent. Hence, the need to ensure fair play by the regulator in the operating environment through pro-competitive interconnection regulations.

Interconnection Issues

IT and liberalization have had a significant impact on telecommunications including the growing requirements for interconnection and interoperability. The most contentious interconnection issues that emerged after liberalization included the need to provide access to appropriate facilities to minimize duplication of resources, the timeliness of providing it, the price structure for interconnection, and the need to maintain network integrity. Also the setting and maintenance of standards as well as guaranteeing consumer protection play a crucial role in sustaining competition and growth in the market.

Interconnection

This is the connection of one service provider to another or a consumer to various service providers. Policy makers view it as the key to creating favorable market entry opportunities and conditions for competition. Lack of efficient interconnection systems may lead to anti-competitive effects as it acts as a barrier to entry, prevents competition, deters future competition, and freezes technology. There are no consistent guidelines that have emerged internationally for interconnection. Each country has tackled the issue of interconnection and evolved its own approach in keeping with its own legal principles and regulatory framework. Many countries use their competition law and policy, though this has not been effective.

In as much as interconnection should be left to market forces, most telecommunications regulators including the CCK retain a residual regulatory role in relation to these issues. And the CCK's commitment to creating a competitive telecommunications environment in Kenya is severely being tested by the standoff with regional telecommunications operators (RTOs) over the interconnection conditions of the licenses.[43]

Special obligations attach to operators with significant market powers to meet request for interconnection points, to publish reference agreements and price lists, to act in a non-discriminatory manner, to unbundle, and cost-orient interconnection policies and to put in place suitable cost accounting schemes.

Tariff Policy

A tariff is a scale of charges for a business or public utility. Telecom operators derive the bulk of their income from two sources. One is fixed charges through connection to the networks and subscription or rental of a line to a local exchange and the other is call (usage) charges. The most important principle underpinning the management of tariffs by operators is that tariffs and charges should reflect both the structure and level of the relevant costs.

The implementation of telecommunication tariffs that follow overall cost trends is a prerequisite feature of a legal framework. Not observing cost-oriented pricing, operators undermine their ability to invest in upgrading and expanding their networks. This not only gives a poorer class of services to existing subscribers but also deprives many potential subscribers of services. The use of price cap mechanism ensures that price changes are kept at arms length from political or other considerations.

The CCK is empowered to protect the interests of all users of telecommunications services in Kenya with respect to the prices charged for and the quality and variety of such services and in consultation with the Minister, it may make regulations generally with respect to telecommunications services including fees and other charges.[44] The competition legislation empowers the Minister to fix maximum prices for certain goods and services of which telecommunications services are included.[45] If the Commission fails to regulate the prices then the Minister may well do it.

In Africa telecommunications tariff is often determined by political considerations rather than by economic needs. Tariff re-balancing is essential in African countries where for example income from international charges is inordinately disproportional to the local charges. These do not reflect market dynamics. Also, increasing availability of call diversion technologies may enable domestic users to bypass some of the high international charges. Therefore, African countries will have to review their tariff policies.

Access

It is essential to facilitate interconnection, which the incumbents view as a competitive threat. This gives rise to the issue of access to network termination points, which involves full unbundling, shared access and collocation. The main bone of contention is full unbundling of the local loop. Unbundling would facilitate efficient utilization of the network resources. In addition to the incumbent's reluctance to unbundle there is the problem of defining the local loop still based on the old telecommunications set-up.

Due to the IT advance, this is unsustainable and is already causing difficulties.[46] Both from an interconnection perspective and the technological perspective, geographic boundaries are becoming artificial and need to be eliminated. The local loop is repositioning from merely being an access point to the customer to a 'window' on the range of providers and services available simultaneously in the market.[47]

The legal framework provides for interconnection of regional telecommunications operators with Telkom networks, terms of which will be mutually agreed though they will be based on principles developed by the CCK.[48] The CCK is mandated to ensure that operators adopt judicious employment of an appropriate technological mix in order to minimize the cost per line. Yet its recent treatment of the entry of the RTOs points to a different view.[49]

The problem of setting interconnection charges is related to the standards that are set and have to be maintained in order to ensure that interconnectivity is enhanced.

Standards

A standard is a technical specification relating to a product or an operation, which is recognized by a large number of manufacturers and users and is important in determining interoperability of networks and technologies. Rules, protocols, and standards need to be flexible and future-oriented to anticipate and encourage innovation and new uses. Moreover, technical standards should not be allowed to preserve markets for existing business by blocking market entry through unfair type approval regulations.

Type approval is the administrative process, which controls market access for domestic and foreign suppliers of telecommunications technical equipment. As such it is a litmus test of whether a country's commitment to market liberalization is genuine or mere lip service.[50] Regulating of type approval is intended to facilitate open network access, consumer and operator choice, interconnectivity, interoperability; compatibility, network security and network integrity. The last two are especially important for radio communication apparatus since they constitute critical safety systems.[51]

The main approaches to approving equipment are based on design and performance standards. The internationalization of standards for telecommunications is mirrored in the evolution of standards for information technologies that are particularly sensitive to interconnection and interoperability of services and products. However, due to the increasing use of the Internet for almost all communications, efforts of standardization bodies such as ITU are now geared to bringing harmony to the industry.

User Rights Protection

Finally, the telecommunications framework in guaranteeing the maintenance of network integrity protects both the operators and the consumers' interests and facilitates operator or consumer choice in terms of network owners or service providers to hook up, access to services as well as the provision of basic services for consumers. With regard to the numbering policy (involves the allocation of numbers), obligations should be imposed on the operators to provide number portability, geographic portability and mobile portability. This reduces consumer expenses as there is no need to change service providers or buy a new terminal for every change of product or service use. Another aspect relevant to the user is the provision of carrier pre-selection options. The trend has been moving towards self-regulation by industry in this last area.

Conclusion

From the above discussion, the limitations of applying general competition rules as opposed to sector specific rules have been highlighted as well as the challenges facing the regulatory environment of the telecommunications industry. The need for flexibility and rules that are more sector sensitive has been shown to be significant in dealing with the issues that are peculiar to the industry. Finally, the rules in an African context may only be effective if applied by an autonomous regulator supported by a strong political will and institutional framework. All these adjustments have resulted from or been exacerbated by the evolution of IT in an increasingly globalized world. And the imperative for change within the sector is still being primarily driven by IT through the need to promote investment in, and usage of IT. Africa should exploit this phenomenon.

Endnotes

[1]Telecommunications is the transmission and reception of signals by any electromagnetic means with radio and cable being the basic carriers.

[2]For instance, telecommunications international arrangements were entered into by states for ease of telecommunications service provisions, as they were state-owned. The ordinary devices of public international law were appropriate. Presently the private enterprises (mainly multinationals) dominate the industry with competition being the means to secure efficiency and the interests of the user. Private law is more applicable to the current environment. For further discussion on the changing international face of telecommunications, see generally, Hague Academy of International Law (1999).

[3]The WTO framework uses a very broad and open-ended definition, see WTO (1998). For telecommunications, its definition encompasses all transport networks and services infrastructure. The coverage includes cable TV networks when these are used in the offer of basic telecommunications services. Kenya's communication legislation (to be discussed later) is deficient in this respect. It is

vague in defining telecommunications.

[4] See generally Boylaud and Nicoletti (2000).

[5] These are also trans-border with the main issues of concern being consumer protection, protection against cyber-crime, choice of law, taxation in e-commerce, and intellectual property infringement. However, a treatment of these is beyond the scope of the chapter, which is limited to the distributive side of telecommunications. See generally Osiro (1999).

[6]WTO (1998).

[7]See Ministry of Transport and Communications' *Telecommunications and Postal Sector Policy Statement,* paragraph 2.1.

[8] WTO (1998).

[9]Article II, (General Agreement on Trade in Services (GATS).

[10]Article I, GATS.

[11]The GATS Agreement provided for Member countries exemptions from the application of this obligation in certain sectors provided it was notified before the entry into force of the Agreement. See GATS' Annex on Article II Exemptions and the Annex on Negotiation on basic telecommunications (BTA). Alternatively, exemption from MFN obligation could be requested after entry into force of the Agreement under Article IX, GATS.

[12]Article XVII,GATS.

[13] For some of these developing countries notably South Africa, Colombia, Taiwan and Mexico, the US claims that they are violating the terms of the WTO Agreements for failing to implement their obligations including the NT and market access measures. See USTR (2000).

[14]Article XVI, GATS.

[15]Forty-four countries (occupying 99 percent of basic telecommunications revenue among WTO Members) permitted foreign carriers to possess telecommunications facilities within their jurisdictions, while fifty-four countries promised to adopt market-oriented telecommunications policy (namely, adoption of the Reference Paper). See Takigawa (1998), p.40.

[16]The US Telecommunications Act of 1996 gave the FCC authority to treat dominant carriers more harshly than the antitrust rules. See Takigawa (1998), p. 38.

[17]Para. 3.4, Ministry of Transport and Communications' *Telecommunications and Postal Sector Policy Statement,* paragraph 2.1. This will last until 2004. Telkom was obliged to sell 26 per cent of its equity to a strategic investor and at least 20 per cent to employees and local and international investors through the Nairobi Stock Exchange. Telkom had problems getting a strategic investor partly due to a KSh.20 billion debt inherited from the KP&TC and also due to a foreign equity limitation of 40 per cent, which they increased, to 49 per cent. However, the government still deferred the sale, which should have been finalized in February 2001.

[18] Kenya Communications Act, ss. 5(5) and 23(2)(b).

[19]Whereas there is only restricted competition in the provision of some services such as mobile services and trunked radio system (because of spectrum availability), there is full competition in the provision of most value added network services.

[20]Restrictive Trade Practices, Monopolies and Price Control Act (RTPMPCA), s. 24 gives the Minister of Finance wide ranging powers to allow a monopoly or duopoly or any such concentration of power.

[21]Controversy plagues the privilege granted to Telkom. And the CCK seems to be aggravating the situation by consolidating Telkom's dominance to the detriment of competition in the industry by applying stringent authorization requirements for new operators who need to utilize Telkom's network. Further discussion below.

[22] ITU has been a principal advocate for the universal access of basic telecommunications services.

[23]The EU's definition of universal service includes basic telephony (or its equivalent) and Internet access. Whether the universal service is defined in terms of geographic accessibility, affordable access or equitable access is important in determining the obligations imposed on operators. Given the low per capita income of Kenya, the CCK is likely to emphasis affordable access rather than equitable service.

[24]E.g. Section 310 of the US Telecommunications Act, which empowers the FCC to deny foreign telecommunications operators authority to operate in the domestic market on grounds of public interest. See Takigawa (1998), p. 36.

[25]Para. 3.3.1 *Telecoms Policy Statement.*

[26] The operators complain that the prices of the licenses were too high resulting in illiquidity. This is affecting their ability to create the necessary infrastructure to apply the licenses' benefits. The other operators, particularly the losing bidders, claim that it is unfair and bad for competition to grant any concessions to these operators.

[27]Directive on a common regulatory framework for electronic communications networks and services. COM(2000)393. Brussels, 12 July 2000.

[28]Communications Directive. COM (2000)393. This is defined as 'services provided for remuneration which consist wholly or mainly in the transmission and routing of signals on electronic communication networks, including telecommunication services and transmission services in network used for broadcasting, but excluding services providing...content transmitted using electronic communication networks and services.'

[29] Ss. 2(1) and 3 of the KCA empowers the CCK to perform these functions under the Act. The Act (s. 84(1)) also provides for the establishment of a policy advisory body, National Communications Secretariat (NSC), which is to serve as the policy advisory arm of the government on all matters pertaining to the info-communication sector.

[30] New Zealand is an example of the latter.

[31]Working Party on Telecommunications and Information Services Policies (WPTISP) (1999).

[32] In fact, their role has increased through the abolition of exemptions of applying general competition rules to the telecommunications sector. For example, the US Telecommunications Act, 1996, expanded the Department of Justice's role in reviewing mergers by removing the ability of the FCC had to exempt mergers of local phone companies from anti-trust review. It has increased also through the forbearance of the sector regulator in some areas such as significant market power, price regulation and reduction in the number of companies or services no longer subject to sector specific regulation. In addition, there is an encouragement for self-regulation in the industry. For instance, in Australia, through the development of voluntary industry codes of practices, technical standards, and in the UK Oftel encourages the same. See WPTIS (2000).

[33]The composition of the CCK board of directors raises questions on its integrity and many believe that government officials heavily influence it in a way that may be in conflict with industry interests. The board of directors includes four Permanent Secretaries (from the ministries of communications, finance, internal security, and information & broadcasting) out of a possible 11 board members, (s. 6, KCA).

[34]There was a scandal linked with the tendering for the second mobile phone provider in the country. The major players and bidders in the industry insist that the CCK had come under undue pressure from politically connected operatives to flout its tendering procedures and disregard other rules and regulations. See also "World Bank shocked by telephone tender scam" *East African Standard*, (Nairobi), 26 August 1999.

[35]These may be substantive, procedural (such as those changing the regulatory system) and institutional constraints. See generally Levy and Spiller (1994).

[36]The FCC applied the incipiency theory in enforcing competition policy within the sector. The regulator could freely stretch the areas of illegal conduct and therefore dominant firms had to refrain from aggressive competition, which got stifled instead of being promoted. Thus the regulatory discretion should not be too wide ranging. In the US the case of *Microsoft* exemplifies the courts' and the competition authorities' endeavor to strike up 'an appropriate balance between the need for checking exclusionary behavior and the need for allowing dominant firms to compete aggressively.' Takigawa (1998) p. 38.

[37]For instance, s. 24, KCA, provides that no person is allowed to operate a telecommunications system or provide any telecommunications services except in accordance with a valid license granted under the act otherwise it is an offence.

[38]KCA, s. 5(1).

[39]Such claims are being made against the auctioning of the 3G licenses in Europe, as their cost has been so high that the possibility of newcomers into the market was greatly limited by the financial constraint.

[40]Trade Related Investment Measures (TRIMs) Agreement as well as the principal obligations of NT. In the US before the enactment of the Telecommunications Act 1996, section 310 of that Act formerly prohibited not only foreign ownership in telecommunications company of more than 20 percent holding in stocks, but it also covered corporations for which foreign persons directly operated as managers. See Takigawa (1998), p. 35.

[41]The *Telecom Policy Statement* (1999), para. 3.3.2, provides that, '...the Government will...encourage Kenyans to actively participate in the sector through equity ownership. Consequently it will be a requirement, initially, that any firm licensed to provide services in the liberalized market segments should have at least 60 per cent of its equity owned by Kenyans.' The provision applies also to investment through the security markets. See the Capital Markets Regulations, Foreign Investors Regulations, Legal Notice No. 3 of 1995, rule 2.

[42]Telkom's problems in getting a strategic investor forced the government to lower the proportion to 51 per cent for the corporation's proposed privatization.

[43]The three RTOs have refused to pay the license fees as they disagree with the terms of their license as defined by the CCK. The RTOs want to have inter-regional connectivity, which would enable them to bypass Telkom's switches; otherwise they would incur additional costs and loss of control. *Daily Nation* on the Web, 'Licensing row puts regional phone firms in limbo', 10 April 2001. Available at *http://www.nationaudio.com*. [Accessed on 10 April 2001].

[44]KCA, ss. 23(2)(a) and 27(2).

[45]See the RTPMPCA, s. 35.

[46]Another problem affects the mobile virtual network operators. These require special access, see Bernard (2000). See also generally, Boylaud and Nicoletti (2000).

[47]Bernard (2000), p.39.

[48]In the case of *DGT &Anor (BT) v. Mercury* (*The Times* 3 August 1994) the court thought that parties to an interconnection agreement should be held to the agreement, but this does not in anyway detract the regulator from his role as an arbitrator between the parties in resolving their interconnection dispute.

[49]The interconnection terms of their license require them to go through Telkom even when it is not economically and strategically viable.

[50]See OECD 1992: 23.

[51]KCA s. 40, CCK may by regulations, prescribe technical requirements to be complied with in the case of radio communication specified. In relation to any other apparatus the technical requirements will be determined by what it will consider to be the minimum lawful interference possible to the rest of the network it is connected to.

CHAPTER 9

STRATEGIES FOR LIBERALIZATION WITHIN THE REGULATORY FRAMEWORK

Sebastian Mathews

Introduction

In the last decade of the twentieth century, Africa has emerged as one of the most dynamic telecommunications markets in the world. A wave of liberalization initiatives across the continent has seen a rapid roll out of a host of communication technologies including fixed line, wireless, cellular, Internet, satellite and data services. The introduction of competitive markets has opened the floodgates of foreign investment and the corresponding inflow of investment has in turn generated a virtuous cycle of multi-sectoral economic growth.

Globally we see an increasing move into knowledge-based economies where the primary currency of exchange is information and the value thereof. The Organization for Economic Cooperation and Development (OECD), a grouping of 27 developed countries, estimates that more than 50 percent of the GDP in major OECD economies is knowledge-based (OECD 1996).

Based on this premise, Africa's economic and social growth thus critically depends on the extent and quality of its Information and Communication Technology (ICT) networks as well as its ability to integrate such infrastructure into the fabric of its societies. An African telecommunications revival represents a concerted and directed desire to actively participate in the global political, economic and social arena. The telecommunications revival represents not just better services but fundamentally it provides the medium to generate, access and distribute economic prosperity.

During 1997 for example, Africa registered a growth rate in fixed telephone lines of 9 percent; its highest since 1985 (ITU 1998a: 2-3). If this rate of growth is sustained then the 1997 estimate of 15.2 million fixed lines should double by 2003 (Minges 2000: 38). On the mobile front, almost twenty mobile cellular networks representing more than three million subscribers, came online across the continent by 1998 (Gullish 2000: 60).

In South Africa alone, by late 2000, the number of mobile subscribers surpassed the 5 million mark, fuelled largely by near exponential growth in the 'prepaid' sector targeting lower income markets. This phenomenal growth of competitive mobile telephony in that country saw the mobile subscriber base overtake the state controlled monopoly fixed line subscriber base in just 6 years. As Figure 1 illustrates, using South Africa as a benchmark, the potential for further telecommunications development in the continent remains high. Such mobile service uptake in spite of significantly higher tariffs illustrates the pent up demand for basic communications access especially in the light of the poor quality of service offered by state-owned fixed line services across the continent. Note that the size of the gap represents one indication of the size of the opportunity for telecommunications development in Africa.

On the Internet front, equally dramatic growth is being experienced albeit from a low base. By the beginning of 2001, all countries on the continent had some form of Internet access. South Africa is unique in that its extensive telecommunications infrastructure and Internet Points of Presence across the country has been its Internet community consistently ranked amongst the top twenty countries in the world in terms of the number of Internet nodes. By 1999 there were almost 400 Internet Service Provider's (ISPs) in Africa with South Africa having at least a hundred whilst at least six other countries, namely Egypt, Morocco, Kenya, Zimbabwe, Nigeria and Tanzania had more than ten ISPs (Jensen 2000a: 48).

Figure 1: Total Teledensity in Selected African Countries (1998/1999 total fixed and mobile penetration)

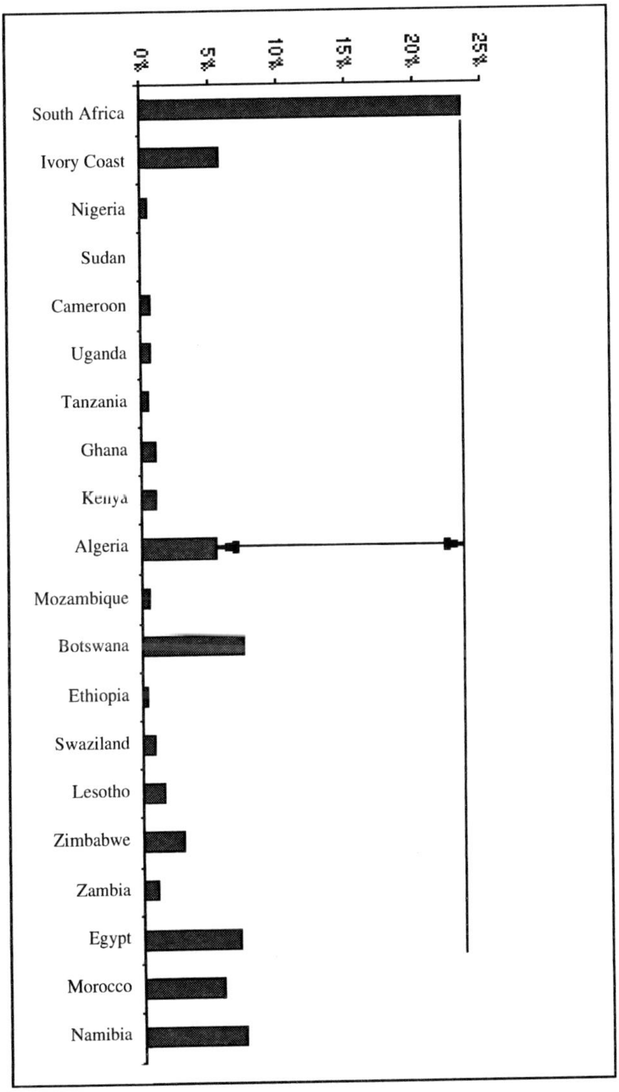

Telecommunications: Africa's Growth Imperative

A significant body of research conducted largely by the International Telecommunications Union (ITU) and the World Bank in the latter half of the twentieth century has brought to bear the increasing realization that an advanced telecommunications infrastructure is a fundamental economic catalyst. These studies analyze the relationship between the gross domestic product (GDP) per capita and telephone density and confirm that telecommunications penetration is a statistically significant predictor of economic growth and vice-versa. A modern and functioning telecommunications system is thus generally accepted today as both a cause and consequence of economic growth. Furthermore, such a system is also considered an indispensable tool for a country attempting to migrate up the economic value chain from an agrarian subsistence based primary economy to a modern service based participant in the global economy.

Telecommunications and the Economic Engine

In developing countries for many years, both the theory and practice of economic development tended to underestimate the importance of telecommunications in economic performance and social development. On the whole, telecommunications infrastructure and services were considered to be less important than energy supply, transportation systems, irrigation projects, or social infrastructure such as schools or health care. Telecommunications was often seen as a luxury when compared to other infrastructure or so called productive investments.

Whilst macroeconomic aggregate numbers may not necessarily be sufficient to assess the actual contribution of telecommunications to development and the level of telecommunications required for a given state of development in any given country, nevertheless strong statistical correlation exists between the GDP per capita and telephone density. Figure 2 illustrates the high correlation between the GDP per capita and fixed line telephone density of twelve African countries and suggests that low penetration may constrain economic growth.

In Africa it is estimated that for every dollar invested in telecommunications infrastructure, three dollars are added to the GDP (Telkom 2000: 3). Further, the ITU inferred in a 1998 study that the lower the gross national product (GNP) per capita of a country, the greater the "contribution" of each telephone to the GNP (ITU 1998b: 9). In effect, the less developed a country is, the higher the influence of one telephone on the GNP. However, since the economic impacts take time to become apparent due to diffusive time lags, these effects are often unmeasured and hence unquantified.

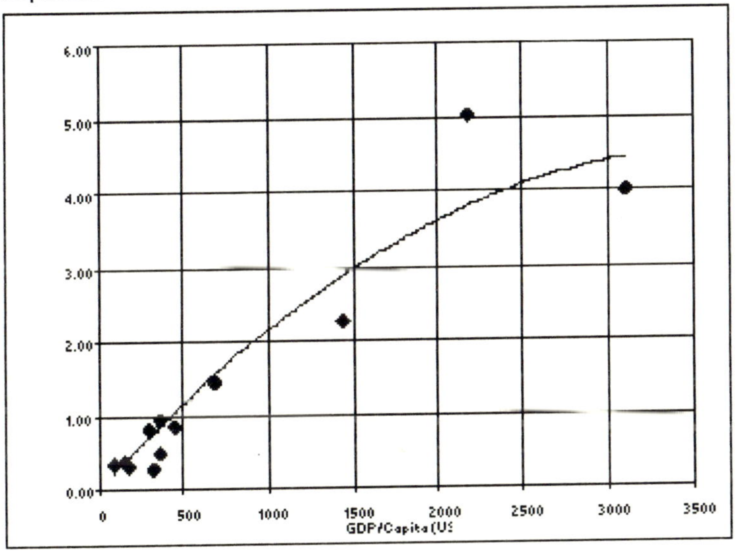

Figure 2. Penetration versus GDP per capita of 12 African Countries

However it is important to consider not just income generation but income distribution as well in assessing telecommunications contribution to economic growth. Developing countries need access to world markets for their products and services, which in turn implies good communications infrastructure. It is not simply that the "haves" are at an advantage, but that the "have-nots" are at increasing risk of social and economic exclusion. Countries that lack a firm ICT infrastructure become marginalized as electronic commerce grows in importance, for example. They are incapable of sharing in the new route to prosperity which electronic commerce affords, and remain dependent on the export of basic commodities, for which the world price is often in decline. As a result Africa's share of world trade has fallen from about 4 percent in 1980 to less than 2 percent in 2000 (Sarker 2001)

Worldwide liberalization and growing globalization of numerous industrial and service activities present not only a major challenge in the form of increased competition but also create new opportunities for the developing and transition economies of Africa. Telecommunications and Internet access can provide a young upstart company the same global reach as a major Fortune 500 company. Similarly telecommunications access can provide a country with:

- freer access to markets of their neighbors as well as to those of advanced countries;
- increased foreign investment as foreign investors are made aware of opportunities;
- and technological, market and production cooperation with foreign-based companies to take advantage of lower production costs and economies of scale.

Evidence suggests that countries that have invested in communication infrastructure have attracted high levels of foreign direct investment as well as domestic investors into other sectors (Emdon 2000: 76). The value of the ICT infrastructure is in the linkage effects to other sectors more than in the infrastructure itself.

Studies by the ITU have attempted to quantify the foreign exchange benefits arising from improved telecommunications infrastructure in Africa. It has been found for example that in some cases, telephone calls made by African business were on average worth over ten times the cost (ITU 1998b: 14). This figure is even higher in rural areas in Africa because of the opportunity cost associated with alternative means of communication, transport and time.

Investments in telecommunications contribute cost savings, increase income and enhance general commercial efficiency but also have externalities in that such investments help maximize the benefits of other infrastructure investments. For example, service industries such as the tourism and travel sectors, which in Africa are almost entirely foreign exchange orientated, are particularly impacted by improvements in telecommunications infrastructure. However, even the agricultural sectors benefit significantly especially for non-traditional and perishable stocks such as flowers and vegetables that are often directed to volatile and price sensitive foreign markets. The entire country profits through higher revenues, higher taxes, and increased foreign investment.

As such the benefit for the economy as a whole exceeds the sum of individual user benefits through its diffusive and synergistic effect in the expansion of the production and exchange of goods and services. Telecommunications investment increases the overall productivity of a country in a way that could otherwise not be achieved.

Telecommunications, Information and the Knowledge-Based Economy

It is also important to differentiate the inherent nature and demand for telecommunications services from other capital-intensive economic development sectors such as transport and power. Economically viable power generation, for example, requires a minimum level of demand side power utilizing machinery, equipment, or even consumer goods and similarly developing transport sectors imply the widespread availability and affordability of vehicles. Such end user level requirements require significant demand side economies of scale for economically profitable utilization of the infrastructure and this is not necessarily guaranteed for a particular nation at any given point in time.

In the provision of a telecommunications infrastructure that includes the end user hand set, an economically viable minimum level of utilization is guaranteed on the assumption that human beings are inherently social beings that consistently require information to interact and hence an information medium to communicate these interactions. Further, a wide body of information theory suggests that bandwidth usage will fill up to meet available bandwidth i.e., human interactions tend to increase on every available communication channel.

In order to achieve sectoral or general productivity gains that lead to economic windfalls, favorable conditions for new initiatives, new enterprises, new products and services must exist. In today's global economy these favorable conditions include ready access to a multitude of information. A modern telecommunications infrastructure is thus a precondition for not only increasing productivity in existing enterprises and for existing lines of products and services but even more so for the creation of new activities and of new jobs.

Applying the Principles of an Information Society to a Knowledge-Based Economy

The principle economic benefit of developing the telecommunications sector lies in the very nature of information. Unlike other services such as water, transport and power, telecommunications can be seen as a physical medium to transport information. A fundamental thesis of modern finance is that perfect information maximizes economic efficiency and utilization of resources whilst eliminating arbitrage opportunities.

Until recently, traditional economic theory has underestimated the role of information and the communication of that information with regards to economic performance and social cohesion. However, the sustainable competitive advantage of nations has always been closely linked to the free flow and exchange of ideas as much as to access to natural resources. Without liberal access to economic and market information, virtual information monopolies exist, which whether enforced by the state or by private groups, entrenches artificial social and economic privileges. The prevailing lack of equal opportunity acts as a brake to the economic performance of the economy as a whole and inhibits successful structural change and development.

Extending this premise further, if Africa is to fully develop its economic potential, then it must ensure that it does not become marginalized and find itself outside the information monopoly of advanced industrialized nations. Whilst this monopoly is not necessarily enforced, it is certainly self-imposed by a lack of telecommunications infrastructure that limits access to global sources and destinations of information (Brands and Leo 1999: 17).

A definition of modern corporate strategy by Monitor, an international strategy consultancy, posits that for firms to succeed they must be able to make an integrated set of choices in order to correctly position their firms within their industry sector so as to gain sustainable competitive advantage. In order to do this they must have access to valid information and knowledge. In other words, in order for firms to make wise economic and social choices, not only must they have better access to information but they must also have access to better information. The accumulation, collation and organization of information constitute knowledge and the quality of decisions made on the basis of accumulated knowledge constitutes wisdom. Hence the success of a firm depends on wisdom, as it is able to correctly make informed choices.

The quality of information can be enhanced through this process of accumulation, collation and organization i.e. the management of knowledge. Hence the provision of a modern information and communication technology infrastructure enables access to better pools of information as well as providing mechanisms to enhance the quality of this information. Figure 3 conceptually illustrates the increase in economic value due to information flow in a knowledge-based economy.

Figure 3. The Economic Value of Information Flow in a Knowledge-Based Economy

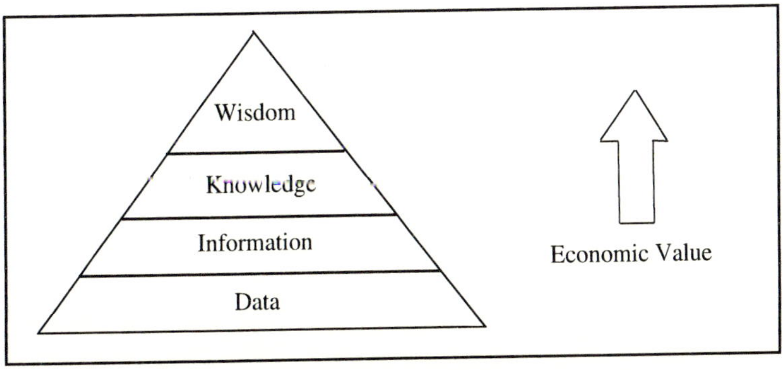

Components of a Knowledge-Based Economy

An information society consists of basic building blocks or components that must be identified, nurtured and matured. Information and thus a well-developed information and communication technology infrastructure becomes the basic construct upon which the modern knowledge-based economy is built. Further the linkages and externalities of such an infrastructure are pervasive and often positively impact other economic sectors. Figure 4 below illustrates the basic components of a knowledge-based economy:

Figure 4. The Building Blocks of a Knowledge-Based Economy

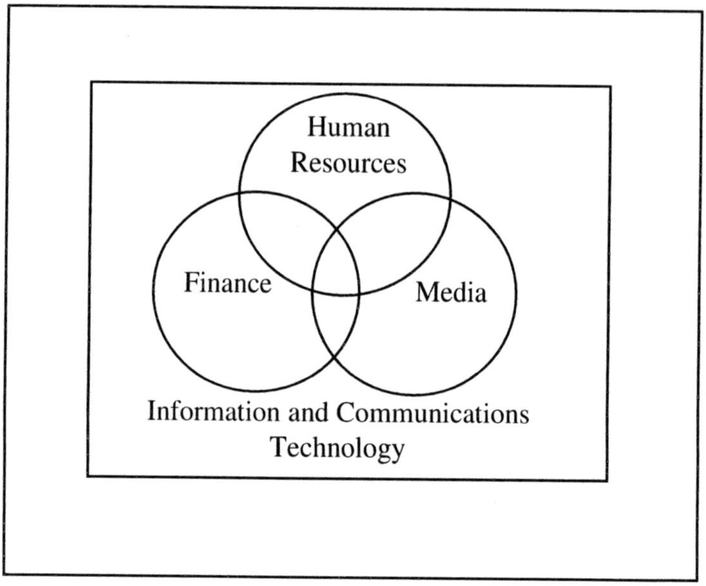

- *Human Resources*: It must be understood that the embodiment of knowledge in human beings has been central to economic and social development. A precondition to effective participation in the knowledge economy is appropriate capacity building of human capital by a targeted direction of educational resources. For example by 2000, India was graduating approximately 125,000 software engineers a year, second only to the USA, a direct precursor as well as consequence of the migration to an information society and a knowledge-based economy. A gap in terms of human resources affects the human capital development required for the planning, implementing and managing contributions to the knowledge-based economy. This also includes the ability to build capacity at policymaking, institutional and operational levels.

- *Banking and Finance*: A modern and well-developed financial sector forms the mechanism whereby value is added to and extracted from information in a knowledge-based economy. Access to investment and venture capital can be considered the fuel for the modern economy. The access to and availability of such capital whether in the form of foreign investment or a strongly developed local financial sector is critical to meeting the cost of building the architecture of the information society. This is in turn creates a virtuous cycle as the cost of capital decreases as transaction and frictional costs decrease with increasing information efficiency.
- *Media*: One must extend past the traditional definition of media, to include the human data, information and knowledge that accumulate in various sectors of society including health, education, media, agriculture and industry. Africa must increasingly generate, market and distribute such content into the global information marketplace. The value of products or services in virtually all sectors is greatly enhanced by the increasing availability of information. For example, art dealers or art lovers in America can be made aware of African art via a website. However if they are further able to purchase online and be guaranteed delivery of the art, then a set of valuable services have been added to the aesthetic service that the artist provides. Such value addition widens customer reach, increases market share and improves terms of trade resulting in higher profits.

Africa's Transition to an Information Society

If Africa is to increase its margin within the global economic pie, it must compete more efficiently in the primary and secondary sectors that it is predominantly engaged in currently. Decreasing the transaction or "friction" costs associated with such activities is thus essential. If the premise holds that access to perfect information eliminates frictional costs then it follows that improved telecommunications access should decrease the costs of production.

Alternatively, in order to extract more value in the global economy many developing nations are increasingly making the transition away from primary and secondary economies to higher value service sector tertiary industries where most of the new activities being created fundamentally depend on information as a primary input. India for example has successfully followed this route and by 2000 India's IT exports of US$ 6.3 billion contributed 15 percent of its GDP. As a direct result of this shift to the information society, more than 180 of the 500 top multinationals outsource their IT needs to Indian companies.

Financial Valuations of Telecommunications in Africa

It is also important to understand that on a commercial level the telecommunications business model is fundamentally different from any other. For example, a most basic accounting concept, that of cost of goods sold, is virtually absent in the mobile telecommunications business. "Airtime" is free and assuming that the annual license fees (being the only "cost" of airtime) are sufficiently low to warrant investment, we can assume that the telecommunication business model unlike virtually any other business is primarily dependent on revenue and operational expenditure as its fundamental business levers. As such, many of the economic assumptions and generally conservative growth forecasts that apply to ventures in other development sectors in Africa do not necessarily translate to the valuations of telecommunications entities. Vodacom, South Africa's first mobile entrant, in the six years since inception has seen pre-tax revenues rise to one billion US dollars by 2000. Zimbabwe's third mobile operator Econet turned an operating profit within just six months of operation in spite of significant regulatory hurdles and government opposition prior to its commencement of operations in 1998.

This phenomenal growth in telecommunications also silences the argument that insufficient value exists in Africa to warrant large-scale foreign and local telecommunications investment. The 1997 divestiture of a 30 percent minority stake in Telkom, South Africa's state owned fixed line monopoly for US$ 1.2 billion to foreign investors was the largest ever inflow of capital into the South African economy. Telkom has committed nearly six billion US dollars for a five-year domestic capital expenditure program beginning in 2001.

The expected Telkom Initial Public Offering of the majority portion of the remaining 70 percent is similarly expected to boost South Africa's economy especially considering that industry analysts value Telkom anywhere from US$ 9 billion to US$ 16 billion. Admittedly the bulk of this valuation arises from Telkom's 50 percent share holding in Vodacom, Africa's leading cellular operator. However, this merely confirms the value that the industry places in the potential of the African mobile market. By 2000 Vodacom had invested nearly US$1 billion worth of infrastructure alone in South Africa since the introduction of cellular telephony a mere six years prior. Vodacom's 2000 reported after tax revenues of US$ 500 million represent a nearly 40 percent leap from the previous financial year and rival earnings from South Africa's largest mining houses.

One of the factors encouraging private investment is the continuing high level of expenditure per line in the region reflecting the intensive usage that users in Africa make of the limited number of telephone lines, particularly for international telephone calls. In 1999, fixed line estimates by the ITU placed the annual average African revenue per line at US$ 1,014 against a capex investment of US$ 354 per line. This conservatively indicates total revenue of at least fifteen to twenty billion US dollars per annum. South Africa's Telkom, for example, has seen yearly reported revenues rise to around US$3.3 billion and has contributed US$ 74 million in taxes to the economy by its 2000 financial year.

As Table 1 illustrates, by early 2001 a range of international operators made strong moves to establish operations in Africa, some with up to eight active mobile cellular investments on the continent. Assuming that mobile operators can overcome the various obstacles to entry that exist in African countries, high level valuation calculations as shown in Figure 5 indicate that significant value can be created over the next few years. Valuation analysis by Umbono Management Consultants calculates future market potential on a conservative basis using the "value per unpenetrated GDP/Pops" method. Potential operator values are calculated on the following assumptions for the sample of African countries presented above:

Mobile Operator	License to Operate
Vodacom	South Africa, Lesotho, Tanzania
MTN	South Africa, Rwanda, Cameroon, Swaziland, Uganda, Nigeria
Econet	Zimbabwe, Botswana, Lesotho, Cameroon, Nigeria, Kenya
France Telecom	Cameroon, Botswana, Ivory Coast, Madagascar, Mauritius, Egypt, Congo, CAR
GTE	Egypt
Investcom	Ghana
Millicom	Senegal, Ghana, Mauritius, Tanzania
MSI	Egypt, Uganda, Zambia, Malawi, Sierra Leone, Congo, Sudan, Gabon, Chad, Burkina Faso, DRC
Portugal Telecom	Botswana, Cape Verdi
Telecel	Zambia, Zimbabwe, Ivory Coast, DRC, Madagascar, Burundi, CAR, Guinea
Telecom Malaysia	Guinea
Telia	Namibia, Uganda
Vivendi	Kenya
Vodafone	Kenya, Egypt, South Africa

Table 1. Investments by International Telecommunications Operators in African Mobile Cellular Markets (2000)

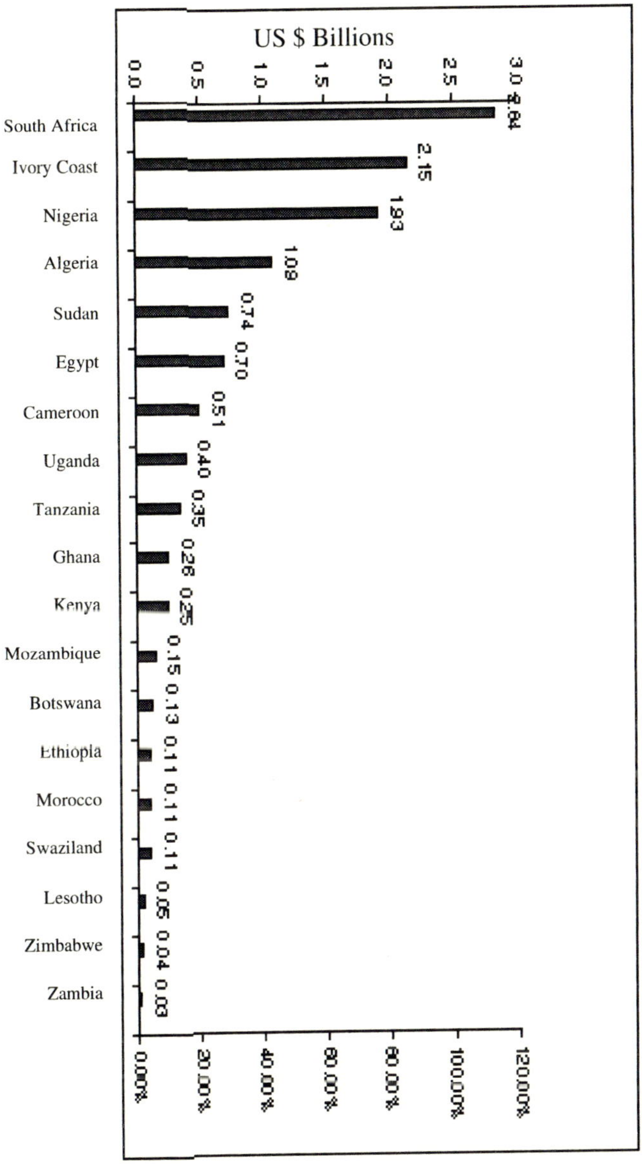

Figure 5. Potential (Currently Untapped) Mobile Operator Valuations in Selected African Countries (2001 US$ Billions, Valuations in 5 Years Time)

- A teledensity/purchasing power coefficient of 80 percent of South Africa's factor for Sub-Sahara, and 100 percent for North Africa.
- Enterprise value of US$1,275 per subscriber (based on 1999 ITU estimates).
- Fixed line penetration growing 10 percent Compound Annual Growth Rate.
- GDP per capita growing at 2 percent per annum.
- No detailed consideration is given to unique circumstances in individual countries.
- Future value of SA market is considered to be highly conservative.

Telecommunications Regulation and Economies in Transition

The development of modern communications plays an important positive role in maintaining and strengthening economic, commercial and social cohesion not only in advanced market economies, but also in developing countries that are in the process of transition towards the market economy and stable democracies in former socialist countries.

Africa in Transition

The dynamic expansion of the private economy through the encouragement of private initiative has seen the creation of new enterprises and jobs as well as attracted domestic and foreign investment. This is part of the transition process that many African countries are currently undergoing as they move away from the socialist economic and political models. It is striking in such transitions how the availability or lack of telecommunications services can make the difference between stagnation and the maintenance of old monopolies and privileges versus the creation of dynamic equal information opportunities both for entrepreneurs and the rest of society.

A rapid and balanced expansion and upgrading of the telecommunications systems in these countries are essential for modernizing their economies, improving their general competitive position and facilitating their integration into the world economy as well as avoiding or reversing the marginalization of large regions or segments of society. Telecommunications has a bridging effect in addressing cultural schisms. The free flow of information tends to bring about social cohesion rather than disruption. Whilst the transition from central state controlled economies to competitive decentralized market economies is an extremely complex process, it is also widely recognized that improving and expanding inefficient and insufficient telecommunications systems is a precondition for successful economic reconstruction and transition (Melody 1999: 11).

Regulation: Constraint or Catalyst

Development polices at both national and international levels must take into account the nature of such changes. The general liberalization and globalization in the flow of information and the regulatory revolution in telecommunications are leading to an unprecedented freedom in the supply and demand for telecommunications services.

Developing countries in Africa face the dual challenge of catching up with rapid technological advances to participate, produce, trade and exchange effectively in the global economy, while providing more people with better, appropriate and affordable access to the opportunities that the telecommunications sector offers. The trade off between the cost of congestion to economic, commercial and social activities and the cost of network expansion influences both policy formulation and operational activities (Emdon 2000: 76). Especially when infrastructure is intended to follow economic growth, the timing of new infrastructure investment can be seen as a trade off between these two alternatives and their costs. Where infrastructure is supposed to lead growth, policy makers and operational managers must in turn carefully manage the possibility of under-utilization, with correspondingly higher costs, if infrastructure expands too rapidly.

In spite of the global examples that point to the inherent profitability of the telecommunications sector given the right balance of government regulation, incumbent telecommunication operators in Africa generally are unprofitable in servicing local operations. Telecommunications entities in Africa have rarely contributed positively to government coffers apart from the "cross subsidization" of services that international net settlements bring in terms of valuable foreign exchange earnings. For a number of countries such as Ethiopia, Senegal and Sierra Leone, ITU figures report net settlement payments from the United States alone to exceed 15 percent of total telecommunications revenue.

The absence of market orientation and responsiveness by public telecommunication operators in Africa results in a lack of planning and investment typically associated with any commercially run organization as well as poor revenue collection, inadequate logistics support and demotivated and poorly trained staff. The aggregate figures on fixed line teledensity in developing countries hide geographical and spatial disparities, concentration of facilities in relatively few urban areas, long waiting periods for telephone connections and generally abysmal service quality. For example, a 1998 ITU report estimated the average waiting time for fixed line telephone in Africa to be 3.5 years (ITU 1998a: 29). State run public telephone networks in much of Africa are poorly maintained and lack sufficient technical capacity that for example only a small percentage of call attempts actually succeed.

A fundamental reason for such poor performance lies in the control and ownership structures of these enterprises. As state run monopolies, they are subject to the normal budgetary practices of governments and as such must "compete" for budget allocations from the state along with other areas of government activity. Since funds are not allocated on the basis of economic criteria as they would be if the entities were competing in the market for capital, under investment is thus common depending on the inclinations and views of the government at the time with respect to telecommunications.

Traditionally in Africa, the overwhelming immediate social and political objectives after independence have tended to take precedence over telecommunications development. Communication has been seen merely as a privilege for the wealthy and not as an intrinsic human social right and an economic necessity. As such the economic and social externalities and linkages associated with improved telecommunications infrastructure have been ignored, at the cost to growth.

Since many governments in Africa do not possess the necessary economic capital to improve their telecommunications infrastructures, there is a growing realization that private sector initiative, capital, technology and skills must be harnessed and directed to this end. However, the participation of the private sector requires the provision of a sufficiently attractive return on investment that is best met in a fully liberalized market. In order to guarantee such a return, a transparent and robust economic environment coupled with strong legal and regulatory frameworks is essential.

The Role of Regulation

Economic growth afforded by telecommunications development and the resultant rising prosperity can also serve to achieve important public policy objectives. Legitimate concerns about universal service, costs to workers in restructuring economic sectors and consumer protection should be answered with concrete steps to ensure that reform does not threaten important public policy goals. While there is little doubt that anti-competitive practices destroy jobs, the positive employment effects of market liberalization will be limited without further reforms to social security systems, labor market reforms, and active measures to upskill the workforce.

The challenges are even greater in countries where the core rules needed to underpin market functioning such as property rights, judicial institutions, commercial codes, and bankruptcy laws are not in place (Singh 1998: 1-2). Markets are highly imperfect and can often be unstable. Failure to provide an institutional and regulatory environment where restructuring and investment can occur can reverse the potential gains of liberalization.

The Role of Government

If we understand the role of government to be limited to the promotion of justice, the preservation of order, and the restraint of wrongdoing, then we can assume that any human activity (economic or otherwise) must be conducted on the basis of a strong legal foundation that provides a robust and flexible framework for operation, promotes a sense of certainty and clear legislative and policy direction, provides recourse in the event of dispute, and advocates the protection of all stakeholders.

It is increasingly becoming clear that the extent and gravity of the telecommunications gap is the result not only of resource and technological constraints, but also of policy errors of omission and commission in both developing countries and former socialist countries (ITU 1998b: 34). In socialist countries the excessive emphasis in the communist economic model of "material production" and thus the general underestimation of the value of all service sector activities led to a neglect of telecommunications development. Combined with the associated political system aimed at controlling and limiting free communications within the population the resultant shortage of phones was thus a sign of both inefficiency of the economic system and the lack of freedom of its citizens.

Regulatory reform requires a stable macroeconomic policy, flexible labor markets, appropriate regulation of capital markets, and complementary structural reforms to provide a supportive environment, and often a context of strong growth, to facilitate adjustments that follow from the reform process. A multi-sectoral approach also boosts gains, since the benefits of sectoral reforms are amplified when competition is vigorous in upstream and downstream sectors of the telecommunications value chain. This calls for independent, competent and effective regulation that is separate from the government's traditional policy making role.

Regulatory Initiatives

Major reform activities have significantly impacted telecommunications regulation worldwide (Melody 1999: 8). The US Telecommunications Act of 1996 fundamentally liberalized local US markets at the most tangible level viz. the local loop. Significantly, the World Trade Organization (WTO) agreement on telecommunication services represented the commitment of 72 countries (which together account for over 90 percent of global telecommunications traffic) to specific programs and timeframes for liberalization. The WTO reference paper also set the ground rules for a transparent and predictable regulatory framework, including the establishment of an independent regulatory body. The WTO agreement was followed shortly by the European Union policy on full liberalization of telecommunications services.

These events have led to a tidal wave of liberalization initiatives in a number of developing countries worldwide. In 1996 alone there were more than 110 telecommunications privatizations in nearly 70 countries. In Africa, this has led to the separating of postal services from telecommunications, granting greater autonomy to telecomm operators, selling stakes in incumbent operators, opening up the market to competition and granting new licenses for basic telephony. The licensing of cellular licenses for example to private operators has seen explosive growth in that sector.

Regulatory Reform

Regulation can boost sectoral efficiency and innovation, enhance overall economic flexibility and potential growth, increase consumer choice and welfare, and increase government effectiveness in maintaining high standards of environmental, consumer, and safety protections. Regulatory reforms have helped ease trade tensions by addressing multiple "behind the border" impediments to international trade and investment.

Such regulatory reform can mean revision of a single regulation, the scrapping and rebuilding of an entire regulatory reform and its institutions, or improvements of processes for making regulations and managing reform. Properly done, telecommunications regulatory reform can also help governments promote other important policy goals such as health, education, environmental quality, security and safety.

The establishment of an independent regulator is the first step in the reform process. Liberalizing trade, by privatizing incumbent state-owned operators, empowering market competition, and reforming government institutions through the separation of policy making and regulation then form mutually supportive elements of this first generation of reform. Figure 6 illustrates how regulatory issues can act as a limiting factor between market demand and technological supply forces, generally serving as a damper between market "pull" and technology "push."

Figure 6. The Regulator as Interface between Technology Supply and Market Demand

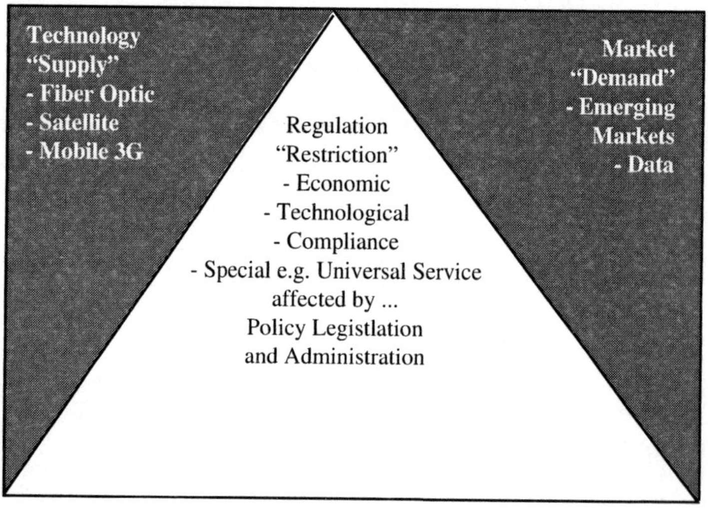

On the supply side, regulatory reform that enhances competition and reduces regulatory costs can boost supply side efficiencies, reduce prices, stimulate innovation, and improve the ability of economies to adapt to change and remain competitive in global markets. On the demand side, regulatory reform boosts consumer benefits by reducing prices for services and products and by increasing choice and service quality (Boylaud and Nicoletti 2000: 15).

In both Europe and the United States, the effects of telecommunications sectoral reforms are still working their way through the economy, but the medium term results clearly indicate the benefits for consumers in terms of prices, quality, and choice are solidly positive. In Japan regulatory reform has seen prices for national long distance and international telecommunications services fall by 77 percent through 1996 and reforms in distribution have lowered prices by one percent annually since 1990.

Regulatory Reform in Africa

In Africa, while the region has generally fallen behind in the reform process compared to other developing countries in Asia and South America, several governments most notably Ghana, Uganda and South Africa have embarked on the liberalization process with varying degrees of success. These countries have already seen development of their telecommunications sectors due to various privatization and market opening initiatives.

Both Ghana and South Africa, for example, preceded partial privatizations with a comprehensive review of the telecommunications regulatory framework, which included the separation of the postal and telecommunications operators. However regulators elsewhere in Africa have struggled to keep pace with the rate of both the telecommunications explosion and liberalization initiatives. Most of the regulatory agencies thus far created in Africa to oversee the telecommunications sector remain quasi separate bodies with limited powers—the sector ministry retaining the authority to issue directives to the regulator, appoint the heads of the agency, and control the agency's funding. The lack of institutional capacity and resistance to structural change in light of changing global economic, political, social and technological circumstances have seen regulators struggling to manage these initiatives and the resultant uncertainty and unclear policy directions have inhibited the potential economic benefits.

Privatization: The First Step

In Africa, liberalization has generally meant the entry of major foreign players into what was an essential public utility industry. These nations have acknowledged the need to maintain and upgrade telecommunications infrastructure. In light of increasing capacity constraints and rapidly aging networks, it was further understood that only foreign investment would bring about the telecommunications enhancements desperately needed.

Prior to these privatization initiatives, most public telecom operators had traditionally neither followed commercial practice, efficient practices, or customer service. They have generally been seen as a source of foreign exchange to government exchequers in the form of net international revenue settlements. By 1997 a total of nine African countries transferred varying levels of ownership and control of state-owned fixed line telecommunications entities to the private sector, raising a total of almost US$ 1.7 billion. The largest slice of this total came from South Africa's divestiture of a 30 percent stake in the local incumbent. In 1996 Ghana sold a 30 percent stake of Ghana Telecom to Telekom Malaysia for US$ 38 million and expects to invest between US$ 300 million to US$ 400 million by 2002 on network expansion. Côte d'Ivoire was able to raise US$ 210 million for the sale of 51 percent of CI-Telecom. Similarly, Cape Verde, with a population of only 372,000, sold a 40 percent stake in Cabo Verde Telecom for US$ 20 million.

It is still premature to fully assess the impacts of these privatization initiatives. International experience strongly suggests that the participation of the private sector through liberalized competitive markets results in network expansion, increased teledensity, increased operational efficiencies, increased capital expenditure and investment, and the introduction of leap frogging technology especially where such liberalization sees the introduction of strategic partners with management control.

In Mexico, Malaysia and Czechoslovakia for example, annual investment doubled by the second year following privatization while in Peru it nearly tripled in the first year alone. In Peru, following privatization of the mainline operator in 1994, annual revenue rose nearly 90 percent to US$ 1.3 billion in two years representing a Compound Annual Growth Rate (CAGR) of almost 38 percent. Teledensity doubled to 6 percent and lines per employee (a common measure of operational efficiency) increased by a phenomenal 162 percent. In Argentina, similarly dramatic results followed the privatization of the mainline operator. Over a five-year period total revenues jumped 140 percent to US$ 6 billion representing a CAGR of 19 percent whilst teledensity increased from 9 percent to 17 percent. In Czechoslovakia, after the privatization of the mainline operator in 1995, total revenues increased by 38 percent to US$ 1.12 billion in the two years following privatization representing a CAGR of over 17 percent. In that period teledensity increased by 6 percent to 27 percent whilst the total lines per employee increased by 23 percent.

However, as successful as some of these initiatives seem to be, the extent and nature of these benefits vary significantly between countries depending on a whole mix of economic variables. It is essential to note that privatization, whilst having the potential to bring in tremendous economic windfalls, is only the first step in the regulatory reform process. The misuse of privatization has seen a number of governments retain such a large share of the proceeds of partial privatizations into government treasuries that the privatized entity is forced to raise further debt from capital markets. On the other end of the spectrum, some governments have undervalued their telecommunications operations and sold their stakes at significant discounts to foreign buyers.

In any event, privatization of incumbent monopolies prior to the establishment of an effective regulatory environment may merely transfer the monopolistic behavior of the incumbent into the private sector, as has been the case in South Africa. Governments that desire only the short term financial benefits of selling a minority stake in state-owned entities may not reap the full harvest that would accrue from a comprehensive regulatory reform process that includes institutional and sectoral restructuring and effective regulation.

The Case for Competition

The next step in the liberalization process, but one that is often the most contentious is the introduction of competitive markets. Several countries in Africa such as Ghana, Uganda and the Seychelles have already licensed Second Network Operators (SNOs) to provide fixed line competition. The WTO agreement, whose initial African signatories included South Africa, Ghana, Senegal, Mauritius and Côte d'Ivoire, commits its signatories to establish open markets within specific time frames. Significantly, this permits telecom operators based in other WTO member countries to build, own and operate independent network infrastructure, offer services and compete in the national markets of member countries. WTO members will then be required to change national legislation and regulatory structures and the dispute resolution mechanisms in line with the General Agreement on Trade in Services (GATS) , which permit any WTO member to potentially challenge and change the regulator of other members in order to reduce anti-competitive behavior. However, the WTO agreement did make provision for "progressive liberalization" which for varying reasons, countries such as South Africa have taken full advantage of in a bid to delay fully competitive markets.

If markets are not fully liberalized, the reform process may be further hindered and perhaps even cease. Valuable foreign investment and technological expertise may be lost as foreign companies go elsewhere. If however markets are fully liberalized, the resulting competitive markets may see little or no profit in providing universal service for example and the public telecom operator may see its value destroyed by foreign competitors.

Another aspect that deserves consideration is the desperate need by many African countries to roll out national networks that provide both demographic and geographic coverage. Traditionally, monopoly operators have performed dismally at this task due to insufficient investment capital and operational skills. But partially privatized monopolies may not necessarily differ in market behavior and practice from state-owned and controlled monopolies simply because there is no competitive imperative to do so. The great telecommunications debate in Africa centers on whether the competition afforded by fully liberalized markets would stimulate the rollout of a national network or merely extract resources from the country.

The pressing questions today include the time frames that must be allowed for monopoly operators to prepare themselves for competition or whether a "big bang" type immediate liberalization would most efficiently and economically serve the country. South Africa has seen its government adopt a prudent "managed liberalization" approach of phased introduction of competition to the incumbent state-controlled fixed line monopoly. This approach has been criticized by would be players in the local telecomm industry as well as certain consumer segments who favor immediate liberalization or at least more rapid introduction of competitive markets. OECD experience suggests that the gains arising from the dynamic effects of market opening result in an extraordinary surge in innovation and faster introduction of new technologies, services, and business practices that multiply benefits for consumers and produce new high growth industries. Direct and indirect effects of sectoral reform help increase flexibility in the labor market and elsewhere. In the United States these effects allowed the economy to adapt more quickly to changes in technology and to external shocks, improved the trade off between inflation, growth, and unemployment, and boosted the US lead in productivity.

Nevertheless there is still no widespread consensus on the nature of policy and regulation that would best answer these questions given the unique nature of the African socio-economic, political, physical, and demographic environments.

Cellular Competition

Very few countries have managed to anticipate the rapid technological changes that have swept through the telecommunications sector. The exponential adoption of mobile cellular telephony, initially thought of as a niche service, has fundamentally altered perceptions of usage as well as caught many countries and regulators unaware. By the end of 1998, there were fifty-five cellular operators on the African continent of which twenty-five were fully privately owned. A number of factors drive this growth.

- Firstly, private international strategic partners back almost all the new cellular networks on the continent with both investment capital as well as management and operator expertise.
- The rapid uptake of cellular telephony on the continent has also been attributed to the pent up demand as well as innovations such as "pre-paid" debit type payment services, which have enabled lower income groups access to mobile technology.
- More significant however has been the almost immediate introduction of competitive mobile markets in over thirteen African countries. Allowing additional market entrants has boosted growth significantly compared to countries that permit only one operator.

This does seem to suggest that while governments preoccupy themselves with the threat of competition in the fixed line environment, technological advances and the market have largely rendered these questions redundant in the mobile world.

Experiences Elsewhere

By early 2001, the Western European telecommunications market was considered to be stabilizing and consolidating after a period of rapid growth. Future annual average market growth rates are expected to drop to 2 percent and the total market value is expected to reach US$ 144 billion by 2003. Primary reasons for these trends include greater tariff reductions as a result of a theoretically fully competitive market in all but two Western European countries. National and local margins are being eroded as prices continue to decline. The resultant increase in call volumes is often not enough to offset the lower tariffs. In the mobile sector, slowing global economic growth and high prices paid for licenses in a number of European mobile license auctions have had devastating consequences for operators and a wave of consolidations has swept through the sector.

However, this has also seen dramatic benefits for consumers as penetration increases with competitors rolling out services into rural areas in a bid to maximize call volumes. The fierce competition has also seen a spate of innovations that have fuelled usage particularly in the data and applications services sector. In order to protect market share, incumbent operators are moving up the value chain. A host of new data applications both mobile and fixed line are poised to enter the market as well as adoptions of newer and alternative infrastructure technologies based on IP technologies, modified cable TV networks, Wireless Local Loop technologies and other technologies to provide new services. Whether this technological lead can be maintained by European operators post the 2000/2001 telecommunications market remains to be seen.

The Asian Experience

Experiences in Asia show that the liberalization process is fraught with uncertainty without the presence of strong legal frameworks. The Indian telecommunications liberalization fiasco continues to be driven by pluralistic pressures from a range of interest groups and political influences. The regulatory authority, the Telecommunications Regulatory Authority of India (TRAI) is weak and ineffective and has seen its authority consistently challenged and ignored by the incumbent state-owned monopoly, the Department of Telecommunications (DoT). By 2001, a full four years after the terrestrial and cellular liberalization program was announced and two years after the licenses were awarded to potential providers, very few actual operations had begun. Virtually each stage of the liberalization process in India has been marred by court battles and scandals in response to inconsistent policy decisions and poorly defined implementation procedures.

For example, a 2001 policy change by the Ministry of Communications allowed the regional state owned fixed line monopolies to offer mobile services at a fraction of the cost that mobile operators can afford (Kripalani 2001: 29). Because these fixed line operators get free spectrum allocation and lower entry fees, they can remain profitable at operating margins that private mobile operators cannot meet considering that they pay up to 95 percent of their revenues to the fixed line monopolies in the form of interconnection charges. Such predatory pricing and discriminatory regulation seem certain to destroy the fledgling competitive mobile cellular industry.

On the other hand in South Korea, far-sighted government policy and effective regulation have paved the way for liberalization and partial competition has seen the establishment of a fixed line duopoly and significant telecommunications infrastructure investment. South Korea now has a teledensity of 43 percent – one of the most developed in Asia, up from 7 percent in 1982 when the market was first liberalized.

Strategies for Liberalization

The Regulator Redefined

Regulatory reforms should be based on competitive markets and require the proactive role of good government in establishing effective, market-oriented regulatory and competition regimes and institutions, as well as social policies that efficiently achieve high standards of protection.

The Regulator as an Agent of Progress

The migration to knowledge-based economies and information-based societies may fundamentally alter the role of the regulator. Such societies will regard as essential the creation of informed independent ICT regulators as catalysts to facilitate the development and application of new, innovative and leap frogging ICT initiatives. The role of the ICT regulator will then become one of an agent of change and progress. At the fundamental level, regulators will become facilitators of information flow rather than the policemen of market practices. The ICT regulator will become a professional "market manager" whose task it is to develop, grow and enhance a nation's ICT sector. The ICT regulator must be able to balance user orientated demand side telecommunications needs against operator orientated supply side technology services.

An independent ICT regulator will be able to contextualize the government mandate to meet economic and social objectives expressed as public policy as well as be able to monitor the changing global economic, social and political climate. The ICT regulator would then identify, recommend and implement the mechanisms that would optimize a country's participation as an information society in the global knowledge based economy.

Regulatory Convergence

At the heart of the emerging ICT sector lies the convergence of technologies. Convergence has many dimensions not just the technological. The pervasive nature of information will see it as a primary input to a number of industry sectors including education, health and banking. The move towards demand driven industry and markets will see a convergence not based on available technology but the needs of various user groups. This will inevitably lead to policy and regulatory convergence as countries and regional blocs seek to best leverage such convergence in order to experience positive economic productive gains and performance.

Regional Level Regulation

The new ICT regulator will develop a pan-regional ICT vision and strategy to maximize the economic potential of trading blocs. This will include harmonizing innovations with other sectors, for example regional electronic stock exchanges in the financial sector. Regional regulation provides opportunities for synergies between trading nations as well as permitting telecommunications entities to maximize their operations. The prevention of duplication and allowance for improved regional communication, standardization and economies of scale in network rollouts has tremendous economic and social benefits. Such regulation also minimizes the chances of isolationist and shortsighted communication policies that individual countries are often prone to adopt. Already significant progress has been made in such regional regulation through the creation of agencies such as the Southern African Telecommunications Regional Restructuring Program, which serves the twelve Southern African Development Community (SADC) nations.

Public Private Partnerships

The perception and role of both government and the private sector in Africa must be redefined. Government must no longer be seen or perhaps more importantly see itself as the sole provider of all services. The private sector both nationally and internationally must be encouraged to take a more proactive role in development in exchange for a market driven return on investment. The regulator plays a key role in the interface between government and the private sector. Such interface mechanisms include:

Emerging Economy Telecommunication Funds

African Telecommunications Funds must be established either on a regional or local level and managed by experienced and competent fund managers with a specific mandate to identify, evaluate and invest in potential telecommunications ventures in Africa. Based on the outcome of such evaluations, such funds would then direct venture capital investment on a strictly commercial basis. By providing a centralized investment pool and world-class fund management skills, risk is mitigated for international and domestic investors seeking investment in high growth sectors in emerging economies.

New Paradigms in Development Finance

International development agencies such as the World Bank and even major regional agencies such as the Development Bank of Southern Africa have come to increasingly realize that participation in the information revolution by developing countries is just as essential as it is for advanced industrialized countries. An increased commercial focus by international and regional development agencies to provide development aid to telecommunications ventures is necessary possibly by taking equity stakes in potential large-scale public sector telecommunications ventures. Development agencies would move away from being viewed as convenient banking facilities providing soft loans and would be expected to play a more proactive management role. For example, WorldTel the commercial arm of the International Telecommunications Union has successfully demonstrated this model for development aid. WorldTel seeks to play a role as a strategic equity partner in telecommunications development projects in developing countries and expects development projects to be run on strict business principles and provide agreed rates of return as well as demonstrate good corporate governance.

Universal Service

Universal service has been widely recognized as an important public policy objective. An argument against introducing competitive markets puts forward the view that market entrants seeking profits prefer to cream high end customers at the expense of providing universal service. This is notwithstanding the fact that incumbent state-controlled or owned telecommunication monopolies in Africa, have rarely if ever met any form of generally accepted universal service obligations. Even after privatization, their ability to meet universal service obligations cannot be assessed without the benchmark of competitive markets.

Further the rapidly changing technological environment has called into question the definition of universal service. Apart from access, affordability is an important determinant in extending telecommunications services to every citizen. By its very nature, universal service implies provision of service to those groups that are uneconomic to serve and require subsidizing. The concept of universal service may have to be redefined and applied in a flexible manner with an economically, socially and regionally valid application of the concept. We may thus have to differentiate between universal service and universal access. Given the gap between teledensity in the developed world and in the developing world, the developed world model of universal service implying high teledensity may not be applicable. Universal access by means of shared services and multiple users may be a better model to follow and more economically justifiable for companies forced to invest.

Another way to bridge the telecommunications gap by developing countries is to adopt the capabilities of emerging technologies. Generally universal service provision has meant the extension of telecommunications to remote and rural areas of countries so that everyone has either the option of a telecommunications device or is in easy reach of one. It is important to differentiate between a telecommunications device and a telephone. If universal access is meant to imply a fixed line telephone then it will remain inordinately expensive to provide such an infrastructure. The introduction and widespread adoptance of mobile cellular technology means that whilst such provision may not necessarily be particularly profitable from a strict accounting perspective, the resultant social goodwill may translate into significant marketing benefit. Further the capital expenditure in such a venture is often far less than a corresponding fixed line roll out. In South Africa, the awarding of a mobile cellular license to Vodacom in 1994 included a universal service obligation of a mere 22,000 subscribers. In six years, the mobile subscriber base in the country has overtaken the fixed line base and there is wide spread perception by the government that the obligations were too lenient. Yet viewed from another perspective, the South African mobile operators Vodacom and MTN with their affordable "prepaid"debit type payment systems targeted mainly at lower income groups have together connected over 4 million lower income groups. Their mobile coverage exceeds 52 percent of the total land surface, an area occupied by over 80 percent of the country's population. It can be argued convincingly that the foundation for true universal service has been laid.

Conclusion

Telecommunications is an indispensable part of any country's participation in the global economy, for supporting international trade and strengthening cultural contacts. Considerable empirical evidence exists that suggests the positive economic, commercial and social benefits of such investment. The recognition of telecommunications as an engine for growth and a key factor in economic development and structural change is becoming increasingly more widespread in the developing world.

The translation of that potential for telecommunications development into actual economic gains in Africa requires a concerted effort by both government and the international and local investment community. The biggest enabler to bridge the telecommunications gap will be government policies. It is critical that government policies are proactive enough to improve access to the information and communication technology by ensuring the build up of adequate network infrastructure.

A progressive, transparent and credible regulatory environment is a key enabler in matching the inherent demand for telecommunications services in Africa with the private sector obligation to provide a commercially attractive return on investment to their shareholders. Considering the wave of regulatory reform initiatives sweeping the continent, the future appears positive.

CHAPTER 10

DEVELOPMENT AND RETENTION OF TECHNOLOGY SKILLS IN EGYPT

Khaled Ismail

Introduction

This chapter addresses the problem of developing skills in the technology sector, and the retention of such talents in developing countries, such as Egypt. Currently, in Egypt, more than 50 percent of the specialists in the technology sector are lost annually either to other sectors or to other countries. Those who remain in the technology sector do not always keep up-to-date with the advancements in technology, and are in many cases deficient of business-oriented acumen relating to product development, marketing, and sales. It is postulated that in Egypt as well as in most developing African countries, a winning model requires partnership between new start-up companies with their counterparts in developed countries, in order to quickly gain and optimize experience in management, quality assurance, and marketing. Close collaboration between such start-up companies and the academic institutions completes the healthy lifecycle for advancing the technology sector by promoting a sustainable university-industrial corporate culture that is mutually beneficial to both academia and industry.

Most of the developing countries to date are characterized as such, largely in part because they "missed" the industrial revolution more than a hundred years ago. The window of opportunities has opened again with the emergence of the technology revolution and its applications. Without a concerted and focused effort, the developing countries will most probably miss the new window of opportunity as they did with the industrial revolution. The impact this time will be even more dramatic, and perhaps catastrophic, since technology today is globally pervasive and indispensable. Participating in the technology revolution has two major aspects to it. One is to become an informed consumer of technology-driven products, such as the Internet, mobile communication, and computers, to name a few. The second aspect is to become a technology producer, by contributing to the design and/or manufacturing of technology-oriented products such as software, Internet services, etc....The former aspect facilitates access to information and allows the developing countries to keep in touch with the developed ones. It can enhance training and educational capabilities by accessing online courses. It optimizes communications in countries typically plagued by the lack of infrastructure (roads, phone lines, mass transportation, etc.). The major drawback of joining the technology as a user is the drain on the already scarce foreign currency resources needed to build the backbone and infrastructure and to import hardware and software products.

Joining the technology as a producer is envisaged to offset the above expenses. In its initial form, such products could be for local use, but ideally they should also be exported in order to generate foreign currency. Becoming a producer would also most likely slow down the brain drain from the developing countries.These concepts can best be illustrated by long-term experiences gained in Egypt, herein used as a prototype developing country at a critical juncture in its development.

General Statistics

One of the most important parameters of success in the field of technology is the ratio of users to producers actively involved in the field, and the quality of both. The role of the universities and institutions is, in part, to continuously generate an abundance of such talents. In Africa, in general, the young generation below the age of 25, amounts to more than 50% of the population, while it is below 30% in Western Europe, indicating that sooner or later, the "aging" Europe will seek help from the "young" Africa in many fields. The number of African professionals currently sustaining western economies is also quite significant.

Let us begin by looking at the statistics for Egypt of those who graduate from high schools every year, and who are qualified to enter higher-level education. Figure 1 indicates that this number is around 450,000 every year, out of which about 13,000 go into technology-related fields (Egyptian Minstry of Education 2000). Since technology is typically on the priority list of most students, those who get to study in this field are selected out of the large group by virtue of their grades, which hence guarantees this sector the elite group of graduates (typically among the top 10% in high school diploma ranking). After finishing their studies, about 5600 of the graduates move on to fields other than technology (Egyptian Minstry of Education 2000).

Figure 1: Distribution of Available Skills

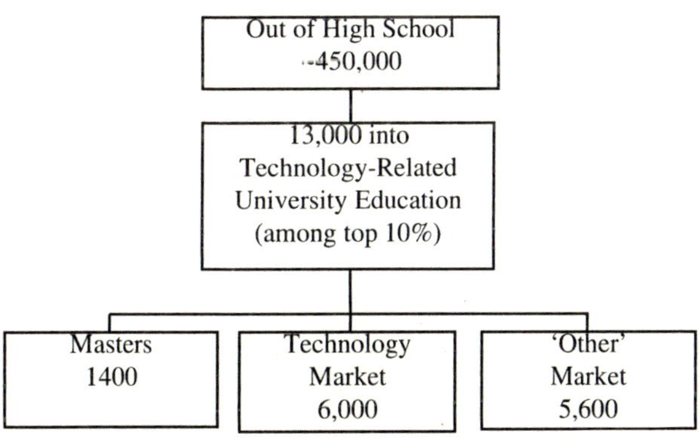

Figure 2 shows those joining the workforce either in Egypt or abroad and those continuing their graduate level education by acquiring a Masters degree. At this level, about 1600 workers leave Egypt to work abroad (typically in the Gulf area; there are no statistics about how many of them continue in the technology sector, but it is believed that most of them do).

Figure 2: Distribution of Available Skills

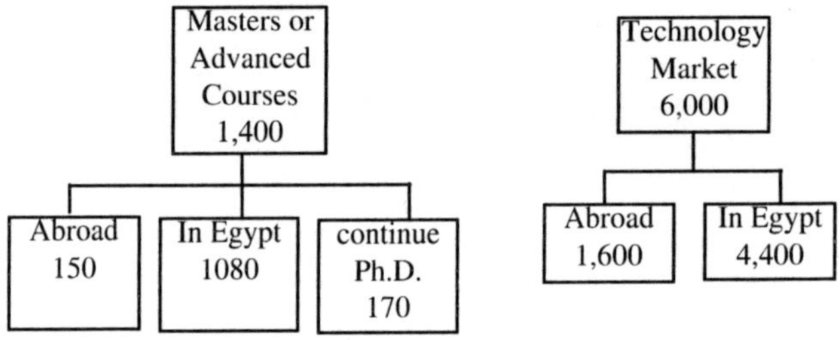

Figure 3: Distribution of Available Skills

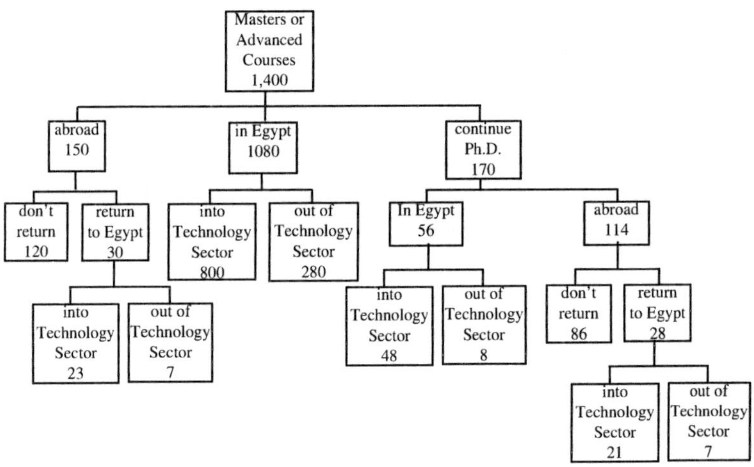

Figure 3 shows the distribution of students going for graduate level and advanced courses annually. The biggest loss is that of about 100 Ph.D. holders in the technology sector, lost annually either to a different sector or to work outside Egypt (mostly in the U.S. and Western Europe).

Figure 4 summarizes the above three figures by grouping the manpower into three categories: one that stays in Egypt and continues in the technology sector, another that gets lost to a different sector and the third that gets lost to work outside Egypt.

Figure 4: Distribution of Available Skills

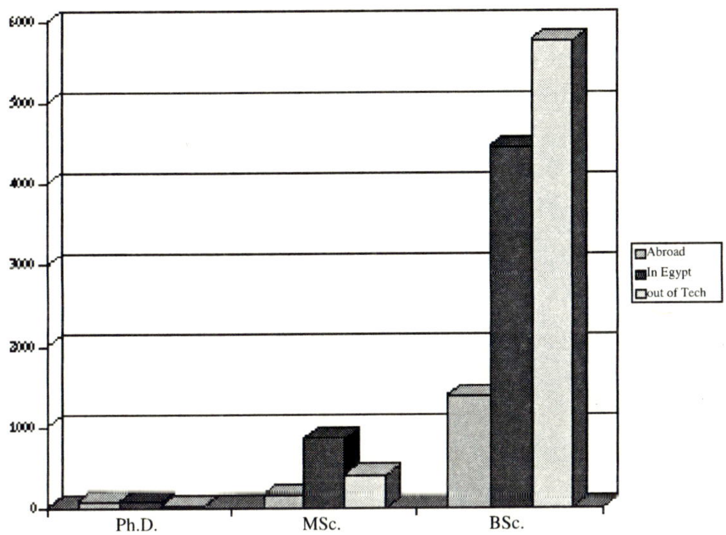

The Hidden Potential

On the academic side, one can clearly identify some strengths, which, if properly harnessed, can become an important ingredient of the advancement of the technology sector in Egypt. They can be categorized into two main strengths, one on the academic side, and the other on the industrial side.

With respect to the educational system, the clear strengths, which can be identified, are the following:

- Technology education is highly selective (only the top 10% high school graduates are eligible).
- A good command of English and quite often a second foreign language.
- A strong undergraduate theoretical education.
- Accelerated growth in the number of public and private universities.

As far as industry is concerned, there are some very favorable features such as :

- Relatively inexpensive and available infrastructure.
- Loaded man-year is far less expensive than in Europe and in the U.S.
- A Central location between the U.S. and the Far East (NY –7, Cairo 0, Tokyo +7)
- Tax shelter for the technology sector (5-10 years depending on establishment).

Identifying the Problem

The problems which deter the advancement of the technology sector can also be categorized as either education-related, or industry-driven. The former category is significantly attributed to:

- Lack of funding that affects primarily the experimental aspect of education, graduate-level research, and the facilities for life-long learning required for professors.
- Projects and graduate research programs that fail to address the most pertinent academic and industrial issues.
- Lack of a work environment.
- Inadequate components of the curriculum that strengthen understanding of the business side of technology.

In the latter category, the technology sector is constrained by the following problems:

- Lack of upper and middle management cadres that have a technology education and possess strong business-leadership skills.
- Lack of entrepreneurial know-how and risk-taking experience.
- International marketing faces difficulties; local marketing is limited to a small consumer market.
- Lack of time specifically devoted to the establishment of total quality management (TQM) in the technology sector.

What is missing?

While Egypt has a well-developed academic sector, that sector has fallen short of satisfying the needs of the Egyptian economy for generating the entrepreneurs who can start up new companies in the various fields of advanced technologies. Thus, there is an abundance of engineers who are academically quite solid, but whose knowledge cannot be harnessed in their fields of interest due to the lack of the appropriate industry and academic infrastructure. This is an enormous economic waste. What is needed is to educate them through a comprehensive interdisciplinary program to become the future entrepreneurs who, based on novel, yet simple ideas can start up new companies that will contribute to the Egyptian economy.

The healthy relationship between universities and research institutions on the one hand, and industry on the other hand has to be fostered through collaborations, whereby the role of industry would necessarily include:

- Corporate financial support
- Creation of internships and endowments that are consistent with ongoing graduate research domains and country-specific needs.
- Creation of attractive job opportunities that are globally competitive and will enhance the retention of Egyptian scientists who would othterwise end up outside Egypt.

Perhaps the role of the universities and research institutions should include:

- Establish a culture of mentoring of well-educated, trained engineers and scientists
- Initiate and conduct research on topics that promote a healthy and mutually beneficial university-industrial collaborative program that is sensitive to national development plans and promotes a globally savvy culture of businessmen and academicians.

The problem in establishing such a relationship is the isolation of both entities from each other. Industry typically does not believe in the capabilities of academia to solve its problems, and does not recognize the immediate return-on-investment resulting from supporting students and research programs at universities. Researchers in academia, on the other hand, stay in their ivory tower tackling problems that have very little relevance to those faced by industry, with a clear focus on publishing in academic journals.

One major breakthrough in this relationship can take place when foreign companies are established in Egypt. Such companies come with prior successful experience of collaborating with academia, and know the value of research. They also transfer their experience in management, quality assurance, and marketing.

The Ellipsis Experience

Ellipsis is a newly established company in Egypt, conducting engineering design for its mother company in the US. Its field of specialization is that of the design of embedded digital communication systems, with particular emphasis on the design of system-on-chip (SoC) wireless local area network solutions. The company currently employs about 30 engineers in the following areas of specialization: communication systems, hardware (VHDL) design and software (JAVA) developers. It has managed to attract several Egyptian engineers holding Ph.D. degrees and working in the U.S. to return to Egypt. One of the most important success factors was the availability of a fast Internet connection, which is readily available at a very reasonable cost (a 512K leased line costs about 10K$/year). According to the Egyptian law of investment, Ellipsis enjoys a 5-year tax shelter, which reduces the financial burden on the company during the start-up phase.

Ellipsis has also established a relationship with Cairo University and the Institute of Technology and Information, allowing graduates of both places to join Ellipsis, and paving the way for joint research leading to M.Sc. degrees for several of the Ellipsis employees.

This model can be replicated for tens or hundreds of companies in Egypt as well as in many other African countries where the basic requirements (skilled human resources and fast Internet connections) are available. This approach would circumvent the typical dependence on government-spoon-fed organizations that are typically doomed to failure, even under the best economic scenarios in Africa.

Summary

In summary, Egypt like many other African countries has enormous untapped resources in terms of manpower. Several real challenges exist and keep preventing these countries from taking full advantage of the technology revolution. The proposed solution is predicated on opening companies that are linked to and modeled on more experienced companies in the more developed countries. Such companies are vital in the creation of synergy between industry and academia. African governments should provide lucrative investment platforms for such companies, providing them with adequate space (creating something analogous to the Silicon Valley), appropriate infrastructure particularly for communication, tax shelters and incentives for funding research at universities, and, most importantly, a mature legal system that protects local and international stock-holders and protects vulnerable African scientists from inappropriate plunder of their intellectual property.

CHAPTER 11

THE FEMINIZATION OF NEW INFORMATION COMMUNICATION TECHNOLOGIES FOR SOCIO-ECONOMIC DEVELOPMENT IN AFRICA

Juliet Evusa

Introduction

During the past decade, there has been a renewed interest in the power of information and communication technologies (ICTs) as an engine to transmit and disseminate information for development in Africa. This chapter examines the current and potential impact of new information communication technologies (NICTs)[1] in Africa, focusing on their gender dimensions and analysis of the extent to which the revolution is serving women's needs and preferences. I will argue that African women need to take an active role as decision makers in the process of using NICTs to accelerate Africa's economic, social, and political development.

The chapter also explores how NICTs can be used to contribute to the emancipation process of African women organized in NGOs[2], grassroots organizations, community based organizations and other social movements. In an attempt to acquire an insight into the social conditions of NICTs and the role members play in the production and dissemination of information, I will present a case study of African women members of NGOs using the Internet as a positive organizing force during the preparation of the Beijing 2000 conference.

The chapter consists of four sections. In the first section, I draw from Hafkin's (2000) historical perspective on the emergence and the convergence of gender, technology and development paradigms in order to develop a clearer understanding of earlier analyses on gender, development and NICTs in Africa. The next section analyzes theoretical frameworks that inform gender, development, and technology. These frameworks seek to build on traditional paradigms to arrive at an appropriate theory for gender, development, and NICTs in the African context. I also include research initiatives that have integrated the complete socio-cultural context in which technological transfer takes place. Section three provides a summary of the results of a practical situation and NICT use and needs of African women following data collected from a preliminary case study: *The role of NICT in providing the socio economic empowerment of African women organized in NGOs*. The last section provides a set of conclusions and reflections.

Historical Perspective: The Emergence and Convergence of Gender, Technology, and Development

In order to understand the evolution of development thinking, this section provides an historical approach to gender, technology, and development. I begin by examining the 1930s era when development initiatives had begun to attract the attention of economists and colonial officials. Not only did these initiatives largely ignore women, Parpart et al. (2000) contend great emphasis was placed on approaches that identified development with modernization by positioning development as a linear process. It was not until Ester Boserup's (1970) pioneer study that the role of women in development received serious attention. Apart from observing that most development projects largely ignored women, Boserup writes technologically sophisticated projects undermined women's economic opportunities as well. It was this study that seriously disputed the argument that benefits from development projects would automatically *trickle down* to women and other disadvantaged groups in Third World nations (Parpart et al. 2000).

Boserup's work, *Women's Role in Economic Development* made an important contribution in stressing that new technologies had played a part in marginalizing and displacing the labor of women (Parpart and Marchand 1995). This opened debates about gender gaps in technology and harnessing women's role in economic and social development. This was a step beyond the generally held view at the time that technologies are gender neutral and that their use are equally available for both men and women (Marchand and Parpart 1995; Parpart et al. 2000). Boserup's analysis, however, did not end the view among governments and many development agencies that technology is gender neutral. At the time Boserup wrote her book, Africa was not in the information technology scenario (Hafkin 2000).

After the decade of the 1970s, the *women in development* (WID) approach was adapted as a theme. Emerging from modernization and liberal feminist theories, this approach aimed at gender sensitive development projects that integrated women in the development process of their countries (Parpart 2000). It was during this era, Hafkin (2000) writes, that technology was generally held as gender neutral; hence, there was no thought of questioning its relevance or potential benefits for African women. However, it was only in the early 1980s when the first real interest in the relationship between gender and technology for development began to emerge in Africa (Hafkin 2000). Although the WID approach had provided a reference that was helpful to development practitioners and government agencies, it included limitations that relied heavily on modernization theory, which ignored the possible contribution of local knowledge (Parpart 2000).

The shortcoming of modernization theory led to the emergence of dependency theory. Based largely in Latin America and the Caribbean, the dependency theorists challenged modernization theory arguing that western dependency was the cause of Third World underdevelopment (Parpart et al. 2000). Developments among dependency theorists paralleled those in radical feminist thinking. Academics and activists influenced by this approach sought "women-only" projects carefully constructed to protect women's interests from patriarchal domination (Parpart et al. 2000; Parpart and Marchand 1995). This approach is referred to as Women and Development (WAD).

Unimpressed by both the WID and WAD approaches, some feminists and development theorists saw the need for an empowerment approach that stressed the need for women's access to productive resources in order to accelerate African economic, social and political development (Parpart 1995). Also known as the empowerment approach, the Gender and Development (GAD) perspective emerged from the grassroot organizational experiences and writings of Third World feminists and has been most clearly articulated by a group called DAWN (Development Alternatives with Women for a New Era). DAWN called for an approach to women's development that recognizes the importance of global and gender inequalities (Parpart 2000; Parpart and Marchand 1995).

The recent arrival of the postmodernist feminist debate in the field of GAD has taken Third World women in new directions. The GAD approach originated from the experiences and analysis of Western social feminists interested in development issues. Proponents of this approach hold the view that since women's status in society is deeply affected by their material conditions of life as well as their positioning in national, regional and global economies, gender relations are not only seen as the key determinant of women's position in society, but also as socially constructed patterns of behavior. Although the GAD approach changed how development policies are implemented and formulated, WID remained the dominant approach of most governments, relief, development, and donor agencies.

The 1980s saw the first concerns about the relationship between gender and science and technology (S & T) in Africa. The International Development Research Center (IDRC), the United Nations Educational, Scientific and Cultural Organization (UNESCO), and the United Nations Development Program (UNDP) worked together with the ECA (United Nations Economic Commission for Africa) to develop science and technology projects in Africa (Hafkin 2000). Their emphasis was centered on encouraging young girls to stay in school and study science and technology. It was during this era that Patricia Stamp's (1989) work became the first publication to examine the relationship between gender and technology in Africa.

Hafkin (2000) identified the first real linkages between gender and ICTs for development with the work of NGONET (Non-governmental Organization Network) in preparation for the United Nations Conference on Environment and Development Earth Summit, in Rio, 1992. This organization's aim was to use ICTs to give women and groups from the South a chance to use an innovative process of information exchange to express their views to a global development forum. In its effort to *demystify* ICTs for women, NGONET inspired the creation of the women's program of the Association for Progressive Communication (APC), which played a similar role for the Beijing World Conference on Women in 1995. The conference provided an opportunity for African women organizations to create a forum for exchange in preparation for the Beijing 2000 conference. My case study is based upon the *flame* listserv set up specifically to stimulate discussion and involve African women NGOs in preparations for the Beijing 2000 conference.

It is the potential of NICTs, particularly e mail's easy and fast way to exchange messages and files that convinced African NGOs to connect online. Newer and more complex Internet applications such as the World Wide Web databases, chatting and virtual discussions add to the network's potential for development and democracy. No longer do women's organizations have to travel to meetings to review gender policies, they can send their lobbyist materials about women's rights by e-mail, organize an electronic forum on human rights violations, and exchange information online. They can also make their contributions and solutions online through discussion groups (electronic conferences), or even on e-mail list servs to a broad audience at the grassroots level. This provides the possibility of women's political networking on the Internet among the international groups as well as at the national level (Kramer & Kramarae 1996). All these experiences add to the very high expectations about the Internet's potential for democracy and development.

Recent Research on Gender, Development, and ICT

There have been a handful of research initiatives that have incorporated the complete socio-cultural context in which technological transfer takes place to reveal what African women seek from using ICTs. Findings from initiatives by Huyer (1997), Kole (1999), Levin (1996), Riano (1994), Lafebvre and Lafebvre (1996), and Rathgeber and Adera (2000) contradict most literature that ignores the technological implementation process and transfer. Lack of gender sensitivity in most technological research has led Huyer (1997) and Kole (1999) to stress the need to understand the gendered nature of the socio-economic, policy and technology systems while assessing and promoting women's access to and the use of ICTs in Africa. In her discussion paper presented at a workshop organized by the International Development Research Center (IRDC) in Nairobi, Kenya, Kole (1999) stressed the need for the creation of a gender-sensitive structure under the assumption that women are often the ones who contribute substantially to the development of African communities, yet, the last decades have shown that females benefit less from development processes.

Clearly, ICTs are as much social as they are technical (Kole 1999). To address these social aspects, it is imperative to choose a framework that recognizes both social and technical aspects. This section offers a theoretical framework that addresses gender issues in ICT. I first explain why I do not use the traditional approach on technology and development. Then I present the social construction of technology approach in more detail that offers useful insights into gender and technology. Traditional models of technology and development have failed to integrate social aspects in their right perspective. These models include theories like modernization, dependency, diffusion of innovation ,and technological determinism.

Modernization and Dependency Paradigms

I will begin with the modernization paradigm. This optimistic view claims that the introduction of advanced technology into traditional societies leads to leapfrogging, that is jumping directly to later stages of development by skipping technological predecessors (Kole 2000; Kole 1999). This means that the *technological push* mechanism is the main driving force of development.

By assuming the positive effect of technology, Kole (1999) contends, this theory fails to elaborate the link between technology and development. It also assumes that economic growth and microeconomic policies are gender and class neutral because they are assumed to automatically benefit all members of society through the *trickle down effects*, and that modern technology is not only superior to traditional technologies but also stands apart from the social context (Kole 1999). This positivistic view on technology and development ignores the fact that in reality, technology and social, political and cultural aspects develop in tandem (Mackenzie and Wajcman 1985).

Limitations of the modernization paradigm led to the introduction of dependency theory. Its proponents, scholars from the Caribbean and Latin America, rejected modernization theory and blamed underdevelopment on the unequal and exploitative economic relationship between the dominant powers in the North and their client states in the South (Parpart et al. 2000; Parpart and Marchand 1995). To overcome this dependency relationship, dependency theorists called for a focus on self-reliance; they believed that technology is not neutral and embodies values, meanings and social relations. These theorists, Kole (1999) argues, hold a pessimistic view towards technology and development due to the fact that modern technology is designed and produced in the North, so that it reflects Northern values and meanings as well as structures of hierarchy and control. To them, separation from the capitalist world system is the only solution for development in the South.

Critiques of dependency theory argue that capitalism does not cause underdevelopment; furthermore, empirical evidence such as the rise of Newly Industrializing Countries does not support the dependency theorists (Kole 1998). Separation from the capitalist system would mean not taking part in any technological development at all.

Diffusion of Innovation Theory

The diffusion of innovation paradigm, developed notably by Everett Rogers (1995), is highly influential in technology transfer research and practice. This approach sees diffusion as a communication process involving the exchange of information about an innovation. Through this process, potential adapters can reach a decision by either adoption or rejection. Like the modernization paradigm, Kole (1999) argues, this theory takes a technological determinism approach by treating technology as a neutral process. Many critical scholars insist that since diffusions on their own do not change social structures, the diffusion of innovation model becomes inappropriate for socio-cultural conditions in developing nations (Kole 1999). Although Rogers (1995) is confident that we have moved from the era of the old paradigm of development (technological determinism), his reference to *laggards* and *non-adopters* reveals a pro-innovation bias, which stems from associating *progress* with *modernization* (Kole 1999). The other bias related to innovation-diffusion is the individual-blame bias. Rogers (1995) describes this as the tendency to hold an individual responsible for a problem in the adoption of new technology, rather than the whole social system of which the individual is a part. This bias tends to overlook social barriers such as access to information and economic barriers among potential technology users. Results from my case study will prove that social barriers are very important aspects that define the use of new technology by its potential users.

Technological Determinism

Singhal and Rogers (2001: 32) define technological determinism as "an approach that considers technology as the main cause of social change." Social scientists using this paradigm perceive technology as the application of invention and innovation that follows from science (Mackenzie and Wajcman 1985). They perceive the entire form of society as being conditioned by technology and believe that technological development in the long run causes organizational changes that have effects on economic, social as well as cultural relations (Wajcman 1991; Mackenzie and Wajcmam 1985). But the question remains: what role does the society play in these changes?

Technological determinists have devoted disproportionate attention to the effects of technology on society by ignoring the more fundamental question of what shapes technology in the first place. A common feature of this paradigm is the failure to analyze the contents of the technological process because technology is treated as neutral (Wajcman 1991; Mackenzie and Wajcman 1985). I intend to challenge such assumptions by demonstrating that technology is affected at a fundamental level by the social context in which it develops. Technology reflects such features of the social and economic environment as the social relations of the workplace, gender divisions in society, social values, policies of the state and religion.

The Social Construction of Technology

Having analyzed theories on technology and development that describe a universal, linear and deterministic evolution of development as a result of the introduction of technology, we now move to a critical view, which address non-economic factors. In this view, technology is not treated as a neutral artifact apart from the social context. Instead, it is conceptualized as a materialized product of dynamic social, political, and cultural processes (Kole 1999). This view is best suited to evaluating the NICT process because it offers the possibility of looking at the electronic process and empowerment as dynamic processes in which different actors play an active part.

Developed by Trevor Pinch and Wiebe Bijker, the *social construction* of technology perspective emphasizes that technological artifacts are open to sociological analysis (Wajcman 2000). Within this perspective, Wajcman (2000) contends, is the concept of *interpretive flexibility* referring to the way in which different groups of people involved with a technology can have very different understanding of that technology. Many scholars and practitioners like Singhal and Rogers (2001) who take non-technological aspects into account in discussing technological and social development portray technology more as an enabling factor than a determinant one.

Feminists have used this concept to alter the meaning and use of technologies where women share the same set of meanings, and attach them to a specific use (Wajcman 2000). To exemplify how the meaning of a technological product changes, Van Zoonem (1992) and Frissen (1992) use the introduction of the telephone in the United States. When it was first introduced, the telephone was constructed as an efficient, rational medium meant for the semi-official purposes of business, doctors, and merchants. These scholars illustrate how many users particularly women, turned the telephone from a rational, business medium into an instrument for maintaining friendship and kinship. As a result of the increased social use, not only did the social meaning of the telephone change, telephone companies started to expand telephone networks to residential areas (Frissen 1992). This led to changes in the production, distribution and marketing of the telephone. Singhal and Rogers (2001) provide us with another example of technologies redefined by their users. In their text, *India's Communication Revolution: From Bullock Carts to Cyber Marts*, they give an example of how the social meaning of the cellular telephone changed. In Bangladesh, ICTs have transformed the lives of rural people and village women. The women have started small-scale enterprises through small loans from the Grameeen Bank to buy mobile cell phones that have been used to provide telephone services to their villages and earn them good income.

The *mobile telephony* presence means that all village residents can receive and make telephone calls obviating the need to install expensive large-scale telephone exchanges and digital switching systems. The Grameen Bank has been successful in providing jobs to poor rural entrepreneurs and connecting the community to the world, and is being replicated in dozens of other countries. Most borrowers are women, and meet weekly to discuss loan payment and other health and development issues (Singhal and Rogers 2001).

The social construction of technology, therefore, qualifies as a good approach that illustrates how technology can take on new meaning and relevance in our daily lives. Although NICTs come to us in the form of finished technical equipments and products, such as video records, terminals, and television screens, user and cultural practices endow these subjects with new meanings and values that in turn reecho in the production process (Mackenzie and Wajcman 1985). To illustrate how social contexts can affect technological use, a thematic analysis of the e-mail conversations of women members of regional NGOs using NICTs for socio-economic and political empowerment was conducted. This case study will demonstrate that African women members of NGOs redefine the gender relations in NICTs by attaching alternative meaning to the contents of the media messages, hence, reconstructing it on the basis of their own values and norms.

Background: Beijing Plus 5 Preparations

At the time of the 5th regional Conference on Women held in Dakar, Senegal 1994, ICTs as an issue and tool for women's agenda received minimal attention at the African level. It is in this context that 14 women from nine countries in Africa got together in September 1999, to develop an African Woman's Network (Flame: African Sisters Online) committed to sharing and developing *Beijing + 5*[3] preparation on the website and to enable participants to build websites for their organizations. What was started as a five-day Beijing workshop triggered the participation of African women NGOs and spread beyond the borders of those who developed it to include women globally. At the time I was analyzing this data (December 1999) the listserv had swelled to include a total of one hundred and thirty subscribers.

Flame is committed to strengthening the capacity of women through the use of ICTs to lobby, advocate and participate in the *Beijing +5* conference preparations regionally as well as at the global level. With a special focus on women and media (in accordance to section J of the Beijing platform for action[4]), the list serv provides a special meeting place where African women and NGOs can get involved in critical online discussions by developing a *space* on which women can continue working for the improvement of their lives. Since then, the context of ICTs and Africa has evolved, under the umbrella of different initiatives for African women in the field of ICTs.

Method

Analysis of documents, African networks, and the World Wide Web reveal that African women are just beginning to use ICTs. To test the assumptions surrounding African women NGOs on the benefits they experience from ICTs and the role they play in the production and dissemination of information, I conducted a survey of existing networks in Africa. This survey enabled me to identified networks that held the potential of communicating to various African NGOs by way of e-mail and the Internet. The *Flamme* listserv was set up under FEMNET (African Women's Development and Communication Network) to keep NGOs updated for the Beijing+5 process. It was a communication link to stimulate discussion and involve women in the Beijing+5 process held in June 2000. After receiving permission from the network facilitator, I set forth my goal of conducting analysis of the conversations of African women NGOs which have acquired computer technologies from Northern sponsors. The *cyberspace* thematic analysis involved conversation of 115 list-serve subscribers in different regions of Africa (North, East, West, South, and the Islands). This analysis was carried from the month of December (about the time the listserv was initiated) through February 2000.

My initial aim was to utilize electronic mail as a tool of gathering data through the distribution of electronic questionnaires to the members of the *Flame* listserv. Although the advantages of using e-mail questionnaires are known to be enticing, my efforts were however, frustrated because of low responses from the participants.

After consulting the listserv facilitator, she advised me to approach the members with caution due to the tendency of African women to feel *over-researched.* This prompted me to change my approach to thematic analysis of e-mail messages. As a participant/subscriber of the listserv, I was assuming a dual position by making contributions to the ongoing discussion on the benefits of the ICTs while observing the flow of the conservations. Participants of the electronic network are members of various international organizations and non-governmental organizations situated in Africa and abroad. The majority of these women participated in their capacity as secretaries, presidents or staff of the organizations. In order to observe information flow from the North to South, I categorized participants responding from African NGOs according to the different regions they were representing. The second category, *African women abroad*, consisted of African women responding from places other than their own countries. Respondents from these categories were mainly students, scholars, and women working abroad with international and governmental organizations. The final category-non-African participants abroad—comprised women from the United States, Netherlands, Cuba, Canada, France and other countries around the globe. It is interesting to note that women from this category were staff members of the donor organizations, as well as organizations with research interests in Africa.

Out of a total of 401 e-mail conversations, I narrowed my focus to 115 messages that addressed the themes surrounding this study. The analysis was based on an examination of various e-mails responses that gave insights of empowering dimensions. I made an attempt to present an unconcealed version of women experiences—in their own words—hence, providing the analysis in a thematic manner. Thanks to the availability of the *Global Link* software, I was able to eliminate the involvement of a French interpreter. Global Link enabled me to translate responses from Francophone African countries and France with a click of the mouse.

Research Findings

Analysis from this case study demonstrates that internetworking has great potential for women's socio-economic as well as political empowerment. Hence, supporting the concept that social aspects are an essential part of the technological process. This, however, does not rule out problems and limitations. Let us first begin with findings revealing how technology can take on new meaning and relevance in our daily lives.

I. Internetworking as a Tool for Democracy

The analysis showed that a total of 52 percent of the e-mail responses embarked on a massive campaign of "cyber-democracy," which included political mobilization, conscientization and sensitization of men and women on the linkages between gender, democracy and development. The findings demonstrate the great potential that Internet use has for civil society development, democracy, and women's empowerment.

The analysis revealed a new kind of communication *space* introduced by participants as decentralized, de-hierarchical and allowed instantaneous registration of many voices and viewpoints. This kind of networking supports Nzomo's (1995) notion of unity in diversity. Unlike traditional media, ICTs allowed the exchange of views, opinions and news that might be impossible in other media under government censorship and control. Participants agreed that NGOs used electronic networks as communication instruments to supply the lobbyist with information to be used in later face-to-face sessions with the government delegates:

> Let us lobby our governments to improve the Telecommunications infrastructures, where there are monopolies, to adopt a liberal form of development and make other players get into the market, including women who have the capital, so as to make ICT's accessible and usable (Regional Gender Organization, Zimbabwe).

Respondents also called for the need to pay special attention to grassroots involvement. The need to make internetworking democratic by involving the grassroots requires repackaging of information that is meaningful and appropriate to the communities. The target groups extend from women with direct access to the network through NGOs or Community Based Organizations (CBOs), to include small and micro-enterprise support groups, resource training and advice centers, health and reproductive rights centers, and other self-help women's organizations.

In order to prevent the marginalization of women at the grassroots level, most respondents saw the need of integrating both traditional modes of ICTs:

> A number of times here in Uganda, there are groups of women who have identified with the acquisition of tape recorders and video cameras as the most urgent

> requirement of their organizations – reasons being that their voices and themselves will be heard and seen by the rest of the world as they usually say. We, therefore, need to work within our own communities, to encourage more women to demand access to the technology from their civic leaders (ISIS-WICCE, Uganda).

Respondents admitted that more traditional ICTs are still more effective in making information more accessible to the grassroots. The need to integrate both modern and traditional ICTs, participants argued, allows the translating and forwarding of electronic data to more appropriate media like fax, radio, video, or print.

Although access among poorer and rural classes is limited, Huyer (1997) contends, a technical combination of old and new technologies will most benefit these women. Combining networks such as print, radio, fax, computer communications and even Internet connectivity, exposes grassroots members to larger sources of information and dissemination networks.

The Internet, therefore, becomes a cost-effective way for activists to get their messages out to the world. With access to communication channels, women can express themselves more and contribute more meaningfully to the democratization process. In Uganda, for instance, the Forum for Women in Democracy (FOWODE) is helping female parliamentarians gain access to the Internet to gather information relevant to performing their roles in the legislative process, with a special focus on legislation affecting women at the community level. Female members of parliament use such information to support their decisions for parliamentary debates and to investigate issues when a new bill is before the house.

Women in Botswana, South Africa, Uganda, Ghana and Kenya have reached major victories in recent times by the way of getting agendas such as Female Genital Mutilation (FGM), women's health and education for girls onto the national debate. Their most common goal has been the empowerment of women at the political level to increase the participation of women in key political and public decision-making. One basic strategy to achieve this goal is using ICTs to introduce the recommendation of gender and affirmative action laws that are based on universal backing. ICTs are being used to sensitize women, who are the majority of voters, on the power of the vote and the merits of casting their votes for committed women. This sensitization is done through electronic solidarity campaigns that advance the rights of women; e-mails have proven an effective tool for mobilizing solidarity and influencing public opinion.

II. Cultural Context: Meaning Attached to New Technologies

The analysis shows that approximately 30 percent of the total responses called for the need to integrate the social, cultural as well as political aspects of the African context with the technology in question. Unlike the developed nations, where the development of Northern ICT is based on individualistic use, women from developing areas organize from the basis of cultural norms and values that are oriented towards the social context. E-mail responses in this analysis validate this claim by calling for the need to build a culture around working online:

> I wanted to second Lillian who suggested that we should discuss more on the challenges women are facing in the use of ICTs and probably suggest solutions that are African (CEEWA, Uganda).

Since the cultural framework of the technological development process defines meaning, technology should be viewed as a cultural product, and as such it can embody norms and values and political interests (Kole 1999). The social construction of technology theory influences the available space of attaching alternative meaning to the technology, this is revealed by the fact that these women were using e-mail as an empowerment instrument. This brings us to the relevance of the postmodernist approach to internetworking within the African NGOs. The social construction of technology process is in harmony with feminist postmodernist approaches that stress the need to pay attention to the concrete realities of Third World women's movements.

When talking about people's needs, it is important to stress the significance of local knowledge, in particular non-technological knowledge. Kole (1998) refers to the alternative meaning of technology as *situated knowledge.* According to this knowledge, Kole (1998) explained, there is need to consider the lived experience of integrating new technology into existing social and technical structures.

III. Feminization of ICT Policies

Without changes in ICT policy frameworks to enable women to access information, political empowerment becomes impossible to achieve. A total of 10 percent of e-mail responses raised concerns about the need to address the legal and policy basis on which ICT infrastructure is developed and work towards feminizing these policies. Absent from debates about policy planning for universal access to telecommunications, African women respondents addressed the need to ensure that universal access is defined in a manner that caters to African women in infrastructure-poor areas. Respondents in this category expressed the need to lobby/influence policy makers to implement policies from a gender perspective with the purpose of increasing women participants in ICTs.

> As far as I am aware, the only African country where African women, led by the Media Women's Association (FAM-Zimbabwe), made a spirited effort to ensure gender was taken into account in the privatization process of telecommunications was Zimbabwe. In other countries that had some civil society lobbying (Kenya, for example), it was extremely difficult to get women's organizations on board (Femnet, East Africa).

In gender and technology studies, Wajcman (2000) contends, the social construction of technology paradigm redefines technology and opens options for interventions so that both men and women have equal access and responsibilities in problem solving. This is made possible by the feminizing of ICT policies so as to increase women's participation in internetworking.

IV. Technology as Gender Stereotyped

This theme describes how gender differences in society influence access to ICTs between men and women. In Africa, women's needs are structured according to their gendered roles and responsibilities, which in turn influence their use and response to ICTs. The social construction of technology theory looks at how the mechanism of technological development is constructed in terms of who designs, controls, finances and produces the technology (Mackenzie and Wajcman 1985). Since women have less access to such resources than men, they are unable to exert much influence over the process of technological development. By attaching male competencies, responsibilities and values to a technology, Wajcman (1991) stressed, it becomes an integral part of the masculine gender identity construction.

E-mail responses (8 percent) confirmed Wajcman's concerns by revealing that women considered the word *technology* to have male connotation. This indicates a high level of discomfort with new information technologies. Since gender differences in society influence ICT access, respondents called for the need to demystify and deconstruct gendered aspects of technology:

> In other cases, I feel that technology has been introduced to women in such a way that appears to be something extremely complicated. This is why it is important to demystify technology and present it in a way that is user friendly (ZRWRC, Zimbabwe).

> Women still believe that ICTs are only for men. These gender differences will continue to prevail as long as they are not gender sensitized. (CEEWA, Uganda).

Research has proven that women are the main economic driving force in developing countries, including Africa, and as the economy becomes increasingly information driven, issues of women's access to produce and disseminate information will be increasingly important, if Africa is to participate in development with the rest of the world (Riano 1994). Therefore, there is need to implement non-gender differentiated technological approaches to facilitate women's access to ICTs.

Summary of Results

Graph 1: Summary of Thematic Findings

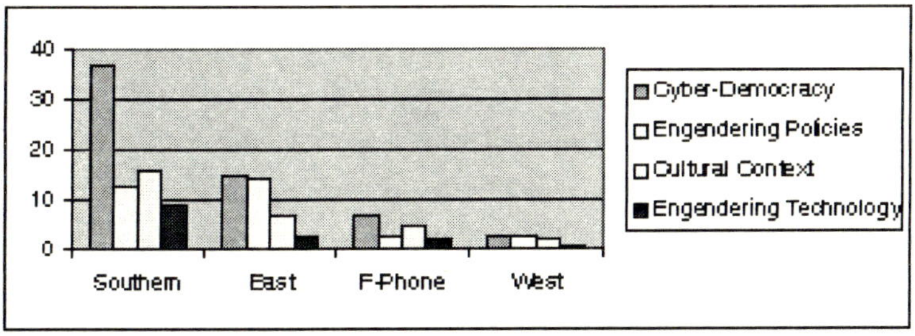

Graph 1 concludes that Cyber-Democracy was the most popular theme among the women NGO respondents; this category scored a total of 52 percent. This was followed by 10 percent of responses calling for the need to sensitize ICT policies to meet the demands of women. Thirty percent of the e-mail responses felt that NICTs should be made relevant to the needs of women in African societies. The last theme, making technology gender sensitive, scored the least (8 percent) in e-mail responses.

The amount of horizontal communication that took place between members of regional NGOs, African and non-African women abroad is an illustration that different groups of people can attach different meaning and relevance to that technology (Wajcman 2000). This supports the social construction of technology theory that technology is an enabling factor and not a determinant one. The respondents not only used the electronic network to address people, they also searched for information of familiar institutes and organizations (for instance the objectives of various non-governmental organizations). Most respondents were partners or contacts of persons who recommended participation. Others were introduced on the same e-mail before subscribing for their personal e-mails.

Internetworking Limitations

Although the results of this case study claim that wiring African countries with the Internet will result in more democratic societies and provide a boost to socio-economic development, such expectations are not always true. Most respondents pointed out barriers and problems associated with Internetworking.

Information Technology Gap

Respondents were concerned with the emergence of new information elitism that will exclude the majority of the populations because of the high cost of computer equipment and low literacy rates. While analyzing the flow of conversation, I noticed that Southern Africa had the highest number of e-mail responses (see Graph 2 below). According to figures from the number of dialup accounts to ISPs, it is estimated that by 2000 there were over 1,000,000 subscribers in Africa. Of these, South Africa was responsible for 650,000, and North Africa 200,000, leaving 150,000 for the remaining 50 African countries (Economic Commission for Africa statistics 2000).

These disparities were further witnessed between urban educated women—in their capacity as women with higher status in NGOs—and women at the grassroots (members of community based organizations) within the same country. Although it is important to note that NICTs may create change in society, pessimists, however, disagree pointing to the dangers and pitfalls of NICTs, such as the likelihood that they may widen the already existing gap between the information rich and information poor.

Poor Infrastructure

Graph 2: Regional Distribution of e-mail responses in Africa and abroad.

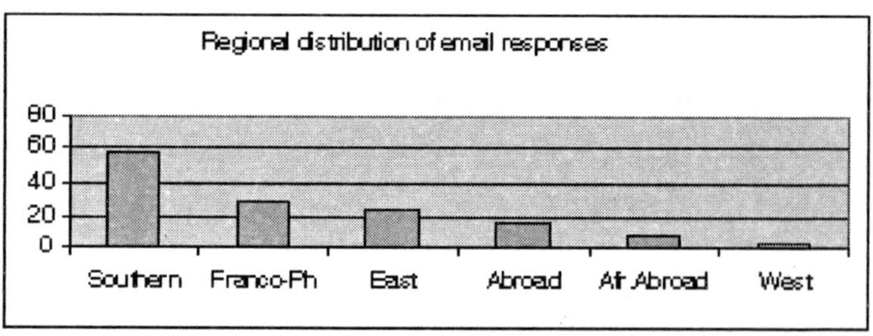

Developing areas encounter low connectivity and poor electronic networks connections, lack of access to computers, or advanced technology and applications and problems involving many aspects of society. Participants attributed this difference to poor infrastructure, high maintenance costs, the low literacy rate among women and lack of trained personnel among other problems. Except in North-South communication flow, South-South respondents experienced electronic network problems due to poor local communication infrastructure.

Low Literacy Rates

The majority of the participants mentioned illiteracy as the number one barrier on the use of ICTs in Africa. Although ICTs pose a danger of widening and deepening the gap between the have and the have-nots, the participants illustrated that ICTs can in fact help eradicate illiteracy. To facilitate the adoption of computer communication technologies, participants anticipated different kinds of training needs.

Language Barrier

After translating the responses from the French-speaking participants, I noticed that there was a communication breakdown between Francophone and Anglophone Africa. I also noted that there was a horizontal communication between countries of the Francophone regions. The only exception was when respondents from either region could speak both English and French. Such respondents became very useful in translating documents into French.

Language is a barrier especially in Francophone Africa, since training packages, software, and electronic conferences and journals are in English. It was also evident that the Anglophone countries dominated the e-mail conversations because 77 percent of the e-mail responses were from non-French speaking regions. Several French-speaking participants complained of the limited availability of information and documents in French. Also troubling is the fact that in many of these African countries the official medium of communication responsible for implementing

ICT policy is usually the language of former colonial rulers—English and French.

In a comparable way, there are distinct needs and problems among respondents from Francophone speaking and Anglophone Africa. Francophone participants suffer more often from language and health problems. Respondents from French speaking countries mentioned concerns over rising epidemics, like HIV/AIDS infection as one of their major concerns. A big chunk of their conversation dealt with the need to use ICTs as an education tool to bring about social change in risky behavior.

Conclusions and Reflections

From the analysis, it is evident that members of women's NGOs are using the Internet as a tool for political and socio-economic empowerment. The use of the Internet during the preparation of the Beijing + 5 conference is an excellent example of internetworking as a positive organization force. The Internet provides a method for working together that significantly speeds communication time and provides a meeting place for those who cannot afford the expense and time to travel and meet in person. This provides the possibilities of international and national feminists using ICTs for political empowerment.

The results show women interacting on the electronic sites with each other by sharing important information about issues that can be integrated into the agenda of women's groups, such as health, political representation, entrepreneur activities, and ICT policies. The daily experiences of these e-mail participants influence their perceptions of the media culture by attaching their own meaning to the messages. These perceptions support the social construction of technology that advocate for the need to examine the contents of ICT by creating a culture around working online, hence demystifying the notion that technology is neutral by making it relevant to the cultural, social and political context of the recipients (Kole 1999). In a sense, the Internet, by its very nature, is a form of alternative public sphere (Kramer and Kramarae 1996).

The notion of empowerment, a feminist postmodernist approach, is our overlapping theme. The adoption of empowerment strategies by grassroots organizations in this study is a practical alternative to the top down development strategies that reminds development agents that development particularly of women cannot ignore the specific situation of life in the Third World (Chowdry 1995). In doing so, Third World women became participants rather than recipients of the development process.

This chapter, however, calls into question the relevance of feminist postmodernist thinking on the concept of democracy. Although the postmodernist discourse is quite in harmony with the vision of African women by emphasizing difference and diversity, African feminists tend to emphasize unity in diversity as a necessary strategy for strengthening the political empowerment of women's movements (Nzomo 1995). The majority of the women speaking with one voice demonstrates the notion of unity in diversity through internetworking for collective political empowerment. This clearly demonstrates the need to create an appropriate theoretical model that will address all these distinct issues so as to make internetworking appropriate to the needs of African women.

Endnotes

1 NICTs (New Information Communication Technologies) in this study denote a wide range of new media. These includes the use of the Internet, e-mail (telnet for countries without access to the Internet) and other computer packages like CDROMs. These technologies are usually complemented by the Information Communication Technologies like the fax, video, radio, and other traditional communication channels like the theater, drama, and other interpersonal communication channels.

2 Note that non-governmental organizations are way ahead of most organizations in implementing effective information technology, especially the Internet, in developing nations.

3 This is a mid-review of the implementation of the Beijing Platform for Action by all stakeholders since the NGO Forum held in Huairou and the Fourth World Conference on Women held in Beijing from 4-15 September 1995. *What is Beijing +5?* (Online). Available: http: //www.flame.org/whatis.html.

4 Adopted at the Fourth World Conference on Women (4-15 September

1995), the Beijing Declaration and Platform for Action identified 12 critical areas of concern for women's advancement and empowerment namely: Poverty; Education and training; Health; Violence; Armed Conflict; Economy; Decision Making; Institutional Mechanism; Human Rights; Media; Environment; and The Girl-Child. *NGO Communication Strategy Proposal.* (Online). Available: *http://www.womensnet.apc.org/beijing/fcom.html* (2000, March 13).

CHAPTER 12

GOING GLOBAL: THE INTERNET AGE AND FREEDOM OF THE PRESS IN AFRICA

Mary F.E. Ebeling

Introduction

On March 28, 2001, the *Daily News*, one of Zimbabwe's few independent papers, reported the brutal beatings and murders of two members of the opposition party the Movement for Democratic Change in Muzarabani. The paper reported that the fatal assaults happened sometime during the week of March 21 and it alleged that the two MDC members were murdered by ZANU-PF supporters. The article detailed the brutality of the attacks by describing the condition of the bodies: one of the victims had his eyes gouged out and both were bruised. During the same week, a MDC MP, Job Sikhala and his pregnant wife, Ellen, were beaten by soldiers in their Chitungwiza home, and fortunately, both survived. *The Herald*, Harare's state-owned daily paper, reported nothing of either story. The British *Sunday Times* reported the same story on March 25, 2001 and members on the NUAFRICA listserv (maintained at Michigan State University), having read both stories online, promptly started a letter writing campaign to protest the killings and beatings by March 30. The letters were addressed to U.N. Secretary General Kofi Annan and Commonwealth Secretary Donald McKinnon.[1] The state-run papers in Zimbabwe were silent, yet through the use of Internet technology, not only was the news of the murders widely published but individuals outside of Zimbabwe were able to take action all within a few days of the beatings and murders.

The digital and information age has transformed media and communications—never before has it been easier to access or disseminate information through networked and wireless technologies, as demonstrated by this story. The information age has been spurred in part by the globalization phenomenon that requires a free, unrestricted and transparent flow of information. While Western governments and nongovernmental bodies have been encouraging this free flow to further the "democratic" enterprise in many developing countries, this new information age generally has been unilateral. Although the problems with globalization are many, ironically, it is this information revolution that benefits African journalists and many are taking advantage of it. By using the Internet, African journalists are able to push repressive government action under the lens of the international scrutiny to rally global response against these injustices. Online activism also makes it harder for some African governments to keep secret their reactionary policies against independent media. Internet technology better enables the media to act as a government watchdog, or what is called the "fourth estate."

African Internet and cellular phone usage continues to grow every year, yet most of the continent's population does not have access to any telecommunications technology at all. And in some places, like Zimbabwe for example, even the price of a newspaper makes media consumption an elite activity. Also demonstrated by the MDC story, then, is the paradox of African journalism on the Internet: A story suppressed by the state can be easily published and distributed through the Internet and, in turn, it can be widely read throughout the world, yet it remains inaccessible to the very people it impacts the most.

Despite continued obstacles in becoming connected either through fixed lines or through wireless technologies to access the

Internet, African journalists have used both technologies to report on events in their countries to reach a global audience, especially in countries whose governments have repressive policies toward press freedoms. Moreover, there are several Africa-based media-rights watchdog organizations that use the Internet to inform a global audience of daily violations befalling journalists and to network with other international media-rights organizations.

Journalists using novel and emerging technologies in innovative ways is not new—the development of earlier technologies, most notably broadcast (especially radio) and fax machines, has enabled the circumvention of official censorship. The Internet has a few advantages over other media, however, because of the medium's dynamism. It allows journalists to publish rich multimedia content globally and instantaneously and it is not constrained by news hole or distribution problems like other media. By looking at issues of free speech activism on the Internet, our understanding of how the Internet is an enabling technology for African journalism in general can be enhanced.

The magnitude of these issues was made clear to me when I interviewed Africans in the United States about African journalism on the Internet, many had stories and anecdotes to share, too numerous for this paper. For every one example I cite, there are hundreds more. I hope that the initial glimpse I present here will lead to further and deeper study in the future.

One note on conceptions of media freedom and freedom of speech: I am not attempting through this chapter to define what media freedoms are or to prescribe for any media what is ethical, especially through a single-minded American lens. I have tried throughout, however, to show examples of what African journalists and media workers consider important and vital to a free press in their own countries.[2]

Background on African Connectivity

Africa represents about 13 percent of the world's total population but has less than one percent of the world's Internet users. Telecommunications is one of the fastest growing technologies in Africa. By the end of 2000, Africa had more than 5.5 million cellular phone subscribers and about 3.11 million Internet users compared with a world total of 407.1 million accessing the Web (NUA 2001). In short, there was about one Internet user among 750 people in Africa compared with about one in three in the United States.[3] Based on these numbers, connectivity is still out of reach for most Africans and the World Bank estimates in 2000 there were about 2.4 computers in sub-Saharan Africa with Internet Protocol (IP) addresses for every 10,000 people (World Bank 2000). Compounding the difficulty for most Africans to connect to the Internet is that they cannot afford a personal computer, much less a visit to an urban cybercafe. In 1999, the World Bank estimated that the per capita GNP for sub-Saharan Africa was US$500. Because most upstream Internet connections in Africa connect directly with servers in the United States, France, Italy or the United Kingdom, costs are very high for users, with usage charges that average about US$2.50 per hour but can reach above US$8 per hour in some countries. Monthly private ISP subscriptions average about US$50, yet the monthly income for most Africans averages about US$42. Moreover, telecommunications infrastructure is still inadequate to meet the high demand for phone lines in most countries, not to mention dedicated Internet lines. One solution to mitigate the exorbitant costs of personal computers and Internet phone-line connections has been the establishment of cybercafes, which provide public Internet access for a per-minute or hourly fee to customers in many large cities throughout the continent. In 2000, there were roughly 167 cybercafes located in Africa, with South Africa having the largest number for a single country at 47 cafes, followed by 67 cafes dispersed among five North African countries (Algeria, Egypt, Libya, Morocco, and Tunisia).[4] The remaining 53 were scattered throughout 17 sub-Saharan African countries. Again, however, because these cybercafes tend to be in large urban areas—most located only in capital cities—and charge high usage fees (hourly rates range between US$1 to $10), cybercafes still remain out of reach for the average African..

Recently, however, more continentally based and locally owned Internet Service Providers (ISPs) have sprouted up, such as Africaonline, M-Web and the newly formed Ecoweb International. It is hoped that because Africa-based ISPs are connected to local servers, enabling Internet dial-ups to be local phone calls, the price of being connected will go down for African users as well as generate revenue for locally-owned businesses. For example, Ecoweb, launched by Econet Wireless Holdings, Ltd., is a very profitable Zimbabwe-based cellular company that has reached into several African countries. The company's financial statement for the year ended June 30, 2000 revealed Econet would provide Internet services for both fixed and mobile networks, through its Ecoweb arm. Econet has faced many challenges from the Zimbabwean government however, especially since the company is the first to successful break the monopolistic parastatal Zimbabwe Post and Telecommunications Corporation's hold on telecommunications ownership in Zimbabwe.[5] Furthermore, Ecoweb's Zimbabwean Web site features headlines from the *Zimbabwe Standard* and the *Zimbabwe Independent*, two independent newspapers that have been under fire by the Zimbabwean government.[6]

With more than 94,000 Zimbabweans subscribing to Econet's wireless services, it is conceivable that in the near future these same subscribers will read these independent newspapers through their cell phones. According to the company's corporate Web site, Econet is already providing CNN feeds through its Nokia phones. Participants at the June 2000 African Internet and Telecom Summit in Banjul, Gambia predict that in the coming years wireless Internet usage will explode in Africa because the technology is relatively cheaper than computers or the installation of fixed lines (African Internet and Telecom Summit 2000).

Controlling Content on the Internet

Governments and legislative bodies all over the globe continue to grapple with controlling and regulating content on the Internet. The Chinese government, for example, has implemented a URL screening system, The Golden Shield Project (Chen 2000), to block Web-based publications considered critical of the government, sympathetic to Taiwan or Tibet, and dissident groups from the national telecommunications network (Earnshaw 2001: 67-69). The project also surveils individual e-mail correspondence.

Controversy erupted in the United States in July 2000 when it was revealed that the Federal Bureau of Investigation was using an e-mail surveillance program called Carnivore. According to the FBI, the Carnivore program enables the bureau to lawfully and electronically investigate criminal activity, including espionage, by asking ISPs (by court order) to install the program on their servers. The FBI claims that the program, which filters through millions of e-mail messages daily, distinguishes ones that are communicating illegal activities. Additionally, the FBI defends the program by saying it is just another application of the wiretapping laws that allow them to surveil criminals.[7] Many groups that advocate for civil liberties and citizens' privacy rights have spoken out against the program, claiming that the privacy and First Amendment rights of millions of innocent e-mail users are being violated by Carnivore (Reporters Without Borders 2001: 123-124).

In spring 2000, a French High Court judge ordered the U.S. Web site Yahoo! to block French access to its auction site that featured the sale of Nazi memorabilia. Yahoo! implemented the order January 10, 2001, after appealing to a U.S. court in December, but the company claims the censoring has nothing to do with the court order (Reporters Without Borders 2001: 47-49). It is certainly understandable that the glorification of Nazism through the sale of Nazi objects is considered dangerous to peace and stability in Europe—13 million people perished under the Nazis' genocidal project—and with the recent phenomenon of genocide in the Balkans, the promotion of Nazism in cyberspace can be seen as fomenting hostilities. Although the Yahoo! auction site does not contain purely journalistic content, many believe this case not only sets a dangerous precedent for creating "borders" on the Internet but also that it promotes self-censorship by Internet publishers (Reporters Without Borders 2001: 47-49).

In a May 2000 press release by the International Press Institute (2000), the organization cited that more than 30 of the 53 African countries have seriously repressive policies against the media.[8] Additionally, there is growing concern over governments' attempts to place stricter controls on the news media. Because African connectivity rates are lower than those in China (20 million users), the United States (116 million), and France (6.4

million), some African governments with poor or repressive media freedom practices are not challenged as much by Internet journalism as they have been by more established media, such as print and radio. But this is swiftly changing as the number of African Internet users increase each year and as journalists' and media-rights organizations become increasingly organized and effective in gaining sympathy from an international readership and political groups. In response, some African governments are already implementing repressive policies toward media freedom on the Internet. In 1996, for example, the Nigerian Communications Commission drafted regulations on Internet content, including restrictions on:

> Any obscene article within the meaning of S[ection].233 (c) of the Commercial Code Act, 1990 i.e. an article the effect of which tends to deprave and corrupt persons who are likely to read, see or hear the matter contained or embodies in it. And any seditious publication within the meaning of S.50 of the Criminal Code Act, 1990 i.e. a publication intended to bring into hatred or contempt or excite disaffection against the person of the President or of the Governor of a State or the Government of the Federation; and to excite the citizens of Nigeria to attempt to alter any matter in Nigeria other than by lawful means; and to raise discontent or disaffection among the citizens to promote ill feelings and hostilities between different classes of the population of Nigeria (Uwachukwu 1997: 121).

Although these regulations were written at the height of the Abacha regime (1993-1998), ensuring press freedoms under the democratically elected (in May 1999) Obasanjo government is still a struggle. There are currently about 20 repressive media laws still in existence in Nigeria. Moreover, during Obasanjo's first year in office there were 162 attacks on the media, including several seizures of media equipment by the police, government, security personnel, or individuals and the death of three journalists. A positive step has been taken, however, by activists and government officials to guarantee open and fair access to governmental information. In 1993, while under the Abacha regime, a coalition comprising the Civil Liberties Organization, the Nigerian Union of Journalists, and the Media Rights Agenda[9] drafted a proposal for the Freedom of Information Bill that was sent for consideration in the House of Representatives at the beginning of March 2001 (Media Rights Agenda 2000). The bill, which ensures open access to official records and information, is undergoing its third reading, and MRA executive director, Edetaen Ojo, said recently that the bill has been set aside for possible approval after public debates are held in Abuja during April 2001 (Ebeling 2000).[10] The Media Rights Agenda is affiliated with Internews International, a Paris-based media-rights organization, and through this affiliation, it is expanding its Internet program to help more Nigerians bridge the international digital divide.

How African Journalist Use the Internet to Mobilize International Support

Journalists are adopting the technology to publish stories, stories that are at times controversial and banned in other media. Many journalists, not only in Africa, are using the Internet to directly confront censorship, as happened with the Operation Tailwind story that CNN reported in the United States. The broadcast, "Valley of Death," reported claims that the U.S. military used sarin gas on American deserters in Laos during the Vietnam War. It was alleged that the military's action was dubbed "Operation Tailwind." When CNN fired producers April Oliver and Jack Smith in June 1998 for their meticulously reported "Operation Tailwind" story, a story the network retracted after pressure from the U.S. military, the ex-producers uploaded the full report and all rebuttals on a server in the United Kingdom, out of reach of U.S. courts (Oliver and Smith 1999).[11]

African journalists are keenly aware of the paradoxical nature of the Internet's influence on journalism on the continent: While the Internet has very clear limits in reaching local readership due to access issues it has a seemingly limitless reach with the global audience. Kabral Blay-Amihere, president of the West African Journalists Association, which represents media workers from 16 nations, said to conference attendees at the Africa Media Forum in September 2000 that the Internet:

> ...[E]nables the journalists' organization to monitor press freedom across West Africa. ...[I] also use the Internet to mobilize international opinion whenever freedom of the press is threatened, putting the government in question in the world spotlight (Freedom Forum 2000: 9).

When repression against Ogonis protesting the environmental degradation by oil companies was at its peak during the military occupation of Ogoniland beginning in May 1994, and leading up to environmental activist and writer Ken Saro-Wiwa's execution in November 1995, activists and journalists used laptops to e-mail their stories abroad to Nigerians living in the United States and Britain (Amosu 2000: 50).

The January 2001 Human Rights Violation Investigation Commission looking into human rights violations by the military during the Ogoniland occupation was closely followed on the Internet by Ogoni refugees in Chicago. One refugee I interviewed who was unable to attend his own hearing read the hearing's proceedings online the following day on Nigeria.com, one of many Nigerian news portals.[12]

The Nigerian press using Internet activism is an example of how technology has empowered journalists to overcome government censorship and reach beyond their domestic readership, in this case, to the Nigerian diaspora and the international human rights community. Journalists and activists were able to alert the rest of the world of the repression they faced through the Internet, sometimes the only means of doing so. And because of the Internet's unique ability to publish and broadcast simultaneously, the response from the international community was swift. Human Rights Watch, the World Council of Churches, and the activist group Essential Action (founded by Ralph Nader in 1982) and its Boycott Shell/Free Nigeria campaign[13] were all mobilized by Ogoni activists' vociferousness on the Internet.

Even when the nascent Internet was emerging from isolated use by university researchers in 1992 when the user-friendly World Wide Web was developed, special interest groups and communities were using it effectively to organize politically either regionally or internationally. Furthermore, in southern Africa, one of the first ISPs to provide e-mail and Internet connectivity in the region, also since the late 1980s and early 1990s, is SANGONet, or Southern African Nonprofit Internet Service Provider. The organization's stated mission is to "promote development and social justice through affordable and accessible communications and information.'[14]

Well before the development of browser software, however, between 1982 and 1987, several communications networks were developed by computer-savvy activists working for social change such as PeaceNet in the United States and the Britain-based GreenNet. Many of these groups organized in 1990 to establish an international network of these NGOs called the Association of Progressive Communications.[15]

Journalists and media-rights organizations in Africa are also using the Internet to campaign on a global scale for the protection of journalists and media freedoms in Africa. In southern Africa, the Media Institute of Southern Africa (MISA), based in Windhoek, Namibia, was founded in 1992, a year after the Windhoek Declaration called for an independent and pluralistic African media (in accordance with Article 19 of the UN Declaration of Human Rights). MISA's primary function is to promote free and open media in the SADC region. The organization's Web site posts daily action alerts that track cases of official media censorship, human-rights violations perpetrated against journalists by governments, groups, or private individuals (including murder)[16], and news affecting the media such as changes in law or media mergers.[17] MISA's "Action Alerts" in turn are republished or linked to many other news or press-freedom sites. Those include the U.S.-based allAfrica.com,[18] in my opinion the best African news portal on the Net, and Digital Freedom Network.[19]

MISA's impact in advocating for the rights of journalists and free expression in the southern African region was evidenced in a statement by Jonathan Moyo, Zimbabwe's Minister of Information and Publicity. In an interview with the state-run *Herald* (Harare's daily newspaper) regarding the ministry's drafting of the Freedom of Information Bill, Moyo said that journalists' unions would be consulted for their input rather than activist organizations that "make noise on behalf of political parties." Later in the interview, he specifically named MISA as being one of those organizations that will not be consulted.[20] This interview came on the heels of Moyo's expulsion of two foreign journalists from Zimbabwe only two weeks prior and the suspension of the accreditation of all foreign journalists.[21] The government has been reacting aggressively against the international as well as the independent press because of their scrutiny of Mugabe's leadership and the ongoing, violent backlash against political opposition by so-called war veterans.

There is growing local and global concern over just how quickly press freedoms are being whittled away in Zimbabwe. There were several highly publicized arrests of journalists working for independent papers in 1999: Mark Chavanduka and Ray Choto of the *Zimbabwe Standard*; and Grace Kwinjeh, Fernando Goncalves, Ferayi Mungazi and Ibbo Mandaza of the *Zimbabwe Mirror*. Mugabe invoked the Law and Order Maintenance Act to arrests these journalists, the same act used by the Rhodesians to arrest him during the liberation war. MISA as well as other international media-rights organizations successfully campaigned for release of the journalists. These arrests sparked a crackdown on independent media in Zimbabwe that has included more arrests, harassment, death threats, deportations and defamation suits against journalists.

Much like the FBI's Carnivore program, the Zimbabwean parliament passed a law in March 2000 allowing the government to order ISPs to install software to filter and monitor communication networks (Amosu 2000). And with the appointment of Jonathan Moyo as the Minister of Information and Publicity in July 2000 after the closely watched parliamentary elections in June, pressure on journalists continues to mount. Numerous death threats have been made against journalists by a small group of war veterans that support the government and have been terrorizing opposition party supporters, nonaligned citizens, and independent papers, including confiscating and burning copies from vendors selling independent papers. The most recent case of a journalist being threatened was the independent *Financial Gazette*'s Bulawayo bureau chief, Njabulo Ncube, in March 2001 by war veteran leader Abel Mahlangu (*Financial Gazette*, March 15, 2001).

Moyo said in a March meeting with the Zimbabwe Union of Journalists (led by Basildon Peta, also of the *Financial Gazette*) and the International Federation of Journalists (IFJ) to discuss their concerns over the dwindling press freedoms that the government could no longer guarantee the safety of journalists and that the government "will never ever tolerate anyone who is contemptuous of our country, its laws and our views," (IFJ 2001a). Although the ZUJ doesn't have its own Web site, the minister's statements were posted on the IFJ's Web site immediately following the meeting.[22]

Prior to his ministerial appointment, Moyo had served on the country's Constitutional Commission, when he began to criticize Zimbabwe's independent media. But even more interestingly, in the early 1990s, when he was a lecturer at University of Zimbabwe Moyo was an outspoken advocate for free speech, writing articles for the independent press imploring the government to improve media freedoms. One article he wrote that ran in the independent paper the *Financial Gazette* on January 23, 1992 was titled "Only [a] Free Press Can Save Our Country from Manipulation," (Kumbula 1997: 175).

The *Daily News* is another independent daily (mentioned earlier) that has been targeted by the government. The paper, accused by the ZANU-PF led government of supporting the Movement for Democratic Change, has now overtaken the state-run Bulawayo *Chronicle* in circulation numbers, essentially putting the *Chronicle* out of business (Rukuni 2000). It is important to note that it was voters in Matabeleland, where Bulawayo is located, who voted in all but two parliamentary seats to MDC candidates.

In April 2000, at the height of election campaigning, the *Daily News* editorial offices were bombed and in January 2001 the paper's printing presses were bombed, believed to be targeted by war veterans. But because the *Daily News* is online in addition to print, the bombings did not prevent the paper in getting the news out to its readership.

Another southern African media-rights organization is the Media Monitoring Project Zimbabwe (MMPZ) based in Harare. MMPZ is affiliated with MISA and Article 19, the London-based international free-expression organization. The organization's Web site features weekly media updates, and because of increasing government repression against independent media in Zimbabwe, the organization is playing a vital role in advocating for Zimbabwean freedom of expression by networking through the Internet with its international partners, such as Article 19 and MISA.[23]

The Freedom of Expression Institute of South Africa[24] is an affiliate of the International Freedom of Expression Exchange and is another southern African media rights watchdog organization. The FEI organizes campaigns to help the South African government draft bills that broaden protections for free speech in South Africa. Although the group focuses on the news and issues concerning press freedoms in South Africa, the FEI also posts regional action alerts on its Web site, making it yet another regional link for media workers in southern Africa and informational source internationally to advocate for press freedoms.

Circumventing Censorship through Internet Technology

In December 1999 the British-based human rights organization Global Witness released "A Crude Awakening," a report (available in print or online) detailing how key Angolan elite individuals, including President Dos Santos, in collusion with international corporations investing in the country's oil economy, have been siphoning oil revenues. The report also exposed how the country's oil revenue has benefited international arms dealers, and helped to maintain the war economy there. Rather than spur development in Angola, the corrupt oil economy has stunted its growth (Global Witness 1999).

The Angolan government, already known for its repressive policies toward the media, promptly banned the publication and warned Angolan newspapers not to quote or mention the report in print. Despite the ban, copies of the report were downloaded, photocopied and distributed in Luanda (Amosu 2000). According to a report by Reporters Without Borders, in 2000 there were only about 4,000 Internet subscribers out of Angola's population of about 12 million (Reporters Without Borders 2001 13). The fact that these relatively few people were able to sabotage an official ban and widely distribute censored material is a testament to what a powerful tool the Internet is.

The banned Rádio Ecceslia is another example of how the Internet has enabled journalists to continue to report despite repression. In July 2000, the radio started broadcasting in Luanda for a few hours everyday from Radio Netherlands via a link with Radio Madagascar. Rádio Ecceslia now has a Web site located on a server in the Netherlands[25] and listeners can download Real Audio files of broadcasts from the site.[26]

On August 4, 2000 the Tanzanian government officially banned the sale or distribution of the book *Mwembechai Killings and the Political Future of Tanzania.* The book details a 1998 riot in the suburbs of Dar Es Salaam that was sparked by a police attack on Muslims that resulted in four deaths. After the riot, police widely arrested Muslims who claim in the book to have been tortured and raped by officers while under arrest. Despite the government's efforts to censor the author, Hamza Mustafa Njozi, and the controversial book, Tanzanians connected to the Internet can download the book in its entirety from the *Islam in Tanzania* Web site.[27]

Outlook for Internet Journalism and Freedom of Speech in Africa

As already outlined in this chpater, the use of the Internet and wireless telecommunications in Africa, although relatively small presently, is growing. In 1996 there were only about 11 African countries connected to the Internet, but by 2000 all 53 countries (and one territory) were wired (Jensen 2000d). With the rapid growth in cellular-phone use in Africa outpacing mobile phone usage rates in other parts of the world, and as more Africans gain access to the Internet every year, journalism on the Internet will continue to be vital in ensuring stories are published despite official repression of the media. Newspapers can be burned, censored, or shut down, but once information is in cyberspace, it is more difficult to suppress. It is too fluid: It can be e-mailed, hyperlinked to other sites, downloaded, photocopied, distributed, and reposted. In some countries, like Zimbabwe or Angola, the Internet has helped bolster the independent media because in spite of bombings or censorship, journalists in those countries continue to publish in cyberspace.

Furthermore, despite Internet and wireless technologies being inaccessible to a majority of Africans on the continent, it has enabled African diasporas and other international readers to stay abreast of news events happening on the continent unlike any other medium. And unlike any other media technology developed so far, the Internet's capability to provide, simultaneously, a local news focus that reaches a global audience, as well as to publish information instantaneously, available to anyone anywhere with a computer or wireless device is revolutionary. The irony of African journalism on the Internet must not be overlooked, however, that is, because most Africans do not have access to the technology, the Internet is a limited source for Africans but it is highly effective in reaching an audience outside the continent, including the African diaspora. This has enabled African journalists to organize more efficiently and effectively, to protect themselves, and to inform the international community about egregious violations of free speech in their countries. Kristina Stockwood, outreach coordinator for the Canadian International Freedom of Expression Exchange, explains:

> Groups in developing countries with access to e-mail send us information right after a violation has taken place. We send that information instantaneously to the rest of the world via the Internet. Just five years ago, we were providing many of those groups with small amounts of money for buying a computer and modem (Roberts 2000).

Some naysayers in the United States believe that there is little potential for the Internet to revolutionize our lives anymore—the hype died with the dotcoms. Others working in traditional media fear the Internet will kill newspapers and broadcast or do not believe that professional journalist standards can be maintained on the Internet. In African countries, however, where journalists risk their safety, and sometimes their lives, to get the story out, there is no room for hesitation—journalists are using the technology to its fullest potential and are revolutionizing free-speech activism.

Endnotes

1 Peter Limb, a bibliographer at the Africana Library at MSU, forwarded the original e-mail letter campaign written by Peter Murphy, secretary of the Zimbabwe Information Centre, based in Sydney, Australia. The ZIC Web site features an image of the raised, opened hand icon of the MDC, evidence of the ZIC's political allegiance with the opposition party. ZANU-PF's symbol is a raised, closed fist.

2 For a superb example of American journalists' arrogance when confronting foreign journalists on their conceptions of what constitutes freedom of speech and censorship, see Horv t. and Rosen (1990).

3 Conclusions of the *African Internet and Telecom Summit*, organized by the Commonwealth Telecommunication Organization (CTO) and the International Telecommunication Union (ITU), Banjul, Gambia, June 5-8, 2000. *http://www.itu.int/africainternet2000/conclusions.html.*

4 Sources of information on the number of African cybercafes vary greatly, however. Reporters Without Borders claim that Algeria alone has about 350 cafes, with the majority of them located in Algiers (Reporters Without Borders, "Enemies of the Internet," March 1, 2001: 11). I've chosen to use Cybercafe.com's (www.cybercafe.com) 2001 numbers because its research is dedicated solely to cybercafes and may be more representative of actual numbers.

5 See for example, reports on the court battles the government lost with Econet: "Zimbabwe Liberalizes Telecommunications," PANA, January 13, 2001 and "Econet Seeks Assessment of Damages from Mawere," *Zimbabwe Independent*, September 1, 2000, a story on a defamatory report on Econet commissioned by the Ministry of Finance. Also see Media Institute of Southern Africa (2000) and Wachira (2001).

6 Econet Wireless Holdings LTD., audited results for the year ended 30 June 2000.

7 See the Federal Bureau of Investigation's Web site: "Carnivore

Diagnostic Tool," no date, *http://www.fbi.gov/programs/carnivore/carnivore2.htm.*

8 International Press Institute (2000).

9 Visit Media Rights Agenda's Web site at: *http://www.internews.org/mra/mra_main.html.*

10 E-mail correspondence with Edetaen Ojo, March 24, 2001.

11 See also, "CNN's Valley of Death: Retraction and Betrayal," *http://media.wmin.ac.uk/tailwind/*

12 I interviewed IkpoBari Senewo, an Ogoni refugee living in Chicago since September 1998. His brother represented his case to the Commission on January 23, 2001. On Jan. 24, Senewo read the story on Nigeria.com, "Saro-Wiwa Invited Soldiers To Ogoniland - Komo," *This Day* (Lagos) Jan. 24, 2001.

13 See Essential Action's Boycott Shell Web site: *http://www.essential.org/shell/index.html*

14 SANGONet is also a member of APC.

15 See APC's Web page for a full history: *http://www.apc.org/english/about/history/index.htm*

16 See MISA Action Alert, for example, the November assassination of Mozambican journalist Carlos Cardoso by a Maputo businessman.

17 To read MISA's "Action Alerts" visit: *http://www.misanet.org*

18 allAfrica.com: *http://allafrica.com*

19 Digital Freedom Network: *http://www.dfn.org*

20 As reported in "New Accreditation Regulations for Journalists," *UN Integrated Regional Information Network (Nairobi)*, Mar. 12, 2001.

21 International Federation of Journalists (IFJ 2001).

22 See IFJ's news release at: *http://www.ifj.org/publications/press/pr/194.html*

23 Media Monitoring Project Zimbabwe: *http://mmpz.icon.co.zw*

24 See FEI's Web site: *http://www.fxi.org.za/*

25 More information is on the Radio Netherlands's Web site at: *http://www.rnw.nl/realradio/features/html/ecclesia000717.html*

26 Visit the site at: *http://ecceslia.snet.co.ao* or download Real Audio files of broadcasts at *http://www.cumbredx.org/cdxsp/cdxsp07170.html*

27 Download the book at: *http://www.islamtz.org/mwembechai.html*

PART III:

Information Technology, Education and Health

CHAPTER 13

THE AFRICAN VIRTUAL UNIVERSITY AND THE PARADOX OF THE WORLD BANK IN KENYA

Maurice N. Amutabi

Introduction

The African Virtual University (AVU), which is a satellite and information technology (IT)-based distance education system sponsored by the World Bank (WB), will be the major focus of this chapter. Through this project, the WB has established campuses in Ghana, Ethiopia, Kenya, Uganda, Tanzania, South Africa, Namibia, Zimbabwe and more recently Senegal. The AVU project is part of the Bank's attempt to grapple with Africa's development problems. It is our argument that whereas the project may have all the good intentions, among them to make Africa catch up technologically, it has serious shortcomings. The chapter will discuss the problems of the African Virtual University project in Kenya at Kenyatta and Egerton universities and its implications for Kenya's social and economic development. It will question the Bank's rationalization or justification for this project in Africa. Whereas technological leapfrogging can be an advantage, it has its own problems especially given that at the same the Bank is responsible for the structural adjustment programs (SAPs) that led to reduced spending in education, which in turn led to the current decay in Kenyan universities that the AVU is supposed to correct. Realizing the importance of technology in development, however, the chapter gives recommendations on how best to utilize and even improve on the idea of the African Virtual University, which is a good concept that was rashly and poorly implemented. The chapter is divided into five parts. First, it briefly examines the ideology of neo-liberalism and capitalist globalization propagated by the World Bank, which underpins projects like the AVU. Second, it looks at the role of the state in education in Kenya. Third, it outlines the AVU concept in Kenya. The fourth and fifth parts assess the AVU and distance education in general and the misfortunes of the AVU project in Kenya, respectively.

The Ascendancy of Neo-Liberalism

The World Bank is the principal source of neo-liberal policy advice to the developing countries. Following the demise of the former Soviet Union the doctrine of capitalist globalization and extreme techno-economic determinism has filled the intellectual vacuum left by the disappearance of socialism. Thus, the Bank has taken to pushing its way beyond its original brief and its neo-liberal policies have been marketed by endless repetition and distorted interpretations of the experiences of the "Asian Tigers", which Africa is supposed to emulate. But these policies are validated neither by historical experience nor by accumulated insights of teachings of the life sciences or economics, rather by the threat of financial strangulation of, and the economic collapse for, the small countries.

For Africa, the problem is not the world economy as such, but the globalization agenda of the international financial institutions, which have systematically and deliberately taken advantage of the debt crisis of the 1980s to institute policy regimes which privilege interests of private capital over the well being of the ordinary people. Societies have been ravaged by structural adjustment programs and contaminated by a virus of uncontaminated greed, nurtured in a culture of social Darwinian winners and losers. The problem lies in the reduced policy space within which countries are now permitted to manage their affairs because of commitments, often extracted under duress, to private and official creditors. The epicenter of these doctrines is located in major universities, think tanks, business organizations, and business journals in the English-speaking world – principally the United States. The globalization agenda is basically for the rich nations from the European Union to the United States and Canada, which regard the entire world as a gigantic playing field from which obstacles to the reach of the European and American business are to be removed like a farmer removes weeds or pests.

The purpose of the institutional reforms propagated by SAPs is to demolish protective economic, social and cultural structures to clear the way for the construction of free market regimes mirrored on the North. The political culture of Europe and of particularly the U.S. is overwhelmingly universalistic and proselytizing. The successful integration of Western Europe and American triumphalism after the end of the Cold War has put new wind in the sails of a neo-imperialist project to impose their particular brand of capitalism and their representational institutions on the rest of the world. Young professionals privileged by the advantage of higher education, including African technocrats trained in graduate schools in Europe and the United States, are attracted to these Northern development paradigms without considering the local factors and relevance for such policies in Africa. It is against this background that the African Virtual University (AVU) is assessed in this chapter.

Kenyatta and Egerton are the two Kenyan universities that were selected as experimental sites for the AVU project. The AVU is a distance education project established to serve sub-Saharan African countries. It uses modern information technologies to increase access to tertiary education in science and engineering throughout this region. Its stated objective is to build world-class scientists, technicians, engineers, business managers, health care providers, and other professionals within Africa, a region that has been lagging behind in the so-called information technological age. By the end of 2000, the project carried out its programs at twenty-six sites. There is no doubt that universities on the continent have had checkered pasts and are replete with problems, both internally and externally generated. It is quite a paradox that these very universities are now providing space for the AVU experiment in Kenya and elsewhere.

Kenyan public universities, like others in Africa are plagued by a plethora of problems. The major ones being the meager resources allocated to them directly by the exchequer, broken down and poorly maintained equipment and out-of-date and poorly stocked libraries, student and staff unrest, unpremeditated closures, brain drain, and over-enrolment and congestion. Others are dilapidated infrastructure, political interference in administration, nepotism and sectionalism, administrative incompetence and inertia, and corruption. Reduced budgetary allocations to the Ministry of Education following the imposition of SAPs from the late 1980s simply made matters worse and eroded the achievements of the early post-independence years. The universities in Kenya were originally established with relatively considerable resources to train the bureaucrats needed to replace colonial administrators and the demands of the emerging civil service after 1963, when Kenya became independent. There is little doubt that they were able to produce a large complement of the post-independence middle classes. The subsequent incapacitation of local universities has often been evoked as the reason for the introduction of the AVU in Kenya.

The Role of the State in Education in Kenya

Whereas many countries in Africa succeeded in quickly expanding education at the primary and secondary levels, they were unable to train sufficient numbers of skilled professionals needed for development. Needless to say, this owed to the failure of the comprador bourgeoisie ruling these countries to mobilize themselves and adapt to the changing needs of their nations. Also, the institutions and structures that they created or inherited from the departing colonial masters were poorly grounded in local culture and society, often oscillating between the western capitalist and eastern socialist models of the Cold War. The so-called indigenous technocrats lacked the commitment and goodwill either because they had narrowly and personally defined agendas, or were inexperienced and appointed on grounds other than merit. For example, in Kenya, the first vice-chancellor of the first university, the University of Nairobi, was not even a full professor. Dr. Josephat Karanja was appointed so as to forestall the appointment of a Luo – there were then only three indigenous Kenyans at the level of full professor, all of them Luos – after President Jomo Kenyatta fell out with his Luo Vice-President Oginga Odinga (Mazrui 1986). The universities of course had their own share of problems that afflicted them in subsequent years.

From the 1980s Kenya, like many other African countries, entered a period of economic recession and stagnation, during which the resources allocated to the universities began to diminish, while at the same time they were required to train an ever-increasing number of students. The demand for education skyrocketed as education became an assured way for upward social, economic, and political mobility. The universities expanded to meet the demand but the quality of education went down because of over-enrolment, congested lecture rooms and hostels, and deteriorating general infrastructure, which provoked student activism and riots against an increasingly beleaguered and authoritarian government that seemed to value quick results rather than long-term investment in human capital. Whatever the reasons and other underlying factors, the net result was the widening of the knowledge gap between Kenya and the developed countries of the North.

In short, by the late 1990s tertiary institutions in Kenya, as in much of Africa, were in a state of crisis, able to give inferior training to a very small fraction of the population lucky enough to access higher education. To be sure, the number of universities grew, although there was no indication that the increase was commensurate with quality. In 1986, Kenya had only one public university and no private university. By 2000, there were six public universities and over ten private universities. Many of the public universities and their campuses were upgraded from lower institutions previously offering diploma training without commensurate improvement or upgrading of infrastructure. If this blind expansion continues, and there is nothing to indicate that it will not, university education in Kenya will not be very different qualitatively from that offered in diploma colleges or even worse. Since the introduction of SAPs, even the World Bank now concedes, the problems of universities have worsened. To quote Edward Jaycox the World Bank President for Africa in 1997: "African universities are facing enormous difficulties, including declining budgets, lack of qualified instructors and outdated academic programs that fail to meet local needs" (USA Today: Technical Report 1997).

Many of these universities lack the basic communication facilities, especially in information technology. The ratio of learners to computers, books, laboratory equipment, lecturers and other aspects is appalling. Thus there is no proper learning and the knowledge gap between Kenya and the rest of the world is going to grow even wider. Therefore, we can almost be certain that the economic and social disparity between Kenya in particular and Africa in general will widen at an even more rapid pace. The Kenyan government has failed to equip universities to turn them into effective instruments of technological innovations, scientific and industrial research and development. The African Virtual University is intended to bring high quality higher education to a large number of students in Kenya using modern information technology thereby producing sufficient numbers of well-trained African scientists, technicians, engineers, business managers, etc. required for economic development. This, it is argued, the local universities have failed to provide.

Interestingly, the local universities have declined since the late 1980s because of lack of funds as a result of the very SAPs designed and imposed by the World Bank and IMF. SAPs led to the introduction of austerity measures that led to the introduction of cost sharing in healthcare, education, and other sectors. They also led to reduced government spending on education hence reduced budget allocation to universities. This was the beginning of decline in quality of university education in Kenya. What therefore makes the World Bank feel that it can succeed with the AVU and yet it has denied the local universities the necessary resources required to catch up? What makes it imagine that the lack of funds afflicting the students in local universities will not affect the AVU? If the World Bank is opposed to increased spending in the social service sector and welfare by governments, then how genuine is it with the AVU, which is still within the realm of education? These are some of the questions addressed in this chapter.

The AVU Concept in Kenya

The AVU was first launched in Kenya at Kenyatta University in April 1998. The AVU is a satellite-based distance education project. It is a network of Internet facilities that uses a multimedia approach to teaching and learning. The AVU learning package comprises of live pre-recorded lectures transmitted by satellite and viewed on a TV screen plus handouts, textbooks, lecture guidelines and programs/schedules on transmission, and other materials, which are transmitted electronically. The AVU network is modeled on the highly successful distance learning networks operational in the North, especially the United States, Europe, Australia, and other parts of the developed world. The materials come from Colorado State University, the University of Massachusetts, and the New Jersey Institute of Technology in the United States and the University College Galway in Ireland.

In the initial phase the AVU was expected to establish partnerships with a number of institutions of higher learning in the countries of operation and others in the vicinity. They were supposed to provide the nucleus of scientific innovation in education, a kind of centers of excellence. These centers and their satellites were expected to offer technology-based credit courses and non-credit seminars using digital satellite technology. The AVU was supposed to adapt the academic resources from existing videotaped curricula of some of the leading universities in the United States and Europe. Other courses have been and will continue to be developed specifically for the AVU by institutions of higher education in Ireland and other European countries. Programs were expected to be broadcast to partner institutions in Africa via satellite from the U.S. and Europe. By the beginning of 2001 these expectations had only partly been met.

According to the World Bank, the main aim of AVU is to tap new information technologies to overcome the many financial and physical barriers that often prevent students from African universities from gaining access to quality higher education. The argument has been that most African universities have become increasingly irrelevant in the rapidly changing world of globalization (Amutabi 1997b; Walters 1997). It should be conceded that African universities often graduate disproportionate numbers of students in the humanities rather than the sciences and engineering, making the AVU project quite plausible. That there is need to bring high quality education to as many people as possible in Kenya cannot be denied. It is not in dispute that there is a great yearning for high quality education in Kenya. The inability of the existing institutions to satisfy this need is evident. Most universities in Kenya are able to admit only a tiny fraction of eligible students. In the year 2000 Kenya's public universities admitted only 8,000 candidates out of a total 48,000 who qualified for university education (*Daily Nation* 23 March 2000).

As has been shown in several countries, distance education has proved to be a flexible, cost-effective means of bringing higher education to those who are unable to attend a national public university. Before the concept of distance education arrived, one of the solutions to the shortage of local university space was to send students abroad en masse. A key question about distance education is how best to do it, that is, the choice of the appropriate technology that will convey learning meaningfully. Bates (1995) suggests a number of criteria, which I have elaborated into the following seven:

- Rationalization: This addresses the question of justification. In this context, we need to find out whether the AVU concept, for instance, is better than the alternative distance, open and cooperative concepts of education existing in Kenya, and also how well the AVU programs are reconciled with local needs, costs, and convenience.
- Relevance: Here, the concern should be on the capacity and potential of the anticipated program to address and meet local needs and demands. For instance in the case of the AVU, one should find out if the working environment of the graduate will be in an Internet atmosphere as simulated by AVU or in deep rural areas where there is no access to the Internet. Will such a graduate who is computer and Internet-dependent be usefully utilized in a working environment devoid of these

appliances and connection? Will his/her knowledge be such that it can be improved to suit local situations or will it be dependent on Northern materials and ideas for the entire life of such a graduate?

- Access: This is concerned with the number of people who will have access to learning and the flexibility or rigidity of conditions of admission; it has to do with the space available and capacity it can accommodate without compromising standards; it also concerns accessibility to a particular technology and learning aids to the learner. It has to do with logistics, that is location of the learning center and its suitability to the learners in general.
- Cost: This has to do with the official fees which the individual learner is charged and other incidental or hidden costs like purchase of stationary, uniform, overalls, individual software like diskettes and learning manuals, charges for private use of other services like e-mails, the cost of transport to the learning place from the homes of learners per day, per month, or per semester. It is also concerned with the cost of installation of a certain technology vis-à-vis estimated long term benefits to the institution and the country; the comparative advantage of the installation with regard to other technologies available on the market and their level of obsoleteness in the near future; the structure of each technology and its relevance to local conditions. The cost of the technology should not be such that the poor who are the majority are locked out because they cannot afford it. In other words, the costs should not be prohibitive to the institution and the learner.
- Teaching and learning: Teaching and learning should always go together. What is therefore critical is that the teaching should be in such a way that the learners can imbibe readily without obstacles like accents, newness to technology, unfamiliarity of vocabulary, etc. It concerns the kind of learning that is needed by society. It also has to do with instructional approaches that best meet these needs.

- Sustainability: This has to do with the period after the funding from the World Bank comes to an end. It is concerned with whether the local universities will maintain the projects with their shoestring budgets. It is interested in finding the best and most convenient technologies for supporting the teaching and learning.
- Spillover Effects: To what extent will this type of education affect the environment, which is the country? Is it capable of influencing development? Can others also learn from the graduates of such a system independent from where they themselves trained? If not, then the system is only establishing dependency structures, which will affect generation after generation.

Assessing the AVU in Kenya

Rationalization

Is the rush for the Internet and information technology in teaching the solution to Kenya's education problems? Or is this part of the scheme by the North to quickly engage and enhance the captive nature of Africa using new technologies? There are more urgent issues that need to be tackled to enhance development other than just university education. Kenya has attained only 65 percent literacy. It means that many citizens are still illiterate. Also, the country is faced with a very high dropout rate at the primary and secondary school level. Attrition has also been very high at the universities especially following the introduction of SAPs. This means that such a basic issue as retention is still a problem in Kenya. The AVU programs are only of benefit to the elite who can afford the costs. In October 2000, Kenyatta University students went on a three-day strike demanding withdrawal of the Kshs.500 (approximately US$7) e-mail fee per term imposed by the university administration. The university eventually closed down. Whereas to the university authorities the e-mail appeared to be a priority, it certainly was not to the students. Even if students saw it as a need it was prohibitive because of the cost. The AVU programs are facing the same problems. The fact that the AVU programs require affiliation to a campus makes them similar to the current university set-up in Kenya. University education is lagging behind not because of inadequate technology but because of cost and lack of resources.

Relevance

Is the AVU curriculum responsive to Kenya's educational needs? Is it relevant to Kenya's labor needs? Will it address or even alleviate the plethora of problems of high quality manpower production experienced in Kenya? Unfortunately, the AVU does not promise to address these problems. As an Internet dependent education system, its relevance to rural Kenya is out of the question. Kenya is 80 per cent rural and much of it is not connected to the Internet. The AVU will only address the needs of a tiny fraction of people in urban areas. Many university graduates are expected to work in rural areas to stimulate development there. Since the AVU is Internet and computer-dependent, many of the graduates are likely to be irrelevant in rural settings where they will not have ready access to the Internet to consult. Thus, before Kenya establishes the relevant infrastructure for massive Internet learning and utilization, and before a significant proportion of the population is literate, the AVU concept will remain largely irrelevant. Those funds are better of being diverted to improve areas of literacy and information technology infrastructure in the existing institutions. Dependency on the Internet rather than on hands-on-learning is already causing problems in the North. A story is told of a student in the U.S. who was asked the name of the President of Rwanda. She asked for a short moment and quickly clicked on a renowned search engine. When the search engine turned up nothing, the student answered, "Rwanda has no president."

Access

According to many experts of distance education, access is usually the most important criterion for deciding on the appropriateness of a technology for distance learning (Walters 1997; Korsgaard 2000). At present the telecommunication infrastructure in Kenya is wanting with regards to the needs of the information age. The network is so limited that it will inevitably restrict access to a large number of learners. The AVU has the necessary equipment, but access remains a problem for the majority to grapple with. This is due to the fact that very few Kenyans can afford the cost of telephones let alone software. Another problem related to access has to do with the location of AVU centers in Kenya. Kenyatta University is located in Nairobi thus will serve mainly an urban clientele and will remain inaccessible to the rural areas. Even the Egerton center is far removed from the rural areas. The use of local rural centers like markets and small towns as satellite centers would ensure greater access to, and impact of, the AVU distance education model. Since the centers have been linked to universities, they still perpetuate an ivory tower mentality associated with these institutions with their bureaucracy, iron gates, watchmen, and rigid selection criteria. The fact that these centers are located in university towns makes accommodation expensive and beyond the reach of ordinary Kenyans.

Costs

The AVU centers at Kenyatta and Egerton have the "state-of-art-technologies" and current computers and Internet servers, TVs, VCRs, and other sophisticated multimedia appliances. All these must have cost a fortune to acquire yet they are likely to benefit only a few select students who can afford the cost. The money invested in these appliances could have improved the current problems in the local universities. The World Bank and other donors are currently covering the initial costs for the technologies chosen in AVU projects. It is not clear whether these funds are in the form of loans or grants. Eventually it is envisaged that the AVU centers will be self-sustaining institutions and that they are to survive by charging fees from their students. The question is, should the money obtained as a loan or a grant be spent on expensive hardware that will not be maintained adequately when the donor pulls out? From experiences in other sectors, one may wonder whether these centers will be properly maintained and sustained in the long term.

In so far as the programs are designed in the North, the AVU is enhancing dependency, which many African governments should be striving to steer away from. It is therefore very important that careful consideration be given to the appropriate choice of technology in terms of costs and effectiveness. Needless to say this does not imply that Kenya should be choosing out-dated technology for distance learning but at the same time it should not grab the latest technology, which may not be appropriate. A careful analysis of the available technologies should have been made taking into consideration the country's existing telecommunication infrastructure, what each technology can achieve given the fields of study chosen, its adaptability to changing needs and circumstances, and its cost and sustainability in Kenya.

Teaching and Learning

Teaching should always take into consideration the background of the learner if it is to succeed. Learning can only be smooth if the learner and the teacher are properly reconciled with regard to their aims and objectives. The AVU offers courses and curricula designed in European and US universities. The instructors, using American or British accents, are often incomprehensible to local students. Also, certain textbooks and teaching materials produced in the North do not have direct application and relevance to Kenya. The courses have obviously been developed with a certain background in culture, technology and relevance in mind, which by and large are those of the North. The AVU should be flexible enough to accommodate the differences that exist in culture, infrastructure, developmental priorities, and political realities so that each country can reap the maximum benefit the project has to offer. Thus, Kenyan students may feel alienated and even confused by these materials if these differences are not recognized. There is need to develop and design materials according to local needs, rather than the general holistic treatment of courses. Unless that is done the AVU's distance education project seems destined to perpetuate dependency in learning and teaching, which can stifle local innovation, invention and discovery.

Sustainability

Will Kenya either through the government or private enterprise be able to continue the project once donor funds are exhausted? This is where the cost criterion becomes important. The cost of the technology chosen should be such that Kenya will be able to afford and maintain it when the World Bank withdraws. Unfortunately, this does not seem to be the case. This is of immense importance since there are many examples of donor-initiated well-meant projects, which have folded after donor funding was discontinued mainly because not enough thought was put into the question of sustainability when the projects were launched. From what has happened at Kenyatta University where students have resisted the AVU facilities like the e-mail service, the sustainability of the AVU experiment in Kenya remains in question.

Spillover-Effects

In 2000, the AVU at Kenyatta University trained 200 teachers in computer literacy, introduction and advanced Internet. They were also taken through a program on multimedia teaching (Aduda 2000). Considering that Kenya has close to 300,000 teachers, 200 are a drop in the ocean. Further, some of the teachers were from rural schools where access to electric power and telephone is non-existent to the majority of schools and computers are beyond reach. Therefore in as much as the AVU objectives may be noble, their effect is of very limited impact in Kenya. Thus, from the onset, AVU is destined to serve a very limited clientele. However, if the trainees are encouraged to train others through similar or other appropriate training, the impact can spread. For instance, high school and primary school teachers can be trained during the summer and given the appropriate technological support, these schoolteachers will be able to teach a large number of students if computers are made available in the schools. The obvious lack of preparedness and/or interest shown by students in the science and technological fields can be overcome. Clearly, the problem centers on the availability of financial resources because the schools will require electricity supply, computers, and upgrading of software. This is the very investment the World Bank and IMF have been discouraging the government to make through SAPs.

It is the private sector that has found short-term training in information technology most attractive and it is them who appear to be interested in the AVU programs and who can afford the costs. When they return to their respective companies and institutions, some of the trainees conduct in-house training and help demonstrate the usefulness of such training through improved output/efficiency/cost saving, etc. This often leads to better remuneration for them and encourages others to improve themselves through distance education. But convincing policy makers that the free flow and acquisition of knowledge and information will substantially help in the economic development and stability of a country and that it will not necessarily be a threat to their power and existence is a big, if not the biggest problem in introducing information technology for distance education in Kenya and many other African countries. Peter Knight (1995) says that Africa is the least connected continent not because of the lack of resources but because of the lack of political concern and will to establish connectivity. He points out that in many countries inefficient monopolistic state telecommunication companies have prevented the free flow of information. It stands to reason that successful and meaningful Internet and multimedia learning cannot go on in such an atmosphere of intolerance and information censorship (Lutta-Mukhebi and Amutabi 2000). It is for this reason that the liberalization of the telecommunications – noted in several chapters in this volume – needs to be welcomed.

Distance Education

There is no doubt that distance education has the potential and capacity to contribute to growth and expansion of high quality university education (Amutabi 1997a). In many countries distance learning has proved that it is indeed flexible and cost-effective: the University of South Africa (UNISA) has the largest number of student enrolment in Africa in its distance education programs. Their alumni include several distinguished personalities in Africa playing very useful roles in the development of their countries (Walters 1997). The British Open University produces 9 percent of all the undergraduates in Britain each year at a cost of 5 percent of the national university-operating budget (Bates 1995). More importantly, distance education enables working people and those students unable to make it to the conventional universities – the vast majority – to get high quality education. The AVU has rightly focused on science, engineering, business and the medical fields. In Kenya distance education has had several obstacles, including the lack of commitment by the government in treating the department of adult education as an important part of education, and the general apathy by the populace and employers who prefer graduates of conventional systems of education than distance learning. This is however changing (Amutabi 1997b).

Distance education especially "telelearning" brings about independence of time, location and distance. The fact that one can learn without being absent from the job is definitely an advantage. The fact that one can get the lessons on videotapes and receive and submit assignments online means less travel expenses. The fact that one can also do assignments from any place and without being confined is also a new freedom that makes the learner to be incharge of his or her learning "in the university without walls." It would appear that many workers who were unable to pursue higher education now have the opportunity to do so.

The distance education courses have also tended to be arts and humanities-based. However, the emphasis should also include agriculture since agriculture is the backbone of Kenya's economy. The country ought to be able to feed itself before anything else. Since agriculture will have a greater spillover effect to manufacturing, transport, infrastructure, etc, it is a critical sector, indeed; any development that the country can aspire to is most likely to come initially from agricultural development. Studies such as improved farm management, afforestation, irrigation and land reclamation, dry land farming, soil science, fish farming, horticulture, resource management, etc. are too important to be ignored for a country like Kenya. Another important area of emphasis should be the improvement of primary (elementary) and secondary (high) school teachers. Distance education can gear the teachers for life-long learning and improvement, attributes which they can pass on to their students who will hopefully, at least some, get prepared for distance education and open learning. Thus a change of culture will start at a fairly young age. The realization that learning is a life-long process is quite critical in a rapidly changing world.

The advantages of bringing high quality education through modern technology are many and varied. Because of dwindling resources existing institutions of higher learning are unable to keep up their libraries and laboratories, let alone to maintain adequate funds for research. Consequently, the academic staff – in addition to low pay – find themselves isolated, overworked with heavy teaching loads, and unable to keep up with developments in their fields. Under these circumstances it is difficult to attract any promising young faculty members. Many of those trained abroad choose not to return and the few who do get disillusioned and frustrated very quickly.

If it had been instituted properly and widely the AVU could have helped in solving some of these problems. The AVU library concept is particularly fascinating. The university instructors and their students can have access to online journals and archived materials, participate in online discussion groups, subscribe to list servs, etc. The staff can keep up with developments through short courses and have access to the libraries in the developed countries through the AVU's electronic library. Interaction with faculty in the developed countries might even lead to joint research projects on crucial problems of development in Africa. This is beneficial to the country, the institution, and the individual. But such benefits are going to come in the long term and hence the need to make sure that the AVU project is sustainable. The greatest advantage of AVU is perhaps its projected aim of preserving African rare collections through scanning and digitization. This will make these rare documents and materials available more easily to online researchers, students, the academic community, and the wider public. The other aim of promoting online publishing of scholarly work by African scholars also is likely to make an interesting contribution.

The Misfortunes of the AVU

The AVU has certain disadvantages inherent in it. It has been seen by some scholars as a vehicle for enhancing the global hegemony of the North in the globalization process (Amutabi 1997b), suspicions fed by the fact that the hardware, software, and the learning packages come from universities in the North. The AVU's misfortunes are worsened by the fact that the same World Bank that has called for reduced spending by governments in education in Africa is the one now advocating for this luxurious and more expensive educational approach.

Like any other distance education program, the AVU intends to save the African governments scarce foreign exchange by keeping the students in their home countries, while allowing them to pursue foreign education. Also, while the AVU was designed to meet growing demands for university education and alleviate congestion at the existing universities and save them money, in so far as the AVU programs are generated in the North, capital transfer still takes place through purchase and importation of equipment much of it already obsolete by the standards of the North. Many students have already shown their inability to pay beyond the pilot phase of the AVU programs. Micro- and macro-economic conditions have increased rural and urban poverty in Kenya. This has reduced the purchasing power of many Kenyans hence their capacity to support higher education privately. Even those who go to private universities in Kenya and those who pursue parallel or evening degree studies at the public universities tend to be the sons and daughters of the wealthy. Thus, the AVU education is limited to the rich, which runs counter to the AVU idea of expanding education opportunities for all.

The AVU program relies heavily on e-mail communication. Whereas e-mail may be quick and efficient, and even cheap, there must be an infrastructure to make it operational. Today, there are many private entrepreneurs providing Internet services in Kenya, especially in Nairobi, but these developments must be seen against the backdrop of very high user-cost and the poor state of basic communications. Unstable, unreliable and insufficient power supply in Kenya, not to mention the 16-month power-rationing program in the late 1990s, is a case in point. There are also the problems associated with unreliable and congested telephone links, unaffordable cost of telecommunications (telephone, fax, e-mail, etc.) services. In as much as the goals of the AVU are good and aimed at leapfrogging African university education, they are overly optimistic and take too many factors for granted. For example, at a very basic level, a telephone line and a personal computer for each lecturer, which are taken for granted in the North, are still a luxuries in many Kenyan universities. It is too much to expect the same lecturer who has no access to a telephone line and a computer to begin using the Internet to send assignments and results to students through e-mail.

The online digital library is a brilliant idea and has the potential to overturn the current intellectual stagnation for luck of current literature in Kenya's university libraries, which are slowly turning into archives. However, the use of the digital library really depends on the easy access, extension, availability and ability to maintain and use a network of computing services. There is still widespread and inadequate computer literacy and facilities in Kenya. Some lecturers in Kenyan universities and their students are still computer-shy, so that using the new information technologies will be a gradual process (Amutabi 1997b). Africa's share of the worldwide telephone network is just 2 percent, and it accounts for only 1 per cent of computers in use in the world. The AVU would have been a great idea if computers were as easily available in Africa as the transistor radios. Even at the AVU centers the computer hardware and software are far much below the speed and sophistication seen in universities in the North.

Conclusion

In conclusion, the African Virtual University is a timely project, which, however, will need careful adjustments to the realities and needs of Kenya if it is to contribute to its development and the welfare of its people. The most important considerations that need to be made include the appropriate choice of technology and its accessibility, the proper choice of what needs to be taught, and the project's sustainability. Already there are tensions at Kenyatta University that led to a strike in 2000 and from which 36 students were suspended and four expelled for allegedly being the perpetrators of the disturbances. This is already casting aspersions on the AVU's acceptability. Participants in any project are the best gauge of its success and failure. The events unfolding at Kenyatta University indicate the latter. The students are supposed to be willing partners for the project to succeed.

Both private and public universities in Kenya are already in different phases of the implementation of their own versions of distance education variously known as "Module II" at the University of Nairobi, "Parallel/Evening/Alternative degree programs" at Moi University, "Sandwich programs" at Maseno University, and "Evening programs" at the United States International University-Africa. Thus the AVU idea is not really novel and its haphazard implementation is indicative of the arrogance of many projects of the World Bank in Africa, including the SAPs. They are often imposed on people without due regard to their feelings and even inputs. Agreed, project implementations and appraisals do not at times require referendums, but courtesy demands consultations with the would-be beneficiaries. There is obvious duplication of purpose between the AVU projects and the current twining of Kenyan universities with European or American universities. Many of the efforts of the universities in the North in Kenya and elsewhere are duplicated by the AVU centers. Compared to the partner universities from the North which come to Kenya as collaborators and partners, the AVU has operated from a position of arrogant superiority.

As long as costs to gain access to education and the Internet are higher in Africa than in the North, money will continue to mark the dividing line between knowledge and ignorance. The World Bank may even increase its centers in Kenya but as long as the cost remains high, its aim of making high quality educational opportunities available to many Kenyans will remain a mirage. In Kenya, income disparities are wide and are bound to widen even more with the kind of privileged access promoted by the AVU. The World Bank also needs to reconcile and resolve the paradox of promoting both the AVU and SAPs in Africa. It is SAPs that led to the decay of education in African universities and the AVU cannot undo the damage. The targets that the AVU envisages cannot be attained with the current levels of impoverishment in Africa where people are barely surviving. It was supposed to enhance access but because of cost only upper classes can afford it. The AVU is, therefore, creating and enhancing societal stratification, because only the children of the rich can afford its costs and are computer-literate.

CHAPTER 14

TELECOMMUNICATIONS AND ITS IMPACT ON HEALTH CARE IN TANZANIA

James Katzenstein

Introduction

It was a warm afternoon and we had been driving for almost three hours on roads that were just wide enough for one car in each direction. On one side of the Land Rover rose an almost perpendicular cliff face and on the other, a sheer drop to the valley floor below. I had been fighting motion sickness for almost an hour and as cars and, occasionally buses passed us in the other direction I closed my eyes and hoped that our driver could keep the car on the road. This was no idle fear as, periodically along the road, signs were erected honoring those people whose vehicles had gone over the side.

At last we arrived at our destination, a village carved into the side of the mountain, and we interviewed the mother of a family whose husband had recently died, leaving her to raise nine children. She was suffering from a goiter and told us of repeated bouts of malaria throughout her family. What public health information she did receive, came from a village health care worker. We interviewed the village health care worker who had been trained some years before but had not received upgraded training since her initial training. The village health worker told us of the time when she had difficulty delivering a baby whose head was too large for the birth canal. In order to save the mother's life, she killed the baby and removed it.

A lack of developed physical infrastructure together with a shortage of trained physicians and nurses contribute to health care problems in the developing world in general and Tanzania in particular. Even where public health information exists, there is no way to efficiently deliver it to the point of greatest need. Health care workers in remote areas frequently do not get the training and education needed to keep their skills sharp and to upgrade their knowledge level, and, when faced with a situation that is beyond their abilities there frequently is no way to get help.

There is also a growing disparity between health care in the developed and the developing worlds. Africans are dying of different diseases than Europeans and Americans and that gap may grow larger as western scientists focus more on man made and degenerative diseases, e.g. heart disease, cancer, and psychological illness. Fewer than ten percent of deaths in Europe are caused by communicable diseases while diseases such as tuberculosis, HIV/AIDS, diarrheal diseases, malaria and measles cause 60 percent of deaths in Africa. Fewer than 3 percent of Africans are age 65 or older while 14 percent of Europeans attain that age (Ratzan, Filerman and LeSar 2000). The separation, both physical and cultural, between the two societies reinforces the different focus in health care. Western scientists with a relatively large resource base focus on health issues in their society while African health care workers and scientists struggle with a shrinking resource base and major barriers to communication and health care information from outside of Africa.

Within Africa there are health care facilities that are capable of delivering diagnosis and treatment on a larger scale than they are currently doing. Lack of physical infrastructure prevents the care from being delivered, even though it is available. In Dar es Salaam, Tanzania's largest city, there are health care facilities available, but people in places such as Morogoro and in Zanzibar cannot access them, because residents there simply cannot get to

Dar. Morogoro is 193 km from Dar and takes 4-5 hours by bus. Zanzibar is an island off the mainland and requires a boat ride. When one is sick with malaria, both trips are impractical.

This chapter describes a project that is beginning to address the issues of increased access to health care and health care information in Tanzania using telecommunications technology and a unique consulting model. The ultimate goal is to build a global network of western and Tanzanian health care organizations and universities that uses a telecommunications infrastructure to solve public health and health care issues in the developing world. The telecommunications infrastructure also provides the means by which health care facilities within Tanzania can provide diagnosis and treatment remotely to patients who currently have no access to health care. The consulting model promotes an environment of equals that bridges the cultural differences and encourages the emergence of problems to address as well as the solutions to those problems. Ultimately, it facilitates the sharing of medical and health care information; allowing medical and nursing programs in Tanzania to access the latest knowledge and insuring these programs remain current.

I will begin this paper with a discussion of the technology of telemedicine and then tell the story of our efforts in Tanzania to prepare the environment. I then will describe the KM Consulting model that facilitates the development of the human systems interface, which converts the telecommunications technology into a telemedicine system. This is followed by a discussion of the appropriate technology for this particular project and the implementation process that is planned. I will close with some thoughts about evaluation of telemedicine in the developing world.

Examining the Technology of Telemedicine

Telemedicine is an evolving science that is both facilitated and constrained by the state of telecommunications technology. Many definitions for telemedicine exist but the one I find most useful is that it involves the provision of interactive health care services where the patient is at a location remote from the provider. Telehealth is a subset of telemedicine but does not include direct patient care. Thus, distance medical and nursing education is a part of telehealth. Telemedicine technology, as it pertains to this study, encompasses an architecture that includes hardware infrastructure at the physician site, the patient site, and the connection between the two. It also encompasses the software and content e.g. instructional CD-ROMs, live training sessions, electronic patient records, etc.

Any conversation about telemedicine must include the state of the telecommunications technology and infrastructure. The goal of the overall architecture is to transmit appropriate information between the two sites, which is accurate and timely and therefore facilitates communication in a seamless way. This may be some combination of live images, stored images and data as well as historical patient information. Telemedicine becomes practical if the interface between the patient, the physician and the health care worker is trouble-free. The physician and the health care worker concentrate on delivering the service to the patient and not on how well or poorly the equipment is working. If the focus is on the quality and reliability of the infrastructure, skepticism builds among the patients and the providers and effectiveness is lost.

The infrastructure at the physician site may contain a computer to store patient information on an electronic patient record and a camera to capture live images of the physician. The computer may connect via telephone lines to an Internet service provider who provides access to the Internet or the physician site may be connected to the patient site via telephone system without the use of the Internet. The patient site may contain cameras and/or computers to facilitate live communications between the patient, the health care worker and the remotely located physician. It may also contain sophisticated equipment to capture high quality images, vital statistics, e.g., temperature, blood pressure and heart rate etc., and real time results of EKGs, ultrasounds, etc. The issues that affect the reliability of the telecommunications infrastructure, e.g., the connectivity between the patient site and the physician site, are bandwidth, reliability, latency, security and ubiquity.

Bandwidth can be thought of as a pipe through which images and data are transmitted. The narrower the pipe, the longer it takes to get data through and since image files are usually quite large, they require a large bandwidth. Increasing bandwidth availability throughout the world is making telemedicine more practical, and the cost of the technology that provides bandwidth has been declining as more users come on line. Reliability impacts telemedicine delivery in two ways. Poor reliability requires investment in maintenance to keep the architecture running, which drives the costs upward and it also drives up the frustration level with respect to both the health care providers and the patients. Reliability in remote areas is crucial to the successful implementation of telemedicine architectures in places like Tanzania because qualified maintenance of telecommunication equipment is scarce or non-existent, and because there is a cultural bias towards the visual rather than the descriptive. One can endlessly explain how the system would function if it had not gone down but the doctors, nurses and patients will not internalize the process until they see it happen. Health care providers will just stop using the equipment if the reliability remains poor.

Latency in telecommunications refers to the time that an image takes to travel from one point to another. As the time increases the quality of the image, if that is what one is transmitting, degrades. This is particularly true with real time image transfer over great distances such as between the U.S. and Tanzania. If the image is being used for diagnosis, e.g. where the remote physician needs to see the patient clearly to make an accurate diagnosis, latency can be a serious problem. Security is an ethical problem and in some areas a legal one. The secure storage and transmission of patient data so that only a select audience can access it is crucial to the effective use of telemedicine anywhere in the world.

The selection of the architecture for the project in Tanzania required a balance. We needed to have an infrastructure that would be simple and reliable enough so that the Tanzanians could operate and maintain it while at the same time flexible enough to usefully extend health care within the country. The vision statement of the project defined that as follows:

> We will provide the ordinary Tanzanian, throughout the country, with consistent quality

> medical care that is comparable to that which he/she could expect to get in the United States or Europe. We will also provide the Tanzanian medical establishment with the opportunity to learn and share the latest in medical knowledge, no matter where in the world it exists (Vision Statement of Project Tanzania 2000).

We wanted to use a hub and spoke architecture to provide a platform for public health information as well as diagnosis and treatment of individual patients. We also wanted the capability of communicating with universities and medical schools in the West, as well as access to web-based public health information and other content available in stored formats, e.g. CD-ROM.

The extent to which telemedicine is used in the United States depends on the availability of alternative platforms, the cost factors in delivering telemedicine compared with its alternatives, the amount of system training required, and the amount of hands on "art" required by the physician when treating a patient. Telemedicine programs have been developed in the U.S. to serve places where there are few or no physicians and where health care facilities are scarce or spread over great distances. As a result, for example, 30% of American telemedicine programs are in prisons where it is difficult and expensive to move patients to health care facilities outside the prison (Primer on Telemedicine, U.C. Davis 2000). Another developing field for telemedicine is home care for the aged, where patient mobility is a problem. Telemedicine allows diagnosis, treatment, record keeping, authorizations, scheduling and monitoring of patients by remotely located physicians. It also eliminates the need for hazardous trips to the hospital and its potential for contamination from exposure to other patients.

In Tanzania, where there is an overall shortage of physicians –1/20,000 inhabitants compared with 1/1,600 in the U.S. a serious deterioration of physical infrastructure e.g., roads, and extreme poverty which limits mobility, all militate against people being able to gain access to health care. Telemedicine and telecommunication technologies may be the best hope of overcoming the infrastructure deficiencies in Tanzania and providing access to health care for a greater number of Tanzanians.

Setting the Stage for the Development of Telemedicine

This project started in April 1997 when I traveled to Dar es Salaam as part of a consulting workshop. I met Dr. Kairuki, the owner and family-head of Mission Mikocheni Hospital, who had reached a 10-year milestone in the history of the hospital. Dr. Kairuki began the hospital with four beds in his house and it had grown into a 130 bed general medical center serving the medical needs of the population surrounding the hospital. He now faced two issues that I felt I could help him solve. The first was the problem that many entrepreneurial family owners face when their enterprises grow large. The doctor was a practicing gynecologist as well as the director general of a significant medical center. He was also active in the medical community and the chief of his rather large family. Getting to see him was difficult because there simply was not enough time for him to accomplish all the tasks he had set for himself. The second issue addressed the long-term strategic plan for the medical center itself. In what direction should the hospital grow in order to provide the greatest benefit for the people of Tanzania, and once that goal was established, how could it be made a reality? The first step in the process was relationship building between the two Americans on the project, the doctor and his senior staff. The Americans needed to understand the Tanzanian culture in which they functioned and made decisions. At the same time we encouraged the doctor to begin building a different, more equal relationship between him and his staff.

As the relationship formed, a consulting model evolved, the theoretical underpinnings of which will be discussed in the following section. Briefly however, the model consisted of a group of multicultural equals called the client consultant system infrastructure (CCSI) coming together in a unique way called structural coupling. Identification of the issues that the group wanted to work on together with the process for solving the issues evolved using a group process called action research. What I found particularly interesting is that the model, the problems to be addressed, and the beginnings of the solution all evolved at the same time and without anyone particularly understanding where the group would ultimately end up.

The group at first wanted more patients treated by the hospital and additional nurses trained to provide added staff as the hospital grew. The overflow would serve at other hospitals in Tanzania and elsewhere in Africa. During the next eighteen months, the group in Tanzania using project management techniques that had emerged out of the first set of meetings, began one of the first private medical and nursing schools in Tanzania. When I returned, eighteen months later, the facility had been built, the curriculum had been developed and approved by the government of Tanzania, faculty had been hired, and the first class was starting nursing training.

During the eighteen-month interlude between visits, I learned about health care in the developing world and began the process of engaging with and understanding deep poverty in Tanzania. When I returned to Tanzania for the month of August 1998, the group began to look at expanding health care within the country and addressing the issue of maintaining the level of health care education and training among the professional staff at the hospital and university. The concept of globalizing the hospital emerged. Globalizing, to us meant that the hospital and the university would couple with comparable medical schools, hospitals and other universities outside Tanzania. As medical education and information in the West advanced, the medical center in Tanzania would be in a position to take advantage of the new knowledge quickly and routinely. We felt that information technology was a key component of a globalization plan but we did not have the infrastructure nor did the medical complex have access to the Internet at that time.

During the next interlude, the medical complex acquired email capability and communication between the Tanzanians and myself became more regular. Additional members with expertise in nursing, information technology and computers joined the group in the United States and we began to put together the plan for an information technology infrastructure. We designed a standard setup for each computer within the hospital that had standard memory capacity and standard software, linked together in an Ethernet local area network (LAN). Our Tanzanian group members identified required hardware and software components in each computer and, together with the American members, developed a list, which I bought at a local computer store. Loaded down with electronic and computer parts and software, I made my third visit to Tanzania. Several months prior to my visit however, the visionary doctor who started the hospital and the university passed away of liver cancer at the age of 59. I believe his death was a tragedy to the family and to the nation of Tanzania.

During my next visit, for the month of April 1999, we painstakingly disassembled, upgraded and reassembled each computer while the maintenance department strung computer wire all over the hospital and university. When we got stuck, we would email to colleagues in the U.S. for advice. A family friend at the University of Dar es Salaam also helped bring the entire network together. We now had reliable Internet access, consistent email communication with partners in the U.S., and standardized computers that could be shared throughout the university.

It was time to test the concept of using telecommunications technology to make the first real-time links between the hospital staff and the American partners in the U.S. We also began to discuss the potential of using telecommunication technology to project health care outside the hospital to remote areas in Tanzania. Our first test was a failure. We used an 8x8 camera that plugged into a telephone and a television. Unfortunately, the lack of sufficient bandwidth and latency defeated us and, although we made the connection, the image was fuzzy and there was no movement. It was however, a beginning. The group developed a plan in which we would train nurse practitioners and several physicians and nurses in telemedicine while at the hospital. We would then open two clinics, one in Morogoro and one in Zanzibar, each staffed by a nurse practitioner who would triage patients and treat those she could treat without help. Those she could not treat herself would be diagnosed and treated jointly with a physician at the main hospital connected by telecommunication and telemedicine technology. My next challenge was to find the appropriate technology that could be adapted to the environment in Tanzania, and to fund the project.

Upon returning home, I began to learn more about telemedicine and to establish relationships with organizations that could provide both funding and technical advice and support. I attended seminars in advanced telemedicine systems at East Carolina University's Brody School of Medicine, and enlisted their aid in designing the system for Mikocheni. I also identified organizations to help with stocking remote sites, identifying and procuring medical content, and providing a framework for electronic patient records (EPR). The project also received some financial support as well as systems support from The Fielding Institute, a university specializing in Organizational Development.

I would now like to describe the consulting model and some of its theoretical underpinnings that evolved out of this action project and which I believe is crucial to the effective implementation of telecommunications technology transfer between the developed and developing worlds.

The KM Consulting Model-an Exercise in Structural Coupling

The consulting model is the means by which the technology and systems were identified and transferred between the partners in the United States and those in Tanzania. It is called the KM Consulting Model and consists of four components which, taken together, result in a unique way to accomplish development work. The four components are: 1) an organization, materialized as a structure called the client consultant system infrastructure (CCSI); 2) the process of coupling (Maturana 1997) and interpenetration (Luhman 1995) by which the consultant becomes part of the system and the means by which the individual members stay connected; 3) the techniques of action research as a process by which members of the CCSI accomplish work; and 4) the context as the overall system in which the CCSI and the research or action project takes place.

The consulting model is driven by a structure that I call the client consultant system infrastructure (CCSI). It is an extension of Susman and Evered's (1978:588) client system infrastructure, which they define as "an infrastructure of ad hoc and permanent face to face groups...developed within a client system to conduct action research." I have extended Susman's and Evered's definition by first adding the consultant to the infrastructure and second, extending the definition to include the use of electronic communications to augment face-to-face communications. The organization that is the CCSI, what Maturana calls an abstract concept that distinguishes a class of objects or relationships from its environment (Maturana 1998), is embodied in a structure as the client consultant system infrastructure that is developing telemedicine research on the ground in Tanzania.

The reason for the success of the CCSI lies in Maturana's concept of structural coupling. According to Maturana, when two or more unities interact with each other in a recurrent manner, over a long period of time, structural changes take place in the unities and in the environment such that structural congruence occurs. "The result will be a history of mutual congruent structural change as long as the unity and its environment do not disintegrate: there will be structural coupling" (Maturana and Varela 1998:75). When applied to a social structure such as the CCSI, structural coupling is the force that allows the participants to achieve, over time, structural congruence with each other and with the environment in which the intervention is taking place. In my view, it becomes the glue that holds the CCSI together as a structure. Structural coupling has to do more with rhythms and patterns and relationships than with structure in the traditional bricks and mortar sense. McWhinney (February 2000, Personal email communication) conveys the sense of poetry that is structural coupling in the following informal introduction to the coupling process:

> Coupling is relating to music, to the drumming of repeated perturbances. Coupling is the cosmic dance of electrons and neutron stars. It is co-vibrating, mutually responding, making love.

To couple requires systems with sympathetic sets of organizing qualities resilience and flexibility, that allows them to join in a give-and-take, to-and-fro, exchanging energy, feelings or information. Coupling occurs when systems resonate: when they are repetitively receptive to receiving and sending, passing back and forth their energies in tune.

Coupling reinforces relations by adaptation; we learn a tune by hearing and singing along. We recursively couple with the "beat" hidden under a cacophony, responding rhythmically until the external pattern has an internal representation.

Coupled systems become a unified system, a whole inclusive of sub-systems, that is more capable of responding to complex rhythms and unexpected violations than were the component parts. The solo dancer joins another to form a coupled pair; the pair finds three more to form the square for a quadrille and then joins the assembly to parade the hall. So atoms become molecules, and molecules proteins, and proteins perform a million dances on the way to become cells and the cells synchronize their functions to become living systems and differentiate them to become animals. Such construction is called structural coupling (McWhinney February 2000, Personal email communication).

The fourth element of the KM Model is the process by which the CCSI identifies the needs of the environment and from which the means of satisfying those needs emerge. This is the process of action research (AR). It is important to note that the process of AR and the structure of the CCSI are linked in a dynamic relationship that allows the resulting system to not just exist but to accomplish something. Maturana (1987), in discussing biological entities, says that the purpose of entities is to exist and maintain themselves autopoietically. They continuously invent and reinvent themselves in response to perturbations from the environment solely for the purpose of continuing to exist. In the consulting environment however, the pressures of the workplace demands that the CCSI accomplish something.

Ernest Stringer (1996) defines AR in terms of an explicit set of social values as an integral part of AR, and envisions a collaborative approach to investigation that seeks to engage subjects as equal and full participants in the research process. He states that AR has the following characteristics:

- It is democratic, enabling the participation of all people;
- Equitable, acknowledging people's equality of work;
- Liberating, providing freedom from oppressive debilitating

conditions;

- Life enhancing, enabling the expression of people's full human potential.

Figure 1 The KM Consulting Model: AR as process to accomplish work

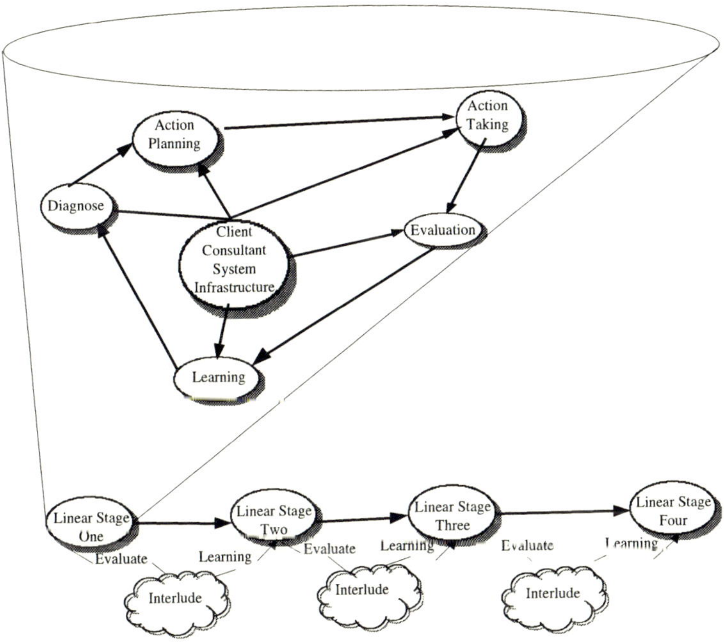

Susman and Everend (1978) emphasize the cyclical nature of AR. They call AR a cyclical process with five phases: diagnosing, action planning, action taking, evaluating, and specifying learning. Stringer (1996) speaks of a continually recycling set of activities in which participants examine the system, explore, analyze, and interpret the system, and then act upon the system. An evaluation takes place and participants recycle the process again.

In addition to the cyclical phases of Susman's model, I have added a component that allows the project to proceed linearly through time. Thus, rather than having discrete projects with distinct end points, we have one ongoing project with parts that are being worked on at various times using Susman's cyclical phases. Figure 1 illustrates this process.

A novel aspect that emerged from this research was the interlude that occurred between the linear phases of the project. I define an interlude as the time frame that falls between periods of activity in which the client consultant system infrastructure is not working as a unit. Interludes are particularly crucial within the international context, and played a key role in facilitating the AR in this study. During the interlude, the members of the CCSI were working remotely from each other without specific knowledge of what others were doing, or, even if others were doing anything at all. I recognized the significance of the interlude following my first visit when, to my frustration, my letters and emails to the Kairuki family were ignored. I was forced to proceed with the evaluation of the first visit alone, working with other sources to integrate my own personal learning with contributions from other scholars and action researchers. The interlude allowed for a period of reflection in which the participant-consultant withdrew from the system and examined what has been accomplished in the role of the researcher. John Mingers (1995:10) states:

> All explanations or descriptions are made by observers who are external to the system. One must not confuse that which pertains to the observer with that which pertains to the observed. Observers can perceive both an entity and its environment and see how the two relate to each other. Components within an entity, however, cannot do this, but act purely in response to other components.

An interlude allows the researcher to temporarily withdraw from CCSI member status where he/she can observe, explain, describe and analyze what has occurred during the time when the consultant was inside the system. The consultant-participant becomes a researcher-observer and then can perceive both an entity and its environment and see how the two relate to each other.

Finding the Appropriate Technology

Telecommunications technology is rapidly developing as a technology, but in the United States, telemedicine technology is being integrated into current medical practices with great difficulty (Field 1996:76). Physicians have had to change the way they consult with each other, the way they schedule patients, and the way that patient data is stored and accessed. The technology has also had an inward hardware focus rather than an outward customer focus. "Most product specifications and proposed technology solutions reflect the perspective of the technology vendor rather than the user of the product" (Field 1996:76). As a result, users of the technology have not found it particularly user friendly. According to the Committee on Evaluating Clinical Applications of Telemedicine, practitioners have experienced problems with convenience, reliability, quality and integrity of equipment; lack of time to learn the use of complicated hardware requiring extensive training and highly technical user manuals; equipment purchase decisions based on grant and other financing requirements rather than appropriateness; and lack of flexibility with proprietary systems (Field 1996:76). From a human perspective, the mix of clinicians, e.g., doctors and nurses, clinical support people, engineers, computer specialists and administrators, and managers and consultants have added to the complexity of a successful telemedicine implementation. Each technical discipline speaking its own language can give the whole process a "Tower of Babel" quality that is further complicated by language differences between countries when a project moves away from the U.S.

The criteria for selecting telemedicine technology for the project in Tanzania is focused on minimizing or avoiding the above pitfalls while, at the same time insuring the technology, from an operating perspective, is as invisible as possible to the clinicians and the patients. The criteria for selection is as follows:

- The technology is usable off the shelf with a minimum amount of customization.
- It does not require a high level of sophistication to use or to

maintain.

- It is compatible with equipment from different manufacturers so that as the system expands it does not have to be reinvented.
- If a component fails, it can easily be replaced.

When one component is replaced by newer technology, the new technology must be compatible with the older components.

The telemedicine component of Mikocheni's information technology architecture is computer-based and will build on the Internet and computer capability that has been developed earlier in the project. High capacity computers have become commodity items in the United States and together with inexpensive high quality cameras and software such as Microsoft NetMeeting, teleconferencing capabilities can be implemented using off the shelf technology. Maintenance and operations are well within the capability of the staff at Mikocheni or can be easily obtained using the resources of the multinational client consultant system infrastructure (CCSI). Software for the electronic patient record, CD-ROMs and web-based medical data bases for content are likewise off the shelf technology that can be readily integrated into a telemedicine system.

The system for delivering telemedicine diagnosis and treatment consists of interdependent components and processes that have to interrelate to each other to deliver satisfactory results. The CCSI will, in conjunction with the physicians and nurses at Mikocheni and the technical consultants in the U.S., use the action research process to develop the protocols by which the patient will be seen and the physician and the nurse at the remote clinic will interact. The patient record will be generated using off the shelf software but the content of the record and the protocol for what goes into it, who will have access to it, how it will be changed, etc. will also be developed by the CCSI. There will be other issues that emerge from the ongoing interaction among all who are involved in this project, but the process for solving issues is firmly in place and the solutions, like the problems, will emerge from the process.

The Implementation Process and the Next Steps

The implementation begins with the purchase of computers and their modification to use small cameras that provide 20 frames per second real time video. The computers will initially be linked by direct cables and then via the intranet at Mikocheni, where the concept of telemedicine can be demonstrated to the nurses, physicians and administrators. An electronic patient record will be loaded to the computers and will be tested and debugged prior to detailed discussions with the staff and administration concerning the format, content and protocols that are needed to make the EPR effective in the Tanzanian environment.

Next, the CCSI, the doctors, nurses and administration must decide the most effective way of delivering telemedicine. Specifically, issues such as under what circumstances telemedicine will be used, what information will be transmitted, and how quickly can that be accomplished. How much leeway will the remote site nurse practitioner be given? Can public health information be delivered to the remote site for the benefit of all the local inhabitants? These and other questions must be answered and then translated into a telemedicine system that will be overlaid on the technology to produce a functioning telemedicine network.

The CCSI will oversee the development and implementation of training programs for nurse practitioners, nurses and physicians in order to insure consistent levels of training for all participants. I am anticipating a practice period during which the physicians and nurses will become accustomed to using the telemedicine system and when unanticipated problems can be addressed prior to establishing remote sites. During this period, plans will be drawn up to establish the first remote site. At that time, questions of available bandwidth, access to a remote facility, and stocking the remote facility with equipment and drugs will be addressed.

An evaluation of the available medical and public health content from identified sources in the U.S. will begin the process of augmenting and upgrading skills of doctors and nurses at Mikocheni. The CCSI will facilitate in-service programs for both doctors and nurses and will work with the university to augment classes in nursing as well as develop a plan to keep the university's medical library stocked with the latest information.

The advantages of this project design are that the technology is off the shelf and readily available. It is also cost efficient and within the limited budget available to the project team. The equipment is reliable and can be maintained with limited technical capabilities available in Dar es Salaam or with limited help from western project partners. As the scope of the project expands to include additional remote locations, the additional equipment requirements should be minimal. The major advantage in the design of the human systems interface is that it is designed and implemented by the CCSI, which is locally driven with help from western partners. This is an African project and the human systems, having been designed by Tanzanians, instead of being externally imposed, will be culturally appropriate to the environment.

Further evaluation of the effectiveness of the concept of telemedicine in a developing country such as Tanzania should also be addressed. The effect of telemedicine on the level of health of a community compared with alternatives is crucial to the long-term viability of telemedicine technology in Tanzania. How clinical applications of telemedicine affect the quality of care and its outcomes is the central evaluative question (Field 1996:165). This central question leads to questions of access to care, its costs and cost effectiveness, patient and clinician perceptions as well as clinical outcomes that becomes the framework of a series of research questions that should be asked and answered within the context of this project.

Conclusions

Telemedicine has the potential of impacting health care in Tanzania by providing access to care for people who, because of a damaged national infrastructure, cannot get to a clinic or hospital. It also has the potential to provide access to the latest health and medical information to the medical establishment in Tanzania and to the general population through the use of Internet technology and CD-ROMs.

The KM consulting model forms the basis for developing the human systems which, when overlaid on the technology forms the telemedicine system. The model is crucial because it empowers the local medical establishment within an internal organization, the client consultant system infrastructure, and allows the issues and solutions to emerge from within rather than being imposed from without.

The long-term effectiveness of the telemedicine system in Tanzania is yet to be evaluated but it seems clear that the technology of telemedicine should be as simple and as invisible as possible, consistent with the needs of the clinicians and patients who use it. It should be a means to an end rather than an end in itself.

Finally, the long term viability of telemedicine as a vehicle for gaining access to improved health care and to the latest in medical and public health knowledge, will be judged by its effect on the level of health within the communities that it serves, compared with the alternatives.

CHAPTER 15

WOMEN'S NGOS AND SOCIAL CHANGE IN UGANDA, ZAMBIA AND ZIMBABWE: HOW INFORMATION TECHNOLOGY CAN ENHANCE WOMEN'S HEALTH IN AFRICA

Muadi Mukenge

Introduction

In 1999, the Pacific Institute for Women's Health launched a three-year capacity building initiative, *Women Connect*!, that combines communications and health strategies for the improvement of women's lives by working through women's non-governmental organizations in Africa. The purpose of the project is to strengthen the communication and advocacy skills of women's rights organizations, a strategy aimed at the broader goals of women's empowerment and helping to build civil society. Specifically, *Women Connect*! assists women's organizations to use all forms of media to communicate and advocate for the causes they feel are important in their communities – for example, women's reproductive and sexual health (including family planning and HIV/AIDS prevention and care), inheritance rights for women, and the reduction of all forms of violence against women. The project also works with groups to connect them to email and the Internet so that they can network with each other and with other groups around the world, sharing problems and solutions. Between May 1999 and early 2001, *Women Connect*! has held training workshops and provided project grants to 26 NGOs in Zimbabwe, Zambia, and Uganda. The impact has varied among the groups, and this paper will share some key highlights. We will not present conclusions about *Women Connect*! since projects are on going. More defining data will be available once a final evaluation is completed in October 2001.

The *Women Connect!* intervention was designed by Doe Mayer, professor of film and television production at the University of Southern California's (USC) School of Cinema-Television, as well as a consultant for the Pacific Institute. The project grew out of a pilot called Women Linking. Women Linking was funded by the Annenberg Center for Communication at USC and worked with a single NGO in Uganda, Action for Development. It was from that experience that a broader initiative with a larger number of groups was conceptualized: *Women Connect! Women Connect!* was funded by the Bill and Melinda Gates Foundation, with continuing support from the Annenberg Center, and is administered by the Pacific Institute. Professor Mayer serves as Project Director.

Project Rationale

Developing countries have seen the emergence of NGOs, or civil society organizations, as a driving force behind legal reform, social service provision, and advocacy for social change and sustainable development. The 1994 International Conference on Population and Development (ICPD) in Cairo identified women's NGOs as playing key roles in the empowerment of women, the improvement of women's health, and disease prevention. Reproductive health challenges such as HIV/AIDS, unintended pregnancy, maternal morbidity and mortality, and gender-based violence demand the involvement of civil society organizations with strong community ties. The Center for Communication Programs at Johns Hopkins University has documented how using media has brought reproductive health into public discourse in places where taboos previously prevented it (Priotrow 1997).

For example, social marketing campaigns strive to sensitize target communities about family planning and using condoms, and women's NGOs are increasingly launching campaigns of their own for the same purpose. We believe that by strengthening their communications capacities, women's groups can more effectively impact their communities and their nations and regions. Technology is one strategy to achieve that goal.

At a time of growing unease about the global gap between those with access to information technology and those without it, Africa is quickly becoming an attractive environment for exploratory projects that harness communications tools for development. These endeavors in communications are still small, and there are many challenges. Frequent power failures and overloaded telephone lines make it difficult to connect to the Internet. The vast majority of Africans do not have access to a computer. However, it is increasingly clear that a well-placed computer, like a communal well or an irrigation pump, is becoming another tool for development. Furthermore, Africa's rapid adoption of Internet connectivity and satellite use, and the growth of Internet service providers have contributed to increasing access to the new technology. The technology optimists include the UN Economic Commission for Africa, the Leland Initiative of USAID, SangoNet, Abantu, Gender Links, ACACIA, and a few other organizations that have launched pilot IT (information technology) projects in selected African countries. The Pacific Institute has taken this idea, combined it with traditional and mass media, and now works with women's NGOs on a long-term basis to apply these strategies to address socio-economic realities, especially those affecting women's health. IT *is* important for Africa, as indicated by one trainee: "Thanks for taking the time to come and ensure that women's organizations in Africa do not get left behind."

Our Approach to Communications

Women Connect! has provided technical assistance in developing communication strategies, using traditional and mass media effectively and linking groups electronically. We take the view that it is only by strengthening an organization's entire range of communication capacities that it will be able to clearly define its own mission and become truly sustainable. *Women Connect!* activities address several objectives: 1) to strengthen media campaign strategy skills by working with groups to define their messages, audiences and media choices, and to develop organizational capability to evaluate and sustain these efforts; 2) to provide technical assistance in utilizing the Internet to access needed technical information, strengthen partnerships, stay linked internationally, and learn from sister organizations; and 3) to develop strategies for strengthening and expanding reproductive and sexual health and rights outreach.

Health Observe and Project Philosophy

An integral purpose of *Women Connect!* is to enable women's organizations to participate more effectively in the provision of information and services about women's health, including reproductive and sexual health, safe motherhood, family planning, and prevention and care related to sexually-transmitted diseases, including HIV/AIDS. In order for *Women Connect!* to be successful, we have to respond to the way African women address this agenda. Health is defined in a broader context, and includes the absence of violence and the presence of other factors that affect a woman's ability to be productive. In this sense, land and inheritance rights also become health issues.

Summary of Accomplishments

Our feedback from the field informs us that some of the project objectives are gradually being met, but not without challenges. Some challenges include, first, the length of time needed to learn, fully comprehend, and apply the communication strategies; second, high staff turnover among participating NGOs; and third, budget constraints caused by high inflation, which have limited the capacity of *Women Connect!* project grants. There have also been positive developments. For example, in January 2001, the Women's Action Group, a leading NGO in health education, legal counseling, and policy reform in Zimbabwe, sent us the following memo after receiving funds for additional communications work from another donor we identified:

> We owe the success to you guys at *Women Connect!* The grant you gave us went a long way towards achieving our communication goals.

Another participating NGO that offers skills training and education, the Zimbabwe Women's Bureau (ZWB), used its Internet connection to download information on health issues and disseminate it in its monthly newsletter. ZWB has raised its members' awareness in such areas as reproductive health and rights, home-based care for AIDS patients, female genital cutting, and male responsibility in reproductive and sexual health. In Uganda, Sandra Okoed, Program Officer at Akina Mama wa Afrika, had this to say about our work:

> The impact of the small grant and *Women Connect!* training on our lives ...and work, will be increased communication and information access to facilitate decision making in the areas that are crucial to women's health, especially reproductive [health], education, adult literacy and creation of a vibrant civil society.

Here are other examples of the impact of the *Women Connect!* initiative on its 26 partner NGOs:

- Two NGOs have become the first women's groups to launch an Internet café in Uganda and Zimbabwe, focusing on training for women and women's organizations. One café will also train women Parliamentarians.
- Nine NGOs are conducting media campaigns and producing media materials through participatory research, planning and implementation. Highlights include a grassroots newsletter that addresses community needs in reproductive health knowledge and services, and a community publishing initiative that involves villages in producing media on AIDS, domestic violence and rape.
- Eight NGOs are newly connected to the Internet and have e-

mail accounts. Along with six other groups, they are receiving advanced training on the use of e-mail and the Internet.

- Eight NGOs are acquiring Internet research skills to locate, analyze and adapt health information for repackaging into traditional media.
- Eight NGOs have developed websites to share program objectives and strategies for addressing complex realities in a developing country.

Summary of Program Activities

Women Connect! has completed the following program activities to date:

- Providing training in communications and health strategies to 30 women's groups in Zimbabwe, Zambia and Uganda.
- Providing individualized technical assistance with the aid of project facilitators in each country.
- Providing small grants of $3,000-$5,000 each (totaling $121,500) to 26 participating women's groups to carry out year-long communications projects.
- Developing two resource guides on communications strategies and electronic connectivity, including relevant websites, listservs and discussion groups.
- Providing NGOs with over 30 different publications on health, advocacy and communications strategies.
- Facilitating the sustainability of grantees' communications projects through new donor linkages.
- Facilitating networking among grantees in an effort to maximize efforts and share strategies and resources.

When *Women Connect!* was conceptualized in 1998, we chose to work in Africa for several reasons. First, we knew that women's groups played an important role in promoting opportunities for women to advance their social, political and economic rights. Second, women's groups had limited capacity to utilize media and information technology in strategic ways. In fact, the Information Age is still in its infancy in Africa and many groups do not have Internet capability at all. Third, women are under-represented in technology training and use. One observer remarked: "In developing countries women will face additional barriers of lack of income, lack of time and lack of training all of which restrict levels of usage" (Morna 2000: 22). Finally, we felt that communications could go a long way in helping women's NGOs achieve their organizational goals through networking with groups that worked on similar issues.

The landscape is gradually changing among participating NGOs in the three capital cities where we have worked. In general, NGOs within the same city now know about each other's activities in more detail and have identified areas where they overlap and can collaborate. They communicate with e-mail frequently and attend strategy meetings that, in the past, were missed because of an inefficient postal service. In Zimbabwe and Uganda, groups are using e-mail to plan and implement advocacy campaigns. They have used online advocacy to bring attention to arrests of women activists, traditional customs that violate girls' human rights, and biased media portrayals by foreign agencies. Some NGOs now have significant access to current information on health topics, including reproductive and sexual health, whereas many other service organizations without access to the Internet depend on materials that are several years old. This updated information is being channeled to NGO branches in rural areas. A major lesson grantees have learned with the IT projects is the time needed for technology adoption, especially with regard to developing vibrant websites. NGOs generally hire firms to create a website, yet are challenged to maintain it and attract readers since they are limited by time and expertise. They now appreciate a phased-in approach to using the Internet, so that initial efforts can focus on e-mail communication, which is less time-consuming and generates feedback quickly. This includes utilizing listservs and discussion lists to share programs, find partners or solicit support.

As the NGOs conduct research to determine the needs in the field, they are generating important information from target communities about attitudes, knowledge and practices around health issues, including HIV/AIDS. In fact, perhaps the most interesting data from the *Women Connect!* grantees has been the collection of the voices of rural and urban women. Some indicate that they do not trust health clinics or that one is stigmatized as diseased if seen there and therefore do not visit them for services. Youth and teachers recruited for an AIDS awareness initiative dropped out because of the stigma in being called "that AIDS group." Disabled women are calling for health workers to be trained on their particular needs, especially in accessing information on HIV/AIDS and facilitating access to funding. Women's groups continue to stress that unless a woman is given the means to achieve economic empowerment, she cannot negotiate safe sex and thus reduce her risk of contracting HIV. The above realities must serve as a lesson to development and international health professionals, as well as implementing bodies such as ministries of health.

Project Methodology

Needs Assessments: African Voices on Health

The needs assessments in Zimbabwe, Zambia and Uganda were carried out by local researchers who possessed some familiarity with the NGO community, communication strategies and health. Together with a local networking NGO, we identified and surveyed a pool of 15-21 women's groups in each country and selected 10 groups per country for the workshops based on the communication needs identified in the assessments and the groups' capacity to implement new activities. All the organizations included health as one of their program areas. However, reproductive health was placed in the context of a broad range of concerns that impact a woman's life and, therefore, her health. These concerns include safe motherhood, access to the political arena where she could influence policies that impact women, protection from all forms of violence, access to property, educational access and socio-economic advancement. Many NGOs are working on strategies to combat the HIV/AIDS epidemic, taking both prevention- and care-based approaches, as well as the prevention and treatment of other sexually transmitted infections.

Training Workshops

We designed our initiative with the participation of a local networking NGO that was able to link *Women Connect!* to other grassroots women's groups. In-country activities consisted of 1) on-site visits to NGO facilities; 2) an executive seminar for NGO directors; 3) a four-day training workshop in media and electronic connectivity strategies for NGO staff conducted with assistance from local co-trainers; and 4) proposal planning discussions for the small grants.

The Small Grants Program

We introduced the small grants program at the end of the workshop and gave NGOs about six weeks to submit a proposal. To date, we have awarded $121,500 in small grants in the three countries. Grantees submit a mid-term and year-end report.

Facilitating Local Technical Assistance

Women Connect! identified resource persons in project countries with expertise in appropriate content areas to assist the funded NGOs in carrying out the small grants projects. In addition, *Women Connect!* hired a local consultant in each country to oversee and facilitate the progress of all the groups on a monthly basis.

On-site Monitoring Visits

Women Connect! staff visited grantees twice every year to discuss progress on the small grants projects. The purpose of the visits was to identify problems, successes, opportunities for collaboration, and technical assistance needs. Because each NGO has different needs, structures and target audiences, these visits were an important component of our strategy.

Sustainability of NGO Projects

The Pacific Institute grantees have emphasized the significant role of the small grants in allowing them to implement the tools introduced in the workshop in a substantive way. Our small grants project support extends over one year, and we took the initiative to identify additional donors who could continue that support once this first phase ends. Thus far, the Global Fund for Women has re-granted eight of the ten *Women Connect!* groups in Zimbabwe and Zambia that applied for additional communications funding. Participating groups in Uganda will likely be considered later in 2001. Our desire to explore additional financial support for the NGOs stems from the groups' own dedication and project results; they want to capitalize on the momentum of activities from this first year. They have a year's experience in doing more effective planning, research, implementation and evaluation than previous efforts. Some express the desire to empower other sister organizations with the same skills. Despite several challenges, select NGOs have shown some preliminary results from the use of communication strategies. Below we provide examples.

Zimbabwe – Using Technology to Advance Women's Health

Women Connect! started working in Zimbabwe at a time of significant social and political change. In mid-1999, the women's movement was gaining momentum after a series of progressive rulings on women's rights, but their struggle also continues as advances are challenged by legal setbacks. At the same time, the country is experiencing economic shocks and political tensions. The accompanying instability and violence have impacted women heavily, as record numbers are assaulted and raped without the means or access to treatment for infections if needed. Women's NGOs have seen their work multiply – in all sectors – but face budget and staff shortages due to donor pullout. In such a difficult environment, it has become even more important for disparate NGOs to network and to combine their efforts to maximize their resources and impact.

Emboldened by the media strategies training with *Women Connect!,* five grantees took on small grants projects that include, first, a grassroots newsletter that addresses community needs in reproductive health knowledge and services; second, a community publishing initiative that involves villages in producing media on AIDS, domestic violence and rape; and third, a media campaign on women's rights launched during an annual advocacy event called "16 Days of Activism." Four other groups completed technology projects. They are using e-mail and the Internet to network with affiliate organizations, to research information on health, and to repackage data for distribution to their target audiences. In all these examples, the groups have demonstrated that they are indeed vehicles for increased access to reproductive health information for women. The Pacific Institute's work in Zimbabwe ended in April 2001, but five of the nine groups have received new funding for continued communications activities.

Case Studies

Zimbabwe Women's Bureau (ZWB) is a well-established NGO with facilities that include a training center for community-based organizations. It operates activities in 13 geographical regions of the country. Focus areas include poverty alleviation, home building and improvement, small business promotion and business management training. The training team develops projects based on perceived needs at the grassroots level. With Pacific Institute training and support, ZWB undertook an IT capacity-building initiative. They now have e-mail and Internet capability and have received training in their uses. The result has been a marked improvement in productivity and management of organizational activities. ZWB has gone further with IT than initially planned, taking the initiative to download health information from the Internet and sharing it via their local language newsletter with target communities in both provincial and rural districts. The newsletter has addressed issues of nutrition, sanitary food preparation and storage, and physical health. Using downloaded information, significant awareness has been created on reproductive health and rights, prevention of infectious diseases, male responsibility, home-based care for AIDS patients, impact of HIV/AIDS on women, harmful traditional health practices and informed consent. They have discussed the impact of rape on women, as well as the legal recourses. They continue to carry out health communications work with additional funding from the Global Fund for Women.

The Women's Action Group (WAG) is a strong NGO known for its programs in legal counseling, health outreach, and policy reform. Its activities extend to districts throughout the country, and it publishes extensively in three languages. Particularly noteworthy are their health education booklets on topics such as cervical cancer, pregnancy, menopause and breast exams. WAG showed an enthusiasm for IT since they felt limited by conventional information sharing methods. Even internal communications were extremely slow, since staff occupied separate buildings and were reluctant to physically carry documents to colleagues. With Pacific Institute training and support, WAG networked all computers and installed e-mail and Internet for all staff. Staff also received training in how to use e-mail and the Internet. WAG is now communicating more frequently with sister NGOs, especially in planning major advocacy initiatives. Staff have streamlined office procedures and are receiving on-going training on Internet research skills that will assist them in developing future health outreach programs.

Zambian Women for Change – Innovative Approaches to Combating AIDS

The Pacific Institute decided to work in Zambia because we were encouraged by the existence of a network of women's organizations that were making attempts to utilize media and information technology more effectively. Despite this resolve, the ability of women's groups to utilize these tools remained extremely limited and sporadic. However, the women's movement in Zambia is quite active, and it has worked diligently in combating a government that has not been responsive to its activities at times. In fall 2000, after several years of lobbying for a National Gender Policy, the women's movement witnessed its adoption by the National Legislature. The challenge now is to facilitate its implementation, and women's organizations have key roles to play. The benefit of the legislation to women's health is significant. Provisions have been outlined to study the prevalence of particular medical conditions in women, to build additional and appropriate facilities, and to facilitate women's access to medical services. Such measures are needed in a country where HIV infections affect girls at five times the rate of boys in the 15-19 age group (Nanda 2000: 14).

Both the NGO sector and government have worked to develop HIV prevention messages and care programs for those suffering from AIDS. The efforts have begun to make a dent – even if slight – indicated in the decrease in the national average number of adults infected by HIV. *Women Connect!* trained staff from eight NGOs, and six of them have been funded for small grants projects. Research by NGOs into the HIV education needs of their communities revealed suggestions to disseminate prevention information through traditional leaders, at bible study groups and through community plays. The grantees finished their activities in July 2001.

Case Studies

The mission of the Zambia National Association of Disabled Women (ZNADWO) is to promote the welfare of women with disabilities and to be a voice through which to channel their concerns. It offers seminars and workshops, and promotes education, training and employment possibilities for women with disabilities. With Pacific Institute funding and technical support, ZNADWO established an Internet connection and is offering training to four staff members (a technology team) on e-mail and Internet skills, as well as strategies for conducting health research. Their e-mail connection is a major help, since most ZNADWO members cannot drive to participate in activities with other women's groups and advance notices can help them plan transportation in advance or simply discuss business online.

As part of their project, ZNADWO is repackaging health information from existing materials into a simpler format appropriate for the low literacy found in disabled people. Some will be transcribed into Braille. ZNADWO has been diligent in carrying out the project objectives despite having a volunteer staff and limited resources. The only NGO without a computer at the time of training, they wrote a proposal and secured funding for a computer from a donor *Women Connect!* identified.

ZNADWO is particularly interested in reaching women who are victimized precisely because of their physical limitations. Says Francisca Muyenga, the director:

> Disabled women are disadvantaged because they are excluded from most job opportunities. Disabled women end up depending on men in exchange for money, and men abandon them after sex. The women live in makeshift huts, some only with a tin roof.

During Muyenga's recent visit to a particular rural area, the disabled women she met said that HIV/AIDS education is very much in need and that the increase in infection rates is due in large part to poverty. "When someone has no resources, and no education, you can give them all the lessons but if poverty is not eradicated, then it makes no sense," they said. However, national AIDS outreach often does not include disabled people. Muyenga has met disabled women who do not know what a condom looks like. She takes male and female condoms to them when she makes visits throughout the country. She goes to health clinics to negotiate free services for disabled women who present a ZNADWO membership card. In January 2001, Muyenga convened a workshop on HIV/AIDS specifically for disabled women — including deaf and blind women. The group is considering a variety of media channels to reach disabled women with prevention messages. Many disabled women have no radios, so they cannot hear health messages via that media. ZNADWO is considering providing radios in each branch office where weekly radio listening clubs can be scheduled with disabled women.

Uganda – Implementing Media Strategies for Improved Impact

Small grants activities in Uganda are ongoing until September 2001, and it is premature to gauge how projects will develop. It is here that NGOs have taken the initiative to integrate traditional media and information technology in substantive ways. Seven of the eleven grantees are downloading information from the Internet and repackaging it into other forms of traditional media for their target audiences. Projects are unique, since each NGO works with different target groups, in different regions and with varying choices of secondary media. Ugandan NGOs have a long history in the country and co-exist with a government that is at the forefront of including women in parliamentary and other roles in government. However, their presence has not erased the socio-economic barriers to women's advancement overall. Some significant challenges are: 1) the inability to successfully lobby for the Domestic Relations Bill, which would extend protection to women in regards to marriage, property rights, child custody rights, among others; 2) the 12-year civil war in the north of the country, where countless women continue to face sexual violations and lack of access to education and health services; and 3) indifference to the plight of women who suffer domestic abuse. Effective communications strategies – including networking through information technology – are relevant to women's groups working in such a challenging environment.

Case Studies

Midwives are but one example of women's health providers who face a range of challenges in their every day work. One of these is facilitating an environment where women in society can make informed choices about their reproductive and sexual health. Robinah Kato is a member of the Uganda Private Midwives Association (UPMA), and tells of warnings she receives from men and husbands for providing information on contraceptives to women. She is constantly threatened with lawsuits and bombarded with other forms of threats. Is she afraid for her life? She answers:

> Of course I am! I have been confronted by people [saying] that we midwives practicing family planning kill the eggs of women in their wombs by giving them contraceptives, especially pills and injections. On many occasions I have been told not to give contraceptives to people's wives without the approval of their husbands. I have been seriously warned and threatened to be sued. But our work must continue...I have had a lot of experiences with spouses and members of different families to know that what I do is very much needed in our community.

The Pacific Institute training and small grant are being used to train Kato and other UPMA members on e-mail and Internet skills, especially research techniques. Such skills are important, as midwives and other medical providers need current information in order to give the most comprehensive care to their clients. Such skills have also positively impacted staff efficiency. The second project component consists of developing health tips that are broadcasted on three radio stations in local languages.

Isis-WICCE promotes the empowerment of women and information dissemination that leads to gender sensitivity and equal opportunity. It generates and disseminates information about women in Africa in an effort to contribute to the global debates about issues of concern to women. Isis operates a resource center with strengths in health issues and documented experiences of women's lives during times of conflict. The continuing civil war in Uganda is rarely mentioned outside of the country, and a similar silence exists for the women suffering its consequences:

> There has been an increase in all forms of violence against women, including sexual slavery, rape, sexual abuse and forced pregnancies, says Juliet Ogutti, the Documentation Officer. Given that women are also always targeted in situations of armed conflict and other types of conflicts they have come out and spearheaded the need for peaceful means of conflict resolution, peace building and reconciliation.

With Pacific Institute funding and technical assistance, Isis-WICCE will expand its facilities to add an Internet café. How will this initiative give voice to women's experiences? Isis is targeting its information technology training to other women's organizations. Improved communications skills by a broad segment of the NGO community will translate into more effective services and more efficient and effective networking, information sharing and advocacy.

What We Have Learned

Our work on *Women Connect!* has identified preliminary lessons learned related to working with women's NGOs, communications and advocacy. These are trends that have emerged and which will be further analyzed in the final evaluation of the project in October 2001.

Women Connect! was designed to be comprehensive yet strategic. First, we feel strongly that a broad communications strategy can strengthen overall organizational capacity; hence we linked media and information technology. Second, technical assistance to participating NGOs and follow up activities to implement skills introduced through a training workshop is critical to achieving objectives and having an impact. The support of NGO directors is equally important.

There are particular trends emerging for the technology projects, and the excitement of acquiring new tools is mitigated by the reality of the accompanying demands. Each NGO has unique needs and must find its own way to integrate the technology innovation into the organization. Doing so can be difficult, and causes changes and puts pressure on systems, relationships, communication and management styles. Adopting new technology requires a commitment to staff training and linking with technical resource persons. NGOs that aim to run an Internet Café must develop appropriate business and marketing plans to ensure effectiveness and sustainability.

The media initiative has challenged NGOs to utilize multi-media campaigns to broaden their reach, when most of them tend to use specific media. We have explored how they can develop advocacy messages that are responsive to cultural norms, friendly in tone and yet successful in conveying new information. For example, the Uganda Women's Network is using the *Women Connect!* grant to conduct a needs assessment that can get at what types of messages *will* work for an advocacy campaign on a Domestic Relations Bill that has stalled in Parliament for 36 years. Says UWONET director Sheila Kawamara:

> The campaign had largely been one where we talked to ourselves as women or the already

> 'converted'. We realize now that we have pointed fingers and not understood people's problems and realities...There is therefore a strong and urgent need to look at issues from the eyes of the people we are trying to reach.

The *Women Connect!* initiative aims to utilize communication strategies to improve women's health broadly. A majority of grantees are conducting research which generates important and useful data on women's health. This data has surprised the NGOs themselves, as respondents indicate their lack of knowledge on reproductive and sexual health matters. Such information will be instrumental in developing future program initiatives, either through the Pacific Institute or the NGOs themselves. Finally, NGOs have shown a particular interest in downloading and reinterpreting a range of current information off the Internet, especially new health information.

The final objective of *Women Connect!* is to foster networking among participating women's groups. We included networking activities in the project design, however, the NGOs have asked for more frequent interaction. This points to a realization of the benefits of connecting for mutually beneficial purposes.

Conclusion and Future Directions

This paper has highlighted how media and information technology strategies can serve as vehicles for developing and communicating messages on reproductive and sexual health and women's empowerment. Our work in Zimbabwe, Zambia and Uganda are only examples of the efforts taking place in regions facing many of the same public health challenges. With the experience of a consistent working relationship with 26 women's groups over the last two years, we can speak about their accomplishments in changing women's lives. We have seen their challenges and their successes, and we believe more than ever that collaborating with them maximizes our mandate to support women's groups.

The *Women Connect!* initiative has established many fruitful partnerships with African women's organizations and serves as a model for communications capacity building and organizational strengthening and leadership development work. Our final evaluation in October 2001 will reveal important learning for organizations working with women's NGOs and on communications and advocacy. There is particular need for collaboration on communications strategies. There are many lessons and strategies that can be shared among sectors. There are donors who have expressed an interest in supporting multi-sectored approaches to development, and this initiative would be such a venture.

As a result of our experience with grant making under the *Women Connect!* initiative, the Pacific Institute for Women's Health is launching the Action Grants Program to support many of the NGOs that participate in our capacity building programs. The Action Grants Program is an outgrowth of the principal objectives of the international program of the Pacific Institute, which are to improve the reproductive and sexual health of women and girls globally, and to support women's and youth-serving organizations worldwide. The immediate objective of the Action Grants Program is to strengthen the effectiveness, impact and sustainability of Pacific Institute's training and technical assistance to women's NGOs and youth serving organizations.

In conclusion, *Women Connect!* has contributed to creating a foundation for women's groups in Africa to take advantage of communications strategies to improve women's health and empowerment. We know that NGOs need more support for these strategies to take root permanently. Making this happen will help NGOs address the complex social, economic and political realities of life in developing countries.

CHAPTER 16

BOTSWANA'S CULTURAL DATABASE AS A TOOL IN THE FIGHT AGAINST HIV/AIDS: A PROSPECTIVE ANALYSIS

E. A. Biakolo

This chapter presents observations from the first phase of a feasibility study of a cultural database for Botswana. In the second part of the paper, the situation of HIV/AIDS in Botswana will be briefly articulated, especially relative to the cultural and epidemiological aspects of the epidemic. Finally the possible uses to which the cultural database could be put in the fight against the epidemic are analyzed and discussed.

The Cultural Database

The Government of Botswana initiated the idea of a feasibility study of a national cultural database in 1998. The terms of reference included: (a) identifying and listing all core and subsidiary cultural, information, and sports institutions in the country; (b) identifying and recommending suitable structures needed for networking between institutions concerned with cultural programs for ease of implementation of these programs for purposes of information sharing; (c) suggesting suitable management structures necessary to facilitate the accessibility of the databank to the general public and other institutions through initiatives which all stakeholders will have an opportunity to provide information to the databank; (d) identification of practical mechanisms for the establishment of a systematic information base, including all the necessary equipment so as to create a web site on the Internet for information accessing.

This tender process was concluded in February of 1999 when the research team of three – an IT specialist, a cultural journalist and lecturer, and this writer – was commissioned to undertake the study. The team's interpretation of the above terms of reference yielded the following: Under (a), the team suggested that the list should provide information on the names, addresses, telephone and telefax numbers of institutions and groups, the heads or key figures of such institutions, date of establishment, typical activities, their frequency, their role in the socio-cultural domain etc. Under (b and d) the most important networking structure to be considered was the Internet and the use of electronic mail, taking into account the level of access to phones, computers and the Internet. With (c), we were concerned, besides the role of IT managers and a cultural board, with the role of the Government Computer Bureau (GCB), the body charged with all IT matters in the country, especially with respect to the independence of the management of the database. The recommendations that were made to the client would be discussed shortly, but first a word on our procedure.

From the outset, it became clear that the investigation had to be divided into phases, with culture occupying the first phase and information and sports to be dealt with in the second phase. The reason for this was more practical than conceptual: we wanted to limit the kind of data to handle at this early stage to enable us to focus on the more important task of simulating a web-enabled database with the Internet as a platform.

As a preliminary step, a pro forma was designed, in order to collect data. Seven principal categories were postulated, vis: administrator, artist/art group, art and curio shop, craft worker, craft/art industry, cultural venue and cultural institution. The aim was to cast the net as wide as possible for purposes of artistic as well as geo-cultural representation. To the latter end, sample data were solicited from all ten-district councils in the country and at all levels: city, town, and village.

In addition to the pro forma, extensive interviews were conducted with various administrators, cultural and artistic associations and their leaders, and also individual producers. The major purpose was to determine current knowledge of, access to and use of electronic mail and the Internet; relation to other cultural agents and producers; and the degree of consumer access to their products and activities. The personal interview format was preferred to a questionnaire method to ensure on site assessment of the state of IT in the regions of the country far from the metropolis.

Conceptually, the principles underlying the creation of the database, as enunciated in the final report of Phase 1, are the following four. First, the government as custodian of culture has responsibility to create, maintain and update records of cultural activities and information on cultural practitioners and producers. Second, a central goal of government's administrative function in this respect is the facilitation of citizen access to cultural life and products, as well as the creation of the means to enhance cultural productivity, since culture is a creation of the people and belongs rightly and primarily to them. Third, the database is designed to facilitate this access and enhance cultural production. Finally, above all the database is intended to facilitate people-to-people contact, enhance overall education of the Batswana and others interested in Botswana, serve as a research tool and encourage preservation of cultural products. It is also meant to serve as a means of identifying, sourcing and updating information on institutions, bodies, producers and agents of cultural life.

HIV/AIDS in Botswana

In figures released in June 2000 by UNAIDS, Botswana's adult HIV prevalence at the end of 1999 was estimated to be as high as 35.8 per cent, unquestionably the highest in the world. Some 290,000 adults and children were living with HIV/AIDS. Comparatively, the prevalence rate (in percentage) for Sub-Saharan Africa was 8.57 and for the world 1.07. One Botswana government publication puts the matter starkly: "It is currently estimated that the HIV infection rate doubling time in Botswana is two years." (Government of Botswana 1997). By a simple arithmetic the number of people living with HIV at the close of 2001 would roughly be equivalent to a third of the population. Given that the current total population figure is a mere 1.59 million and given the mortality figures from AIDS, we are staring a demographic catastrophe in the face. Since transmission is mostly heterosexual, this situation has inevitably raised questions as to the cause of this rapid spread, a trend, by the way, which is common to Southern Africa.

The Botswana *National Policy on HIV/AIDS* (1998: 2) stipulates that several factors are thought to have contributed to the rapid spread of HIV in Botswana. Among them are the extreme mobility of the population combined with good communication system, and the high rate of sexually transmitted diseases, the presence of which causes a manifold increase in an individual's chance of being infected during sexual contact with an HIV seropositive partner. In addition, sexual behavior patterns, which include having multiple partners, and frequent change of partners, are accepted as the norm, and common in both urban and rural communities. Rapid urbanization (approximately 45 per cent of the population now live in urban areas) has led to the breakdown of traditional mechanisms for controlling social and sexual behavior. Poverty, and relative lack of power in social and sexual relations among women are also of significance.

Other factors include insufficient knowledge of HIV/AIDS, sexual abuse, a fatalistic attitude and social stigma, as well as a range of cultural practices.

We shall explore this latter cause a little more since our concern is with culture. A Panafrican News Agency report of 1 December 2000 with the title "Cultural Factors Hamper Botswana's AIDS Campaign" by Wene Owino (2000) details a number of these factors. Among the more sensational of these is a widespread belief that HIV/AIDS does not exist, leading to the interpretation of the acronym AIDS as "America's Idea of Discouraging Sex." Furthermore, many Batswana, like other Africans, rely often on *sangomas* (traditional doctors), most of whom regard AIDS-associated ailments as products of supernatural agents. Similarly, some males believe that infections come from having sexual relations with women who have undergone certain traditional rituals; while others claim that having sex with a virgin can cure an infected person. In a recent paper, J. Nthoi (2001: 2) argues that there is no static traditional view of HIV/AIDS. Using *sangomas,* traditional healers and diviners, as a case study, he postulates that that there has been a radical change in cultural perceptions of the epidemic. He states:

> On the whole, the attitudes of traditional healers towards HIV/AIDS has undergone dramatic change since the discovery of the first AIDS patient in Botswana in 1985. Owing to the newness and the strangeness of this affliction, traditional healers attempted to deal with it on the basis of their understanding of similar and common afflictions they had successfully healed. At first, they trivialized the affliction by associating it with minor and curable sexually transmitted diseases such as syphilis, gonorrhea and *boswagadi*...Over time, and through exposure to the massive body of knowledge on HIV/AIDS coming to them through the various media of communication (television, radio and internet), the traditional healer has begun to dissociate HIV/AIDS from these common sexually transmitted diseases.

On the other hand, the erosion of certain traditional cultural practices has been thought to fuel the epidemic. In the past, traditional initiation schools were an occasion to inculcate more restrained sexual mores and marital values in the young. The absence of such institutions along with the prevalence of a climate of sexual permissiveness current in the Western world, and which has permeated to Third World countries, has meant a radical change for the worse in sexual behavior. For example, a Family Health Survey published in 1996 showed that almost 60 percent of first-time pregnancies were to women aged 15-19; secondly, single-parent (mostly women-led) families are the norm rather than the exception, all of which point to the prevalence of extramarital sex.

Thus, certain cultural norms and practices have either directly encouraged the spread of the epidemic or created a climate that impedes proper response to it. Further, the abandonment of certain cultural traditions has fostered attitudes and conditions that have led to the spread of HIV-AIDS. It may well be then that the critical question to ask is not whether or to what extent cultural practices have led to the spread of HIV/AIDS but what is being done to foster those positive aspects of Botswana's culture and to discourage the negative ones with respect to HIV/AIDS.

In trying to address this question, the Botswana Government's plan, *Botswana HIV AND AIDS: Second Medium Term Plan* (1997-2003) has tried to improve on the *Medium Plan One* (MTP 1) (1987 –1993). It proposes three aspects of the response, namely: policy development, institutional strengthening and service delivery. As part of institutional strengthening it proposes to strengthen social mobilization and advocacy against harmful traditional and cultural norms and values; strengthen efforts on conflict resolution between cultural practices and existing law on sexual matters; research into practices related to sexual behavior; and strengthen traditional structures (House of Chiefs) and their capacity to understand the desirability of change, and involve chiefs in policy development and discourse.

On service delivery it promises provision of Information, Education and Communication on harmful traditional and cultural practices in relation to women and the girl child's sexual and reproductive health as well as incentives for rural employment creation. Nevertheless, there is evidence that supports the view that the abandonment of certain positive norms has led to the rapid spread of HIV/AIDS. This plan says nothing at all about revitalizing those practices. Moreover, the negative approach of the plan could lead to nothing more than the sort of sensationalism we saw in the Panafrican News Agency (PANA) report earlier. It is also shortsighted, whatever the alternative claims in certain feminist circles, to limit action to women and female children (Lesetedi 1999). Male sexual behavior and attitudes constitute a good half of the problems inherent in cultural norms and practices and ought to be addressed as such. Above all, the plan is awfully short on substantives. It lacks practicability. What forms of the Information Education Communication (IEC) does the government propose? What could distinguish these forms from those already in vogue in the country – outdoors and electronic media advertising/publicity, candlelight marches, musical festivals, conferences, workshops, seminars etc.

It is in the light of the above criticism that the idea of the cultural database represented not merely a significant addition to IEC media, but in fact an impetus towards knowledge and information for change with respect to the HIV/AIDS pandemic.

The Database and HIV/AIDS

The following are some of the non-technical recommendations of the research team. The first recommendation is that a cultural databank is not only good and proper; it is not merely in accord with the recommendations of the Southern African Development Community; it is something long overdue. Above all, it is our considered opinion that a cultural database can be established for Botswana. Second, on personnel requirements and their training, we proposed three important positions, namely, system administrator, web developer, and data author/editor. The system administrator is responsible for all network and connectivity matters, to ensure that the database web site is available at all times. The web developer is a skilled programmer who ensures that (i) the database is designed for web connectivity; (ii) programs the interface for database connectivity; (iii) the web page itself is appropriately designed, and (iv) connects the web front with the database engine. The main task of the Author/Editor is to ensure that the data going into the site is accurate, fair and appropriate. Third, there should be widespread publicity for the databank as soon as a decision to implement is taken. Fourth, it is imperative to reach cultural producers in rural as well as remote areas. This can be done through various means, including the Youth Officers in the district council, the *dikgosi* (traditional rulers) in the various localities and by any other means or agents. The Youth Officer, the *dikgosi* or other agents may also liaise with different sectors such as the schools, the brigades and associations. Finally, citizen access to the databank can be achieved through a host of options and resources. For example, under NDP 8, the Government Computer Bureau (GCB) intends to establish some seven regional data centers in the country. Schools and community centers, within 60kms radius can thus be linked to a Wide Area Network (WAN) of these GCB data centers and thus to the Internet. There is also available individual or group connection to the Internet through private service providers.

From the researchers' point of view, the conceptual foundation of the database as a medium for a people-to-people contact, as an IEC resource, meant that perhaps the most critical recommendation of all was the last one. Under the National Development Plan 8, the government, through the Government Computer Bureau would provide wide area networks in regional data centers. The researchers estimated that schools and communities would be linked via the Internet.

The potential benefits include:

- Providing easy reach of the people by the government not only with respect to the HIV/AIDS question but in various other socio-political issues with mass appeal or interest.
- Providing a solution to the much-criticized top-down techno-bureaucratic approach to the HIV/AIDS crisis by involving all segments of the population (Molutsi and Badade 1999).
- Creating further avenues for a multi-sectoral approach to the crisis.
- Introducing technology at grass roots level.
- Promoting open participation through a discussion forum of the web site fronting the database.
- Creating an environment and opportunities for business, including citizen-empowerment small-scale enterprises in such areas as arts and crafts marketing, tourism and e-trade.
- Empowering women and girls and fostering gender equity and overall equal access.

Thus, a web-driven cultural database could assist in the fight against HIV/AIDS at several fronts. First, as an IEC process, it is perhaps the most direct and participatory form of communication and education; participatory not only in the sense of sharing the resource but even more in the sense of enabling mass contribution and engendering collective decision making. Secondly, the information resources to which it gives access are incomparably greater than any other medium can provide. Thirdly, it can also serve as tool in the struggle in another indirect manner, namely in the battle against poverty. Poverty has indeed been identified as one of the causes of the rapid spread of the epidemic. Poverty-alleviating business opportunities tied to the database process are therefore also a form of struggle against HIV/AIDS. All available resources need to be mobilized for this fight.

Conclusion

Even if these potential benefits did not materialize and the database was simply another information resource tangentially connected with the epidemic, it would still have been worthwhile to create it. The availability to rural and remote areas of such a critical resource is a real index of development and another firm measure in mitigating the rural-urban drift. It is reasonable to speculate that the cost-benefit returns of this effort far outweigh the enormous capital required to initiate the proposed program.

PART IV:

Information Technology, Biotechnology and

CHAPTER 17

INCENTIVES AND OPPORTUNITIES FOR INFORMATION TECHNOLOGY IN AFRICAN AGRICULTURE

James C. O. Nyankori and
Ibulaimu Kakoma

Introduction

At the beginning of the 1990s, the utilization of information and communication technology for electronic networking in Africa was limited to expatriates working with international and non-governmental organizations and a few indigenous Africans exposed to the technology in institutions of higher learning overseas. Most of the networks, were used regularly by only a few individuals or companies for electronic mail and file transfers (Grant 1999). Since then an increasing number of the African population is using electronic networking services in daily personal, academic and business activities in schools, businesses, government, or the overall private sector.

The apparent increase in the demand for information and communication services is indicated by the growing local private investments in information and communication technology through cyber cafes, computer classes, expanded computer curriculum in formal educational institutions, and providing Internet service. Furthermore, private and public institutions are increasingly using the Internet for identification and self-promotion as well as business transactions.

The literature on the nascent information and communication industry in African countries has identified management, cultural and political factors as major impediments to rapid growth (Adam 1999). However, major infrastructure building initiatives and strategic projects to build demand for Internet services have been active since the early 1990s. A more comprehensive discussion of these can be found elsewhere and includes capacity enhancing projects funded by international organizations (the World Bank, United Nations Educational, Scientific, and Cultural Organization, World Food Program, and United Nations Development Program), bilateral cooperation (International Development Research Centre of Canada, Canadian International Development Agency, United States Agency for International Development, Overseas Development Agency, The Dutch Ministry of Co-operation, and HealthNet), and private local Internet Service Providers (Jensen 1996).

Examples of Internet service demand building projects in African countries include the Commonwealth of Learning Distance Education Projects, World Research Institute's Ghana Environmental Resource Monitoring Project, Ethiopian Science and Research Network, Food and Agricultural Organization's Regional Early Warning Network, Commonwealth Secretariat's Eastern and Southern African Management Development Institute Network, Association for Progressive Communications' Women's Training Program, Telemedicine Collaboration, and United States Agency for International Development's TradeNet.

Ten years after these initiatives, the private sector has become a major force behind the growth of the information and communication sector in many African countries. Increasingly, the private sector is involved in providing Internet services, Internet hosting, and many other aspects of the information and communication sector, which provide access to business, public and private institutions as well as members of the public for a fee.

The purpose of this chapter is to evaluate factors influencing the utilization of information and communication services in African countries.

Background

For the present purposes we conveniently classify firms in the Internet industry into two categories: firms that provide basic services, or Internet access providers; and firms that provide value-added services, or Internet service providers[1]. Basic services include terrestrial and satellite connectivity hardware, configurations, maintenance and technical support as well as switching infrastructure, telecommunications network, international bandwidth, and international leased line. The primary clientelé of the Internet access providers are the Internet service providers.

All other value-added services including specific technologies and support services for information storage, retrieval and transmission are in the domain of Internet service providers. The Internet service providers cater for a wider variety of client including government agencies, corporations, educational institutions and other service intermediaries.

The demand for Internet services is for general correspondence through electronic mail, information storage, retrieval and transmission for business, academic, public service and personal purposes by end-users. The end-users usually have access to personal computers at places of work, academic institutions, private homes or cyber cafés.

Market forces drive the information and communication sector in African countries such that investors and service providers expect positive returns on their investments, and users expect the costs of services to be more than offset by the benefits. Characteristics of the market environment of the information and communication services in Africa are summarized in Table 1.

For convenience we developed four major categories. First, infrastructure (number of fixed and mobile telephone lines, international bandwidth). Second, service distribution structure (number of cities with points of presence, Internet hosts, Internet domains and the number of, mobile line operators, fixed line telecom operators and Internet access providers. Third, cost structure: costs of local phone call, international Internet link, local Internet link, Internet dialup, and monthly telephone line rental. Fourth, access utilization: dialup Internet subscribers, telephone waiting time/wait list[2].

The Internet value-added service delivery depends on the capacity of Internet basic service industry, organizational structure, public policy, and the market system. Production decisions by Internet access providers and distribution decisions by Internet service providers are interdependent such that expected and actual responses by final users affect investment and operational decisions on basic and value-added services. The apparent interdependence between Internet service providers, Internet hosts and final end users in the information and communication sector is key to developing a conceptual framework and econometric model for predictions.

The Internet service provider is the link between final end users and the basic service provider such that the Internet service capacity[3] increases as the number of final end users increases, and the capacity to provide Internet service increases as the number of Internet host capacity increases. The interdependency leads to the following propositions.

Table1: Characteristics of IT Services in Africa

Item	Mean	Min	Max
Fixed lines (1,000)	485.1	6.0	5,075.0
Mobile Lines (1,000)	129.8	1.0	2,500.0
Cities with Internet points-of-presence	7.0	1.0	100.0
Mobile Operators (#)	1.8	1.0	2.0
Fixed Line Telecom Operators (#)	1.3	1.0	2.0
Dialup Internet subscribers	36,794.4	200.0	650,000.0
International Bandwidth (Kbps)	17,534.7	64.0	400,000.0
Internet Access Providers (#)	18.1	1.0	250.0
Internet Hosts (#)	6,547.1	1.0	167,635.0

Internet domains (#)	1,156.6	59.0	6,583.0
Dialup Internet Cost[1] ($)	67.2	31.0	130.0
International Internet Link ($/month)	2,425.0	500.0	4,000.0
Local phone call cost ($/hr)	1.5	0.4	2.9
Monthly telephone line rental ($/month)	6.3	3.0	9.1
Telephone waiting time/Wait List (Weeks)	3.0	0.2	7.9

Source: AISI-Connect database - *http://www2.sn.apc.org/africa*
Notes: [1]20 hrs/month.

1. The Internet access capacity is dependent on the technology, the market, and public policy. Equation (i) expresses Internet access capacity (IAPS) in terms of technology, the market and policy. Technology is represented by (international bandwidth (BANDWD) and number of telephone lines (TELINES). The market is represented by the presence of a local stock exchange (LSTOCK), and public information technology policy is represented by (REGLAT).

(1) IAPS = f (BANDWD, TELINES, LSTOCK, REGLAT).
We expect Internet access capacity (IAPS) to increase in response to increases in the international bandwidth and with the number of telephone lines, other things being equal, such that:

$$\text{(1a)} \qquad \frac{\partial IAPS}{\partial BANDWD} > 0; \text{ and}$$

$$\text{(1b)} \qquad \frac{\partial IAPS}{\partial TELINES} > 0.$$

The presence of a local stock exchange market (LSTOCK) and independently regulated markets are expected to be associated with higher Internet access capacity due to the expansionary effects of capital inflow and free market operations. Similarly, the presence of an independent regulator (REGLAT) can positively influence basic services through market based decisions on specific technologies, product and service lines as well as business procedures.

2. The Internet service capacity depends on the Internet access capacity and market characteristics. In equation (2) the Internet value-added service capacity (ISPS) is expressed in terms of Internet access capacity (IAPS), and three market elements: monthly fees paid the providers (MFEE), and geographic coverage represented by the number of cities with points-of-presence (CITIPOP).

(2) ISPS = f (IAPS, MFEE, CITIPOP)

We expect a positive response of Internet service capacity to increases in Internet access capacity, and geographic coverage; and a negative response to increases in monthly fee such that:

(2a) $$\frac{\partial ISPS}{\partial IAPS} > 0;$$

(2b) $$\frac{\partial ISPS}{\partial MFEE} < 0; \text{ and}$$

(2c) $$\frac{\partial ISPS}{\partial CITIPOP} > 0.$$

3. The Internet utilization rate (USERATE) is determined by Internet value-added service capacity (ISPS), cost of utilization (CALLCOST), literacy rate (LITRATE) and the level of personal income (PINCOME):

(3) USERATE = f (ISPS, CALLCOST, LITRATE, PINCOME)

The response of Internet service utilization rate to increases in Internet value added service capacity, literacy rate and personal income are positive, and the response to increases in the cost of utilization is negative, such that

(3a) $$\frac{\partial USERATE}{\partial ISPS} > 0;$$

(3b) $$\frac{\partial USERATE}{\partial CALLCOST} < 0;$$

(3c) $$\frac{\partial USERATE}{\partial LITRATE} > 0; \text{ and}$$

(3d) $$\frac{\partial USERATE}{\partial PINCOME} > 0.$$

Statistical Model

The empirical model for the relationships defined in equations (1) – (3) is specified as follows.We used proxy variables to represent several variables in the conceptual model (1) – (3). We used the number of Internet access providers as a proxy for Internet basic service capacity. Similarly, the number of Internet service providers is a proxy for Internet value added capacity. The number of cities with points of presence was a proxy for geographic coverage and the number of dialup Internet subscribers was a proxy for utilization. Finally, service price proxies for the Internet valued added service provider and end users were monthly telephone line rental and local phone call cost, respectively.

4. $IAPS = \alpha_1 + \beta_{11}BANDWD + \beta_{12}TELINES + \beta_{13}LSTOCK1 + \beta_{14}REGLAT1 + u$
5. $ISPS = \alpha_2 + \beta_{21}IAPS + \beta_{22}MFEE + \beta_{23}\, CITIPOP + u$
6. $USERATE = \alpha_3 + \beta_{31}\, ISPS + \beta_{32}COST + \beta_{33}\, LITRATE + \beta_{34}PINCOME + u$

In equations (4) – (6) α_k and β_{ki} are unknown parameters to be estimated, u is the error term and the rest of the variables are defined in Table 2.

The three-equation model of African information and communication sector was estimated using seemingly unrelated regressions technique. The data used in the analysis are from AISI-Connect Online Database (http://www2.sn.apc.org/africa) for which we give credit to Mike Jensen. The results of the regression estimates are presented in Table 3.

Table2: A key of the Acronyms of the Major Variables

#	Variable	Description
1	BANDWD	International Bandwidths (Kbps)
2	CITIPOP	Cities With Points Of Presence (#)
3	COST	Local Telephone Costs (US$/Minute)
4	USERATE	Dial Up Customers (#)
5	IAPS	Internet Access Providers (#)
6	ISPS	Internet Service Providers (#)

7	LITRATE	Literacy Rate (%)
8	MFEE	Monthly Fees (US$/Month)
9	PINCOME	Personal Income (US$/Year)
10	REGLAT	Regulatory Environment
11		REGLAT1 = 1 There Is Independent Regulator, Else REGLAT1 = 0
12		REGLAT2 = 1 There Is No Independent Regulator, Else REGLAT2 = 0
13	STOCK	Local Stock Exchange Market
14		STOCK1 = 1 If There Is Local Stock Exchange Market; Else STOCK1 = 0
15		STOCK2 = 1 If There Is Local Stock Exchange Market; Else STOCK2 = 0
16	TELINES	Telephone Lines (#)

Estimated coefficients for bandwidth and the number of telephone lines in equation (1) are statistically different from zero at one percent level of significance. Similarly, in equation (2) the estimated coefficients for the number of Internet service providers and the number of cities with Points-of-Presence are statistically significant at one percent level of significance but the coefficient for monthly fee is statistically significant at ten percent level of significance. Finally, in equation (3) the estimated coefficients of cost of local calls and per capita income are significant at one and ten percent levels of significance, respectively.

Table 3: African Information and Communication Sector: Parameter Estimates.

Equation	Variable	Description	Coefficients[a]
1. IAPS	INTERCEPT	Constant	-15.0891
			(-2.04)
	BANDWD	Bandwidth (Kbs)	-0.0023**
			(-8.70)
	TELINES	Telephone lines (#)	0.00013**
			(9.24)
	REGLAT1	Independent Regulator	8.1370

			(0.97)
	STOCK1	Presence of local stock market	5.4067
			(0.60)
2. ISPS	INTERCEPT	Constant	-3918.94
			(-3.66)
	IAPS	Internet access providers (#)	46.5983**
			(2.64)
	MFEE	Monthly fee ($/Month)	-15.5968**
			(-1.70)
	ISPS	Internet Service Providers (#)	3.7917**
			(4.69)
	CALLCOST	Cost of local calls (US$/ minute)	-1125.4000**
			(5.10)
	LITRATE	Literacy rate (%)	-45.9138
			(-0.31)
	PINCOME	Per capita income ($/p.a.)	2.5560**
			(1.95)
		System weighted R^2	0.9820

Notes: [a]t-ratios in parentheses. **statistically significant at ten percent or less.

Notes: [a]t-ratios in parentheses. Bold face coefficient estimates are statistically significant at ten percent or less.

Hypothesis test results from the t-test of significance are reported in Table 4.

Table 4: Hypothesis Test Results for the African Information and Communication Sector Model

Equation	Relationship	Hypotheses	Test Result
1a	$\frac{\partial IAPS}{\partial BANDWD} > 0$	$H_o: \beta_{11} = 0$ vs. $H_A: \beta_{11} \neq 0$	Reject
1b	$\frac{\partial IAPS}{\partial TELINES} > 0$	$H_o: \beta_{12} = 0$ vs. $H_A: \beta_{12} \neq 0$	Reject
1c	REGULATION*	$H_o: \beta_{13}^2 - \beta_{13}^1 = 0$ vs. $H_A: \beta_{13}^2 - \neq 0$	Do not reject

Equation	Relationship	Hypotheses	Test Results
1d	STOCK MARKET*	$H_o: \beta^{2}_{14} - \beta^{1}_{14} = 0$ vs. $H_A: \beta^{2}_{14} \leq (\beta^{1}_{13}$	Do not reject
2a	$\frac{\partial ISPS}{\partial IAPS} > 0$	$H_o: \beta_{21} = 0$ vs. $H_A: \beta_{21} \neq 0$	Reject
2c	$\frac{\partial ISPS}{\partial MFEE} < 0$	$H_o: \beta_{22} = 0$ vs. $H_A: \beta_{22} \neq 0$	Reject
2b	$\frac{\partial ISPS}{\partial CITIPOP} > 0$	$H_o: \beta_{23} = 0$ vs. $H_A: \beta_{23} \neq 0$	Reject
3a	$\frac{\partial USERATE}{\partial IHOST} > 0$	$H_o: \beta_{31} \; \beta^{1}_{14} = 0$ vs. $H_A: \beta_{31} \neq 0$	Reject
3b	$\frac{\partial USERATE}{\partial CALLCOST} < 0$	$H_o: \beta_{32} = 0$ vs. $H_A: \beta_{32} \neq 0$	Reject
3c	$\frac{\partial USERATE}{\partial LITRATE} > 0$	$H_o: \beta_{33} = 0$ vs. $H_A: \beta_{33} \neq 0$	Do not reject
3d	$\frac{\partial USERATE}{\partial PINCOME} > 0$	$H_o: \beta_{34} = 0$ vs. $H_A: \beta_{34} \neq 0$	Reject

Notes: *Difference relationships

We reject the null hypotheses for all the estimated coefficients for regulation (REGLAT1), the presence of a local stock market (STOCK1), and literacy rate (LITRATE). The test results suggest the following.

1. The Internet service capacity is influenced by bandwidths and the number of telephone lines.
2. Internet hosting is influenced by the Internet service capacity, monthly fees and the number of cities with points-of-presence.
3. The Internet host capacity, Internet service fees and personal income influence Internet service utilization.

The unexpected signs for the coefficients of bandwidths in equation (1) and the number of Internet service providers in equation (2) are attributable to economies of scale. With higher bandwidths fewer Internet service providers are sufficient than at lower bandwidths. Similarly, fewer large Internet service provider can provide as much service as numerous small Internet service providers.

Computed elasticities (Table 5) show the relative effects of the explanatory variables on the number of Internet service providers, Internet hosts and Internet utilization. The elasticities show that the number of Internet service providers will change more than proportionately to a-one percent change in bandwidths and telephone lines. Specifically, for a-one percent increase in bandwidths, other things being equal, the number of Internet service providers declines by 2.1 percent. Similarly, the number of Internet service providers increases by 4.9 percent for a-one percent increase in bandwidths.

The elasticities for Internet host and utilization have similar interpretations. The number of Internet hosts increases by 2.3 percent for a-one percent increase in the number of cities with points-of-presence, but declines by 0.13 percent and 0.21 percent for a-one percent increase the number of Internet service providers and monthly fees, respectively.

Finally, utilization rate changes less than proportionately for a-one percent change in the number of Internet hosts, price and income. Utilization rate increases by 0.6 percent and 0.15 percent for a-one percent increase in the number of Internet hosts and income, other things being equal, respectively and, similarly, declines by 0.28 percent for a-one percent increase in the service price.

Table 5: Computed Elasticities for the African Information and Communication Sector Model

Variables			Elasticities
Dependent	Independent	Description	
IAPS	Internet access providers (#)		
	BANDWD	Bandwidths (Kbs)	-2.1044
	TELINES	Telephone lines (#)	4.9153
ISPS	Internet service providers (#)		
	IAPS	Internet access providers (#)	0.1396
	MFEE	Monthly fees (US$/Month)	-0.2184
	CITIPOP	Cities with points-of-presence (#)	2.3663
	Variables		

Dependent	Independent	Description	Elasticities
USERATE		Subscribed Customers (#)	
	ISPS	Internet service providers (#)	0.6362
	CALLCOST	Local phone call cost (US$/Hour)	-0.2859
	PINCOME	Personal income (US$/year)	0.1557

Conclusions and Implications for African Agriculture

Recent growth in the information and communication sector throughout Africa has been market-driven involving considerable private investments as well as policy liberalization and widespread utilization in business, personal as well as public service activities. Our results show that market forces are working according to expectation. The Internet service industry will continue to grow as more telephone lines are installed and higher international bandwidths become more affordable. At higher bandwidths the Internet service industry will undergo structural transformation due to economies of scale, which will raise the optimal unit firm size. The resulting shift to fewer but larger Internet-service-provider firms will involve consolidation, mergers or, acquisition activities among existing firms or market entry by a multinational corporation.

Internet value-added services and hosting which links the user to the global network, depend on Internet access capacity, cost and geographic coverage. An increase in Internet service capacity, geographic expansion to increase the number of cities with point-of-presence, business cost reduction, and higher utilization rate will lead to considerable growth in Internet hosting capacity through existing and new firms. Increases in the number of cities with point-of-presence is the single most important direct source of growth of Internet service capacity.

The results have important implications for commercial agriculture in Africa at production, processing and distribution levels. The Internet has made it possible for African farmers to access a wide range of information sources on diverse topics from task specific information to sector wide coverage. A particularly important item is market information necessary to provide a roadmap on what to produce, how to produce, how much to produce, and for whom to produce. This can enhance African agricultural export markets involving trade on a wide range of agricultural products within and outside the African continent. The problem of access to information and communication technology by individual farmers could be alleviated through strategic programs developed by commodity groups.

An illustrative account of potential benefits of information and communication technology to African agriculture is the Africanfoodonline.com a division of Product Marketing Concepts also known as African food wholesale club. Africanfoodonline.com is accessible at *http://www.africanfoodonline.com/* and provides African food products to the global market through an online store. The product line of the Africanfoodonline.com includes native African biscuits and crackers, baby products, coco yam fufu, cooking oil, cray fish, cubes and spices, egusi, fufu flour, gari, mackerel, milk, palm fruit, palm oil, palm wine, plantain fufu, pounded yam, rice, yams, and smoked fish. Africanfoodonline.com is widening the African food produce market beyond national boundaries to the benefit of domestic production and higher income for small farmers in many African countries.

Net revenue gains from strategic applications of information technology to agricultural production and trade arise from two major sources. The first source of net agricultural revenue gains is production cost reduction possibilities due to more efficient resource allocation and utilization resulting from information obtained through information technology. The lower production cost means higher profit levels at prevailing prices in the short run and as more producers adopt the new production technique, the market supply curve is shifted downwards. Consequently, there will be a new and lower equilibrium market price at which consumers are willing to purchase more. The increased demand generates additional revenue to producers.

The second source is characterized by an upward shift in the market demand function resulting from discovery through information technology of new buyers in domestic and/or foreign markets. The upward shift in the market demand function leads to a higher equilibrium market price at which sellers will gain additional revenue. The combination of increased production efficiency and new markets made possible by information by information technology means higher net earnings in the African agricultural sector.

ENDNOTES

[1] Occasionally the distinction is blurred through vertical integration.

[2] Time between ordering and getting telephone service.

[3] Unit size and number of firms.

CHAPTER 18

REGIONAL AND INTERNATIONAL COOPERATION FOR SUSTAINABLE AGRICULTURAL BIOTECHNOLOGY IN AFRICA

E. Babafunso Sonaiya and
O. Galman Omitogun

Introduction

Biotechnology may be defined as the application of indigenous and/or scientific knowledge to the management of whole (or part of) microorganisms, or of cells and tissues of higher organisms, so that these supply goods and services of use to human beings (Banders, et al. 1996). In Africa, several traditional practices reveal biotechnology at work. Shifting cultivation, bush fallowing, composting, crop friendly trees (e.g., glyricidia) interplanted with yams are all natural processes used by farmers to achieve soil fertility and protection. Farmers that practice intercropping (e.g., of maize, millet, sorghum, beans, sweet potato, yam, banana, cassava and sugarcane) to keep pests infestation low and thus maintain biodiversity are also using biotechnology. Ash and neem extracts are used to deter attack by black ants and termites on bananas and yams. Farmers practice genetic conservation and diversity. To maintain genetic diversity, farmers borrow and exchange crop seeds and usually buy or borrow diverse breeds of livestock to cross with their own stock. In food processing traditionally practiced by African women, natural fermentation to prepare "fufu" from cassava, "ogi" from maize and palm wine from palms is widely used.

Can Imported Biotechnology be Sustainably Applied in Africa?

In Africa, because of rapid population growth, changes in agricultural practices, biotic environments and consumer preferences, there is a need to increase agricultural productivity. Biotechnology offers a solution to this increasing need for food production as it brings new tools, new ideas and new approaches to agriculture. In addition, biotechnology has provided molecular biology tools suitable for rapid and detailed analysis of agricultural species enabling the genetic modification of plants and animals in order to increase resistance to pests and diseases, to tolerate unfavorable growing conditions, to produce more protein and to fix nitrogen more efficiently.

There are transgenic products with novel agronomic traits. Such include many crops that are resistant to the herbicides glyphosate, phosphinotricin, bromoxinil and the sulfonylureas and imidazolinones; crops carrying *Bacillus thuringiensis (Bt)* genes for insect resistance; genes conferring resistance to viral, fungal and nematode pathogens and nuclear male-sterile crops lines used to create inbred lines for hybrid crop production. Future transgenic crops are anticipated to include plants developed as sources of modified enzymes, industrial feedstock, biopharmaceuticals and antibodies and antigens for vaccines.

In developed countries, before these products are released, they go through a regulatory framework based upon scientifically and well defined tests. Acceptance of transgenic plant products by the public depends upon the economic benefits that the products bring to all participants in the value chain, including trait developers, seed companies, traders, distributors, ingredient manufacturers and consumers. Across the industrialized countries, legislation has been deciding whether or not genetically engineered products will be labeled as such while ethical debates continue in Europe, Japan, and the United States. It is, therefore, important that legislators, politicians, industries and consumers in African countries receiving these products recognize the urgency of harmonizing legislation with other countries on the use of these transgenic products. A key role of the research community must be to accelerate information into society, especially via schoolteachers and extension workers so that the principles and practices of biotechnology are better understood.

As recommended (FAO 2000), the question of sustainability must consider the following elements:

- The factors determining or influencing the appropriateness of the different biotechnologies e.g., their environmental impact; their impact on human health; their status with respect to intellectual property rights; their status with respect to biosafety regulations and controls; the degree of access to the biotechnologies; the level of capacity building or human resources required to use them; their financial cost; their impact on food production and food security.
- The relative costs (financial, social, political or otherwise) of the biotechnologies versus the relative benefits (productivity, food security or otherwise).
- Whether they are more (or less) sustainable than existing conventional methods for food production and agriculture, given the realities of life in developing countries.
- Whether some of the biotechnologies are more (or less) appropriate than others.
- Whether some biotechnologies are more (or less) suited to certain regions in the developing world than others.

It is the position of this chapter that the human resource base is the vital link in the whole process of sustaining agricultural biotechnology in Africa.

Agricultural biotechnology and biodiversity in Africa

Africa is the center of biodiversity for crops such as sorghum, pearl millet, finger millet, tef, Abyssinian oats, Brassica juncea, pigeon pea, pea, Niger seed, cowpea, sesame, faba bean, ensette, coffee (arabica), cotton, kenaf, castor oil and safflower. West Africa is identified as the center of origin or diversity for crops such as fonio, African rice, pearl millet, sorghum, yam, cowpea, Bambara groundnut, bottle gourd, melon, kola nut, oil palm and date palm. Banana and plantain production in Nigeria contributes 5% of the total global production. These two crops are some of the important crops contributing to food security in West, East and Central Africa (FAO 1996). Sub-Saharan Africa is an important source of animal genetic resources with a wealth of domesticated animal diversity. The share of Africa in the world's population and number of breeds is: for cattle – 12 percent and 15 percent, respectively; goat - 22 and 10 percent; sheep - 11 and 8 percent; pig - 2 and 4 percent; ass (donkey) 20 and 8 percent; horse - 8 and 8 percent; dromedary - 17 and 24 percent; chicken - 6 and 9 percent; duck - 1 and 13 percent, turkey - 2 and 19 percent; and domesticated goose - 2 and 8 percent. Other domesticated animals, which are not yet documented, include rabbits, grass cutters and mouse deer. The number of breeds or species of domesticated animals that are endangered are: cattle - 9, goat - 1, sheep - 3, horse -1, dromedary - 1, chicken - 1, turkey - 2, Muscovy duck - 1, domesticated goose - 2, guinea fowl - 4 and ostrich - 2 (Scherf 1995).

Forests cover approximately 26 percent of the world's total land area (FAO 1999). Forests provide a wide range of goods and services such as timber, fiber, fuelwood, food, fodder, gum, resins, medicines, pharmaceutical products and environmental stabilization. Inherently high levels of variation and extensive natural ranges characterize most forest tree species. This high level of genetic variation needs to be maintained to ensure present-day and future adaptability to changing environmental conditions. Despite the availability of a large number of genera and species, less than 500 have been systematically tested for their utility for human beings and less than 40 species are included in intensive selection and breeding programs.

The most important trend in forestry in developing countries is the progressive reduction in the area of forest due to changes in land use. While forests are being lost, there is growing demand both for environmental services and for wood and wood products, which they provide. Whiteman and Brown (1999) predicted that wood demand is expected to increase by 25 percent from 1996 to 2010. This demand will increasingly have to be met by forest plantations, and with decreasing land areas available for forestry, plantation methods will have to be increasingly intensive. This will necessitate better tree improvement programs in which biotechnology may play a role.

The widespread adoption of improved or exotic plant varieties combined with other pressures such as deforestation and soil erosion pose a threat to plant genetic diversity. To avert this danger, both ex-situ conservation in gene banks and in-situ conservation in farmers' fields are required. In ex-situ conservation, stored material is usually grown out at regular intervals to ensure its continuing viability and adaptation to the changing external environment. For those species that grow slowly, have low fertility, or do not remain viable under current storage condition, tissue culture can be an effective alternative way of propagating selected genotype. However, the sheer size that collections can attain will limit the size of genetic diversity that can be maintained. Biotechnology offers new methods of assessing genetic diversity and the relationships between different populations or samples. DNA fingerprinting is more sensitive in genotyping a large number of samples at a time than the biochemical analysis by isozymes but it is still not being practiced in Africa because of high cost of consumables needed for the analysis. For livestock species, cryoconservation of embryos is still in its infancy. Much research is still needed before a system of gene banks for animals, equivalent to those for crops, becomes feasible. Their establishment and maintenance require considerable investment in facilities and skilled staff. Since benefits of such investments are felt only in the long term, gene banks tend to get low priority from most African governments. This underscores the important part played in this field by the international research institutes of the CGIAR, which collect and store landraces and other genetic materials of most of the world's most important food crops.

Some biotechnologies are already widely used in Africa. All countries use traditional fermentation process especially alcoholic and lactic acid fermentation and most have adopted some modern large-scale processes, for example, enzyme fermentation for brewing beer. Low-end biotechnologies such as inoculates that can reduce inorganic fertilizer requirement and biocontrol agents that reduce chemical pesticides are newer technologies that have not been fully adopted. Nigeria, at the International Institute of Tropical Agriculture (IITA) and the National Center for Genetic Resources and Biotechnology (NACGRAB) and some other sub-Saharan countries have acquired plant tissue culture on an experimental scale. In addition, biofertilizers and a range of biological control agents, notably *Bacillus thuringiensis,* are now being produced in some African countries (e.g. South Africa, Kenya, and Zimbabwe). The application, as opposed to the production, of enzymes, diagnostics and DNA markers is relatively easy. The widespread use of these technologies can be achieved through close collaboration with laboratories, institutions and commercial organizations that have the necessary expertise and production equipment.

Genetic modification is the most complicated and expensive of the modern biotechnologies. Furthermore, it carries controversial risks, not only for the health and safety of immediate users but also for society as a whole, since the environmental impact of organism released in new forms cannot be fully ascertained. The environmental impact of the available biotechnology in most African countries is not monitored by any effective regulatory agencies. It is as important to do this as to educate the masses objectively on the advantages and disadvantages of agricultural biotechnology. Musanje (2000), pointed out that the mass production of *Bt* might be used to control blackflies in lima beans. Blackflies are vectors of river blindness. Another advantage is the mass production of "Round-up-ready" soybean seed, which effectively controls weeds without destroying the soybean plant. A major disadvantage is that while biotechnology products are customer driven in the developed countries, they are, generally, driven by interest-groups in the Third World countries. Corporate interests offer to Third World countries biotechnological products, which are patented and ensure total control of the farmer (customer).

Most African countries lack or have insufficient trained scientists in the basic disciplines required for genetic modification. The institutes of the developed world have the required research capabilities but they rarely focus on the crops and problems that are of most relevance to developing countries. The CGIAR centers, on the other hand, are situated, in most cases, in the developing countries, and have long had mandates and research activities geared towards meeting the needs of small scale farmers. The challenge for the national agricultural research systems (NARS) is to exploit the capabilities of these centers adapting the materials and technologies they produce to suit their own needs.

Genetically modified organisms (GMOs) are defined as organisms that have been modified by the application of recombinant DNA technology (where DNA from one organism is transferred to another organism). The term "transgenic organism" arises where a foreign gene (a transgene) is incorporated into the genome of the organism. The majority of the field trials and cultivation of GMOs were carried out in the USA and Canada. It is estimated that roughly 40 million hectares of transgenic agricultural crops were grown commercially in 1999.

The release in Africa of GMOs or genetically modified organisms needs further research and regulation. Wingfield (2000) reported that some of the recent studies done in South Africa suggest that there is a significant increase in crop yield for "rural" farmers using *Bt* maize. She pointed out that there is a very real need for independent studies done in African countries by rural farmers to determine the real advantages of transgenic crops for rural farmers. Since crops that are likely to hybridize with native flora are different in different parts of the world; and since different diseases are important in different countries, rather than a moratorium on "GMOs" what should be promoted is the appropriate monitoring and evaluation for each circumstance. Each transgenic crop has its risks and benefits and so needs to be assessed separately. However, there is currently little human resource capacity for this kind of assessment in most countries of Africa.

Many industrialized countries have imposed strict controls in this area with risk assessment conducted on a case by case basis. The biosafety aspects of genetically modified trees in particular need careful consideration because of the long generation time of trees, their important role in ecosystem functioning, and the potential for long distance dispersal of pollen and seed. With the exception of India and the Philippines, no developing country has legislation on the subject although Zimbabwe, Kenya and a number of Latin American countries are taking step in this direction. In Nigeria, the Biosafety Guidelines were prepared in 1994 and the legislative draft is now ready for the consideration of the National Assembly. But, once regulations have been introduced, it will be difficult to monitor compliance due to lack of expertise and infrastructure. Extensive staff training, more experience and regional cooperation are urgently needed in this area.

Accepting that modern agricultural biotechnology is already in Africa, it is necessary to search for ways to sustain and direct its use towards beneficial, profitable and acceptable ends. Africa needs to create jobs (especially graduate employment), to create capital and to foster social development. Agriculture and agro-based industries are the largest employers of labor in most African countries. Even though aggregate farm output may be increasing, productivity per unit input has not improved or has done so marginally. The advent of modern biotechnology presents a special opportunity and a special problem for African agriculture. Lettington (2000) alluded to the recent studies by the USDA and the National Center for Food & Agricultural Policy [available on http://www.econ.ag.gov/whatsnew/issues/gmo/] which produced inconclusive and variable results for GM crop yield, cost reduction and profit factors depending on individual crop, year, location and planting conditions. For example, GM soybean crops, Lettington reported, had yields that were approximately 6-7 percent lower and required pesticide use that was higher than for nonGM crops — the advantage being that the weed management process was less complex.

A paternalistic approach to developing countries will cause widespread harm. African countries have to be free to make informed and objective decisions on all areas of technology. The establishment of a clear code of ethics in monitoring, would be a *sine qua non* in evaluating what problems must be solved, what obstacles need to be overcome and what is needed for sustainable agricultural biotechnology in Africa?

Conditions for Sustainable Agricultural biotechnology in Africa

One problem is the long-standing separation of basic and applied research functions between the universities and the national agricultural research system. The disciplinary capabilities for sustainable agricultural biotechnology are scattered unevenly across the two types of research institutions. A more important problem is the lack of ties between research institutions and industry and the productive sector in general just as Trigo and Jaffe (1994) reported for Latin America and the Caribbean.

Many countries in Africa have policies directed at supporting research and development through the funding of projects, training of researchers and creation of specialized research institutes. These policies and efforts represent the first phase of the development of biotechnology i.e. the establishment of basic scientific capabilities. Active research and development are essential for understanding the basic science involved in monitoring international scientific trends and events and in training researchers and science policy makers. The objective of this first phase must be to enable a critical mass of people to master the most important biotechniques such as tissue culture, genetic engineering, DNA fingerprinting, hybridomas and immunochemical techniques.

In order to go further and so allow the groups, centers and various biotechnology-oriented Government Organizations (GOs) and NGOs to survive and flourish, a link has to be created between this first phase and the productive sectors. This link consists of technological entrepreneurs capable of relating the knowledge generated or gained to real problems of the productive sector. Okafor (1994) indicated that if the manpower required for the biotechnology industry in [Africa] was available, the necessary private sector would be more readily developed. The lack of entrepreneurs has obviously hindered the growth of biotechnology. For example, researchers (Adesina 1976; Adesina et al. 1979) at the Obafemi Awolowo University (OAU, Ile-Ife) identified the protein sweetener thaumatin in the sweetberry, *Thaumatococcus danielli*, which is 1,600 times sweeter than sucrose. There was no further development locally and the gene has now been cloned in Europe and the U.S. and is used for industrial production of thaumatin, which sells for $5,000 per kg. In the 1970s, medical and pharmaceutical researchers at the OAU and the University of Ibadan (Sofowora and Isaacs 1971) identified an anti-sickling agent in the fig *Zanthoxylum xanthoxyloides (Fagara spp*) and produced pure isolates of the agent. The more generally applicable dental properties of this fig have now been commercialized in China and Europe while the anti-sickling properties still await commercial application. This shows quite clearly that there is a great need for biotechnological entrepreneurs in Africa.

The establishment of sustainable agricultural biotechnology in Africa will require the development of human resources through mass training and re-orienting of agriculture and biology graduates to serve as change agents and entrepreneurs in agricultural biotechnology. To be sustainable, agricultural biotechnology must not only be environmentally friendly; it must also be socio-economically viable and attractive. There are some socioeconomic, environmental, legal and policy concerns. These include risks associated with the release of GMOs, restricted access to biotech products and techniques due to tightening of intellectual property rights, substitution of the exports of African countries (e.g., cocoa, vanilla, vegetables oils), and the widening gap between developed and developing countries in exploitation of the new technologies. Tissue culture factories have already displaced some botanical products. The social and economic effects of such displacement of botanical production in African countries by plant tissue culture production in industrialized countries are likely to be very disruptive to African economies.

In 1989, there was an international symposium on the "Role of Biology in Resolving the Food crisis in Africa" in Yamoussoukro, Cote d'Ivoire. The Yamoussoukro delegates agreed that training in the relevant basic subjects is important in order to acquire biotechnology and to adapt it effectively. They also agreed that a regional collaborative networking system is required to meet the research as well as training needs. There are now some organizations promoting biotechnology development in Africa. These include the Senegal-based African Biotechnology Research Network and the Foundation for African Development through International Biotechnology (FADIB). There are institutions, including the CGIAR centers, engaged in research in hybridization and molecular genetics in Nigeria, Kenya, Ethiopia, Zimbabwe, Ghana, Cote d'Ivoire and Benin. The UNESCO commissioned the production of a direction for biotechnology research on the continent. The UNDP and FAO's IBPGR, iDAD and AGR have initiated several projects to assess genetic resources in the continent. There have been highly successful internationally coordinated research networks on maize (WECAMAN), cowpea (RENACO) and root crops (EARRNET and SARRNET in East and Southern Africa) as well as on small ruminant (SRNET) and livestock policy (ALPAN). Sonaiya (1997, 1999) described the fortunes of an information-sharing network on smallholder family poultry (INFPD). There are also regional coordinating bodies for agricultural research (such as CORAF in West Africa, ASARECA in SADEC countries) which are stakeholders in agricultural biotechnology development.

ABERP: A Human Resources Development Approach

The first and most important task in any national or regional effort in the development and application of biotechnology is to create an awareness of the technology where none exists and to increase awareness where it already exists. The current level of awareness of the capabilities and implications of the new techniques of biotechnology is very low in most parts of Africa. The challenge is to build cooperation between scientists located in different countries in the region and to foster good working relationships between governments, the organized private sector and multinationals. A key element in this effort is the rapid development of human resources capacity of the African countries in order to enable them to participate meaningfully in the determination of the future direction of biotechnology for the benefit of their peoples.

A program based on the principles of Knowledge Exchange and Learning Partnerships (IITA 2000) can help to meet these challenges by initiating or enhancing cooperation between centers of biotechnology expertise in various African countries. A virtual, multilateral, coordinated and collaborative Agricultural Biotechnology Education and Resource Program (ABERP) is proposed as an arrow head for this rapid build up of human resource capacity and entrepreneurs in Africa.

The idea is to develop courses for postgraduate (Certificate, Diploma and Masters) and professional training in agricultural biotechnology management using public and private institutions (universities, national and international research institutes, and commercial organizations) already established locally and abroad. ABERP will emphasize the adoption of modern information, communication and knowledge technologies in order to create changes in attitudes and practices as related to agricultural biotechnology. ABERP is focused on innovation in two key areas that are indispensable for sustainability of agricultural biotechnology — human resource development, and institutional linkages. The emphasis should be on management ability — information accessing and analysis and dissemination, policy development and analysis, problem solving and commerce development.

ABERP manpower development should focus on training and re-orienting of agriculturalists, biologists, biochemists, pathologists, physiologists, geneticists, breeders and food scientists as well as lawyers, socio-economists and journalists, amongst others. A possible target could be to train 5,000 professionals each year for the next five years. In order to achieve this rapid build up of human resources capacity and agricultural biotechnology entrepreneurs, emphasis must be placed on distance learning and Internet-based virtual education in order to allow professionals to be trained *in-situ.* Electronic, computer based, satellite-linked workstations located in carefully chosen centers in African countries will be employed or developed. Input materials to these systems will come from research centers locally and abroad, educational institutions, commercial and not-for-profit NGOs as well as governmental and international organizations. This networking arrangement between national centers in the region as well as links with international agencies and centers in other parts of the world is extremely important to maintain relevance and currency. Databases and electronic libraries linked by satellite are essential. Coordination could be achieved through a UNESCO/UNU Chair of Agricultural biotechnology Development in Africa.

There is a clear need for coordination because of the myriad of the input sources and the collaborative nature of the required programs. The coordinators must be rigorous in guiding the auto-tutorial and directed learning processes towards the goals of obtaining change agents and entrepreneurs. Participants of the programs must be guided to acquire abilities and proficiencies: in information (mostly electronic but also non-electronic) accessing and analysis; in dissemination of such information through various selected mass media; in regulatory and promotional policy development; in the analysis of the social and economic implications of agricultural biotechnology policies for all stakeholders; in problem anticipation, identification and resolution particularly in the area of sustainable agricultural biotechnology development; and, most importantly, in development of commercial opportunities for agricultural biotechnology, e.g., setting up companies for biotechnology products such as seeds and planting stock, soil enhancers, pest management systems, veterinary products, nutriceuticals as well as agro-services.

Training will be at various levels depending on entry qualifications: Certificate courses for lower entry qualifications or short-term courses (1 - 4 weeks); Diplomas for medium term courses and Masters for longer term courses. The databases, libraries and instructional materials will be utilized as resource centers for agricultural biotechnology promotion within each country and in the whole region through linkages with governmental and educational institutions in each country. Similarly, the teaching faculty of the program who are staff at the participating private and public institutions and organizations, naturally become an interdisciplinary think tank for agricultural biotechnology policy development and analysis. The graduates of the program will be able to manage agricultural biotechnology information for private and public organizations, develop relevant biotechnology policies in government service, manage or assist in projects for national and international organizations and develop new enterprises that blend biotechnology with agricultural production, processing and marketing at the national, regional and international level.

Impact and Beneficiaries

ABERP is envisaged to create greater public awareness of the numerous advantages of agricultural biotechnology; promote the availability of agricultural biotechnology information throughout the country from the electronic libraries linked by satellite; enhance the development of human resources capacity and agricultural biotechnology entrepreneurs; engender a sharper focus for applied agricultural biotechnology research within the universities and relevant research institutes; and finally result in increased productivity and quantity of food and raw materials, reduction of poverty, faster wealth creation and higher human development. Among the expected beneficiaries of ABERP are educational institutions including research stations; private and public sectors of the economy; African entrepreneurs and manufacturers; farmers and peoples of Africa.

A proposal for the Nigerian ABERP (Sonaiya and Omitogun 2001) is already submitted to the Nigerian Ministry of Science and Technology under the National Genetic Engineering and Biotechnology Program Action Plan (2000-2004). Individual African countries should develop their own proposals and, by cooperation and collaboration, achieve the regional purpose of ABERP.

Conclusion: Technology Extension Resources and Activities at the OAU

With the collaboration of the International Center for Theoretical Physics, Trieste, Italy, the Obafemi Awolowo University has developed an academic ICT system designed to: remove isolation of scientists; establish small area computer networks and their connection to the Internet directly in order to provide e-mail, capacity building for technical personnel and training of users within the institution. From mid-1995, when the feasibility study was carried out, to June 1996 when the system commenced operation to now, a lot has been done. The system currently has TCP/IP protocol, uses Linux Operating System, 8 Subnets with Spread Spectrum wireless radio linking subnets to the main center and, most importantly, full Internet facilities on campus. The University has campus-wide local area network (LAN) based on wireless technology with the 8 nodes served by a VSAT. The OAU provides the coordination for the continental Cooperative Information Network (COPINE) and has a fully functional internal Information Technology Coordinating Unit (INTECU) based in the Vice-Chancellor's Office.

With these facilities, the OAU participates in several international collaboration projects. Examples are: Climatological Electronic Conference (June 1999), and the on-going Satellite-based Ph.D. program in Mathematics. The OAU also has the expertise and experience to function as part of a Virtual Laboratory for international scientific research collaboration. Experiences include: Whole Earth Telescope (WET) Research; and Vitamin A and its role in Infection (1996-2000). We have a Central Laboratory, which has analytical, photomicroscopy and general equipment section as well as the ability to access real laboratory equipment from remote terminals located in other laboratories within and outside Nigeria. These facilities are being offered in technical assistance to more than 6 Nigerian universities and colleges.

More specifically, the Obafemi Awolowo University has a general biotechnology focus (Ife Biotechnology Program, IBP) which draws members from the science-based faculties (agriculture, health sciences, pharmacy, science and technology) as well as the humanities (e.g., philosophy), law and social science. The IBP is directly connected to the office of the Deputy Vice-Chancellor (Academic) in his capacity as Chairman of the University Research Committee.

The OAU Faculty of Agriculture has a strong Agricultural

Biotechnology program directed, for now, at plant, animal, and fisheries technology. The Faculty is fully committed to the course of Agricultural Biotechnology education. The Faculty has recently acquired, through a Rockefeller grant, the essential electronic agricultural library (TEEAL) and has completed the complete wiring of every office and lecture theatre in the Faculty buildings for Internet connectivity, e-mail, electronic library and LAN (called neighborhood network within the Faculty). An expatriate associate professor of agricultural biotechnology was recruited to provide leadership for the Agricultural Biotechnology program and a core of postgraduate students has been assembled.

CHAPTER 19

PRACTICAL APPLICATIONS OF BIOTECHNOLOGY TOOLS IN DIAGNOSIS AND MONITORING CASSAVA MOSAIC GEMNIVIRUSES IN EAST AND CENTRAL AFRICA

P. Sseruwagi, G. Okao-Okuja,
R. Obonyo and J. P. Legg

Introduction

Cassava (*Manihot esculenta Crantz*) is one of the world's most important staple food crops (FAO 1999). Cassava production in 1998 was estimated at 158.6 million metric tons, which was grown on c. 16.2 million hectares of land. Africa, Asia and South America accounted for 65%, 21% and 14%, respectively of the total estimated area (FAO 1999). The crop is particularly valued for its tolerance of drought and can produce an economic yield even on exhausted soils unsuitable for the production of other crops (IITA 1990). A number of production constraints, however, hamper the production of cassava in Africa and Uganda in particular, the most important being cassava mosaic virus disease (CMD) which causes up to 100 percent yield loss, especially in highly susceptible varieties (Otim-Nape et al 1994).

CMD is caused by cassava mosaic geminiviruses (CMGs) from the family *Geminiviridae* and genus *Begomovirus* (Swanson and Harrison 1994). The disease is generally expressed by irregular yellow or yellow-green chlorotic mosaic of the leaves, leaf distortion and stunted growth. CMGs are spread by the whitefly vector, *Bemisia tabaci* (Genn.) (Homoptera: Aleyrodidae) and disseminated by man through planting infected cuttings (Bock and Woods 1983). CMD has developed from being a disease of only minor significance as it was described 100 years ago in what is now modern Tanzania to being a major production constraint to cassava in the East and Central African region (Sseruwagi et al. 1999; Legg 2000). Yield losses attributable to CMD have been quantified and are dependent on cultivar grown, environment, type and stage of CMD infection. In Uganda about US$60 million is lost annually due to CMD (Otim-Nape and Thresh 1998) and more than US$1,200 million per year in Africa (Legg 2000).

CMD has been recorded in East and Central Africa for more than a century. It was not until the early 1980s, however, that the viral aetiology of the disease was described, and only in 1994 that the first information on the Africa-wide distribution of CMGs recognized as causing CMD was presented. Prior to 1994, CMD was only monitored using visual assessments of disease severity and incidence. As the impact of the CMD epidemic in Uganda became evident, a biological explanation was sought for the big changes that had occurred in the hitherto relatively benign pathosystem. The expansion within the region of an epidemic of unusually severe CMD, apparently originating in Uganda encouraged the reinvigoration of the research effort to diagnose and characterize viruses associated with CMD using new biotechnological diagnostic tools such as the polymerase chain reaction (PCR) and restriction fragment length polymorphisms (RFLP).

This chapter discusses the practical applications of these molecular diagnostic tools in the diagnosis and monitoring of the CMD pandemic in East and Central Africa. The challenges faced by African agricultural scientists while using these technologies are also highlighted.

Biotechnology in Africa

The development and adoption of biotechnology by African agricultural research systems have been slow, despite the tremendous benefits these technologies could have in helping to alleviate the enduring problems of hunger, malnutrition, and disease in Africa. This is probably because of the high expense of these technologies, the requirement for highly trained personnel in their use and the lack of direct application of these technologies to solving plant disease problems facing many African crops. In the last decade of the twentieth century, however, some progress has been made in the use of biotechnology to diagnose and monitor the spread of CMGs in the East and Central African region.

Biotechnological Tools

A number of molecular tools have been used to diagnose CMGs and these include: Polymerase Chain Reaction (PCR); and Restriction Fragment Length Polymorphisms (RFLP). For purposes of discussion, this chapter only describes the PCR method, since this is the most commonly used technique at the International Institute of Tropical Agriculture Eastern and Southern Africa Regional Center (IITA-ESARC) virology laboratory.

The polymerase chain reaction, or PCR, is an incredibly powerful tool in molecular biology enabling large quantities of DNA to be amplified from very small starting quantities. This has facilitated those areas of research where it is difficult to extract enough DNA from a sample for direct analysis. PCR is based on three main components: complementary DNA primers, the enzyme DNA polymerase, and temperature cycling of the reaction. Using these three components new DNA strands are synthesised with each reaction cycle, theoretically doubling the quantity with each cycle. A typical PCR reaction will continue for about thirty-five cycles.

The DNA primers (Table 1) are DNA strands, typically eight to twenty nucleotides in length, with a defined sequence of nucleotides. A complementary sequence is found in the genome of the plant of interest to which the primers can bind. The DNA polymerase then adds nucleotides to the primers and extends it, using the plant genome as a template, so that a double strand is formed showing the most diagnostic amplicons (predicted products). This process is repeated with each cycle. The process that enables this repetition is temperature cycling, which involves three steps; denaturation of double stranded DNA, binding of primers to single stranded DNA and extension by DNA polymerase. In the first step, the temperature of the reaction is raised to 94 °C, which denatures the double stranded genome of the plant into single stranded DNA. In the second step the temperature is lowered to enable the primers to bind to the plant genome. In the third step the temperature is raised to 72°C, which is the optimum temperature at which the particular DNA polymerase works. Therefore, by using the different primers, PCR has enabled the detection of the different virus strains affecting cassava.

Practical Applications of Biotechnological Tools

Biotechnological tools including PCR and RFLP techniques have enabled the diagnosis and characterization of cassava mosaic geminiviruses (CMGs); monitoring the spread and distribution of the different CMGs in the East and Central African region; identification of CMD-worst affected areas; forecasting future spread; and development of control strategies based around the deployment of host plant resistance.

Table 1: PCR primer pairs for CMG diagnostics

Primer name	Sequence (5' to 3')
ACMV	
1. ACMV-AL1/F	CGGAATCCCTAACATTATC
2. ACMV-R0/R	GCTCGTATGTATCCTCTAAGGCCTG
UgV/EACMV-Ug	
3. UV-AL1/F1	TGTCTTCTGGGACTTGTGTG
4. ACMV-CP/R3	TGCCTCCTGATGATTATATGTC

EACMV	
5. UV-AL1/F1	TGTCTTCTGGGACTTGTGTG
6. EACMV/CP/R	ACTCTATGRGTAATRCCYGA
Universal	
7. Universal F	KSGGGTCGACCTCATCAATGACGTTRTAC
8. Universal R	AARGAATTCATKGGGGCCCARARRGACTGGC

Note:

R = A or G, K = G or T, Y = C or T:
a = Adenine, G = Guanine, C = Cytosine, T = Thiamine

Diagnosis and Characterisation of CMGs

CMGs have been characterized and the genome described as composed of two circular molecules of single-stranded DNA (Harrison 1987). They occur in pairs, each measuring 30 x 20 nm (Fig. 1) with a waist at the mid-point of the long axis and have a sedimentation coefficient of 76 s. CMG particles contain a protein of molecular weight c 0.9 x 106 (Harrison 1987). Diagrammatic representations of the CMG particles in Figure 2 further details out the differences between the different CMGs.

Diagnostic tests using enzyme-linked immuno-sorbent assay (ELISA) done during the early years of the epidemic in Uganda had suggested the causal virus to be ACMV. The serological tests used relied on different CMGs having different external features i.e., differing protein coat conformations. In 1996/97, newly developed nucleic acid based diagnostic methods (PCR and RFLP) allowed comparisons to be made in the genetic material (DNA) of virus

Table 2: Distribution of CMGs in East Africa, 2000

CMGs	Distribution in East Africa
ACMV	All affected areas except Kenya Coast
EACMV	Kenya Coast
ACMV/EACMV	W. Kenya and N.W. Tanzania

EACMV-Ug	Most of Uganda, N. E. Rwanda, W. Kenya, N. W. Tanzania and E. DRC

isolates from different locations in Uganda. These demonstrated clearly that a novel CMG was associated with the CMD epidemic, and that this appeared to be a recombinant form, in which the central portion of the coat protein gene of ACMV was spliced into the DNA-A of EACMV. DNA based diagnostic tests using the PCR procedure were developed to facilitate distinction between ACMV, EACMV and the novel variant referred to as the Uganda variant (UgV)/EACMV-Ug (Figure 2) and these were also able to detect mixtures. Since 1997, these tests have been widely used for the detection of virus samples collected during extensive diagnostic surveys, and have been used to map the spread of the novel virus in the region (Figure 3).

Monitoring the Spread and Distribution of CMGs in East and Central Africa

PCR has enabled the monitoring of the spread and the distribution of the different CMGs, in East and Central Africa (Table 2) and hence the mapping of the pandemic disease (EACMV-Ug) in the region (Fig 3).

In 1995, the first report was made from Kenya of severe CMD typical of the Ugandan epidemic and between this time and 1998, the epidemic expanded across much of Western Province in Kenya (Legg and Okao-Okuja 1999; Legg et al. 1999). The pattern of spread was consistent with that experienced in Uganda, and therefore, using PCR, EACMV-Ug was diagnosed from CMD diseased cassava plants in affected areas. Reports of the occurrence of UGV/EACMV-Ug were also made from near Juba, in southern Sudan in 1997 (Harrison et al. 1997) and from Kisangani in eastern Democratic Republic of Congo in 1998. This evident regionalization of the problem led to it being referred to as the CMD pandemic (Otim-Nape et al. 1997). The first record of pandemic expansion into Tanzania was made in 1998 (Legg 2000). CMD spread down the western shoreline of Lake Victoria in Tanzania between 1998 and the present day has occurred more rapidly than the 20-30km reported from Uganda (Otim-Nape et al. 1997; Legg and Ogwal 1998), and by June 2000, the "front" of the pandemic was reported from southern Kagera region and the western corner of Mwanza region on the Lakeshore. In western Kenya, the movement of the pandemic has apparently been slowed by the physical barriers of the Winam Gulf and the Kano Plains in the Kisumu area (Legg 2000). Severe CMD and EACMV-Ug have nevertheless been reported from southern Nyanza province. Pandemic CMD was recorded from the north-eastern corner of Rwanda in July 2000 and in view of the small size of the country, it seems likely to affect all cassava-growing areas until 2002-3. Diagnostic surveys conducted in the Bukavu area of eastern Democratic Republic of Congo at the same time demonstrated the occurrence of EACMV and ACMV, and mixtures of the two in single plants, but EACMV-Ug was not found and there was no evidence of the rapidly spreading severe CMD characteristic of the pandemic. Unconfirmed reports have also been received of the presence of EACMV-Ug as far west as Congo-Brazzaville, but virus diagnoses will need to be validated and supported by field data before it can be concluded that this represents a westwards extension of the pandemic (Legg 2000).

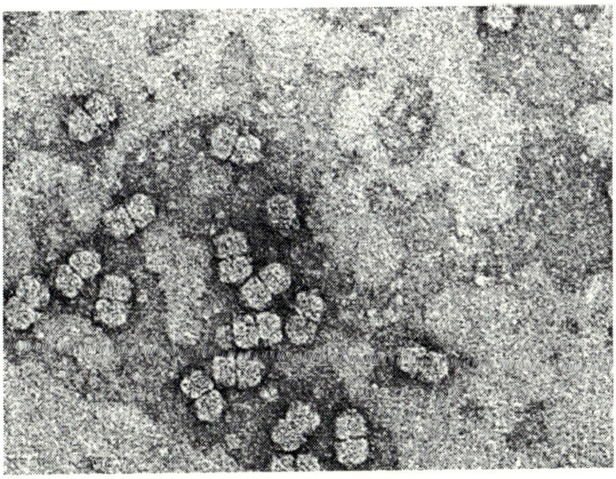

Figure 1:Geminivirus particles viewed under an electron microscope. Bar represents 100 nm.

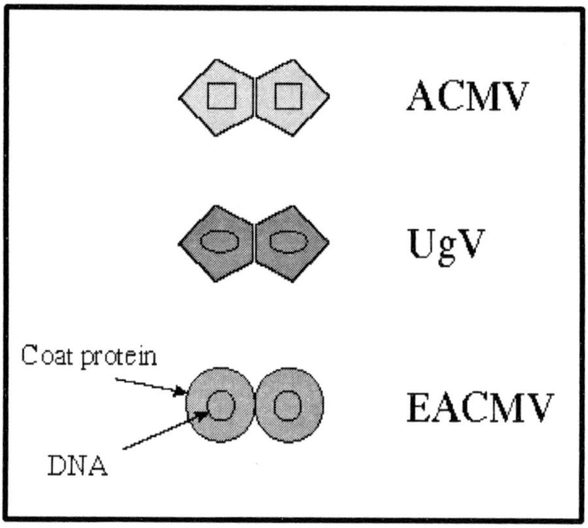

Figure 2: Diagrammatic representation of conformation of Cassava mosaic geminiviruses.

Forecasting Future Spread

Forecasting of future spread of the CMD pandemic is of vital importance to any serious agricultural program seeking to control the further spread of the disease in the East and Central African region. This requires that regular diagnostic surveys be conducted to establish the extent of spread of the disease. Over time the rate of spread will be established and therefore future spread could be ascertained. Assuming for example that the CMD pandemic continues to expand to countries neighbouring Uganda, at an average rate of 20-30km (Otim-Nape et al. 1997; Legg and Ogwal 1998), annually in all directions, a zone can be identified which is likely to be affected in the ten-year period from 2000-2010 (Legg et al. 1999). Important cassava-growing zones included within this band include south Nyanza in western Kenya; Mwanza, Shinyanga and Kigoma regions in Tanzania, the entire areas of Rwanda and Burundi, and an extensive area of eastern Democratic Republic of Congo. Using this information, therefore, control measures can be deployed, emphasizing the critical importance of regional coordination in developing and implementing these control initiatives.

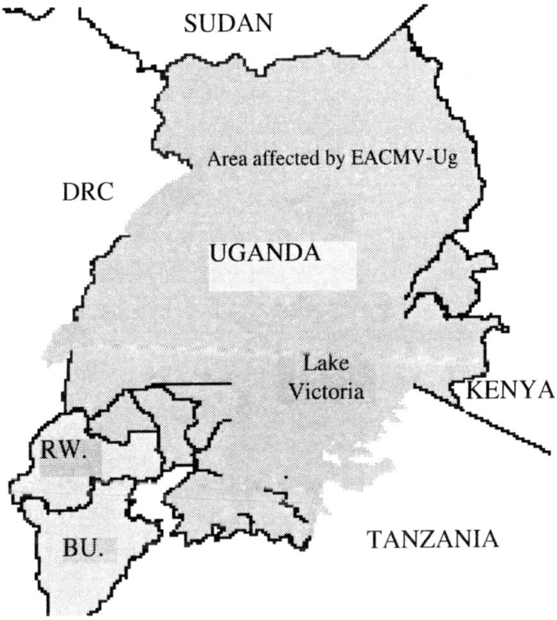

Figure 3: Areas affected by the Pandemic of Severe CMD (EACMV-Ug) in the Lake Victoria Region of East Africa (yellow shade) and the threatened areas of expansion into Rwanda, DRC, Sudan, Kenya and Tanzania (Legg 2000).

Regional Control Efforts

Significant success has been realized in controlling the effects of the CMD pandemic through the identification, multiplication and farm distribution of resistant varieties, the initial set of which were derived from the International Institute of Tropical Agriculture's breeding program based in Nigeria. By 1999 it was estimated that in excess of 100,000ha was under cultivation with CMD resistant varieties in Uganda, and this area continues to increase. By contrast, control programs are at a much earlier stage in both western Kenya, north-western Tanzania and Rwanda, whilst little action is currently being taking in other countries either already affected or immediately threatened by the pandemic. A regional program bringing together Uganda, Kenya and Tanzania, was initiated in late 1998, and comprised the following key components:

- Forecasting spread and targeting control: through monitoring and diagnostics surveys incorporating assessments of CMD incidence/severity/infection type, CMG distribution and whitefly abundance.
- Germplasm exchange/introduction: open quarantine in Kenya and Tanzania for the introduction of CMD resistant germplasm from Uganda; tissue culture introduction from IITA-Nigeria to Tanzania; and regional plant quarantine facility, Muguga, Kenya, to establish elite CMD resistant varieties adapted to the mid-altitude environment of East/Central Africa in tissue culture, virus index and supply to strategic sites throughout the region as yet unaffected by the pandemic.
- Germplasm evaluation: development of a "fast-tracking" approach for the evaluation of CMD resistant germplasm through on-station and on-farm stages. Kenya and Tanzania; "hot spot" evaluation of new germplasm. Tanzania; and farmer evaluation through technology transfer centers (TTCs), where evaluations are combined with training and the provision to farmer groups of small quantities of released CMD resistant varieties for multiplication.

- Training: agricultural workers and farmers in cassava pest and disease management. Both through formal training courses and in-field practical exercises.
- Networking: consultative project planning and implementation process through national and regional steering committees.

Challenges Facing the Development and Adoption of Biotechnology in East and Central Africa

Although some progress has been made, especially in Uganda in the diagnosis and monitoring of CMGs using molecular biotechnological tools (PCR and RFLP), a number of challenges still limit the full development and adoption of biotechnology by national cassava programs in the region. Among these are: the highly prohibitive costs of these technologies, the requirement for highly trained personnel in their use, the requirement for regular and consistent electricity in the laboratories, the lack of direct application of these technologies by scientists in the identification of the different viruses in the field, and the hazardous nature of some of the chemicals used, e.g., ethidium bromide. If the full benefits of these technologies are to be realized in the development of Africa, there is need to train more personnel in molecular diagnostics and to sensitive African governments to invest more in the development of biotechnology both at national and institutional levels. Otherwise, molecular biotechnology will continue to play a key role in forecasting future spread, identifying worst affected areas and targeting control strategies based around the deployment of host plant resistance. These and the many technical and logistical difficulties facing cassava workers will ensure that the challenge of the CMD pandemic will remain one of the greatest to face East and Central African agriculture during the first decade of the new millennium. Only a coordinated, sustained and generously financed regional control program will ensure that the successes realized in Uganda are replicated elsewhere, and that cassava's continued role as a key to the region's food security is assured.

CHAPTER 20

INTEGRATED SUSTAINABLE BIOREMEDIATION IN THE INTEGRATED NIGER DELTA

Charles Onyekachi Akalezi

Introduction

There are presently rising concerns in Africa about whether environmental constraints will limit development and whether development will cause serious environmental damage in turn impairing the quality of life of this and future generations. A number of environmental problems are already very serious and require urgent attention. Humanity's stake in environmental protection is enormous, and environmental values have been neglected too often in the past. Two ways in which environmental problems can undermine the goals of development relate to, first, environmental quality. The quality of environment health is now a key measurement of development. Potable and easily accessible water and clean air are environment indicators, which are now used to check the veracity of economic indictors such as GDP. This is what development is about: the benefits from rising incomes are offset by the costs of healthcare and the reduced quality of life through pollution, this cannot be called development. Second, there is the issue of soil degradation. Degradation undermines productivity. Degraded soil, depleted aquifers, and ecosystems that are destroyed in the name of development have jeopardized the prospect for an assured future for coming generations.

The severity and multiplicity of impacts from pollution and industry along the West African coast are perhaps best illustrated in the Niger Delta. With over 500,000 ha of mangroves, the Niger Delta is regarded as one of the most important wetland systems in West Africa. The Delta is also the center of Nigeria's petroleum and natural gas industries. Criss-crossed by oil and gas pipelines, and riddled with wells, this fragile aquatic environment is subject to chronic operational spills and leaks from oil and gas mining operations. Natural subsidence in the Delta is also accelerating due to these mining operations, exposing settlements in the area to the adverse effects of sea level rise. The Delta, which accounts for nearly two thirds of the Nigerian coastline, is situated in the highest industrial discharge zone in the west and central African (WACAF) region. The wetland zone from Lagos Lagoon to the corridor between Nigeria and Gabon accounts for the bulk of the effluent load of the 18 coastal countries between Senegal and Angola. Approximately 43 percent of the biochemical oxygen demand (BOD), 36 percent of the suspended solids, 83 percent of the oil and grease and 60 percent of the chemical oxygen demand (COD) of the region are dumped into the zone.

As environmental awareness gains ground particularly in Nigeria, activities in the ecosystems become more obvious. One of the most recent examples of this, is the relationship between the Shell Petroleum and Development Company in the Niger Delta and the Ogoni community of Rivers State. This problem has now exploded across the Niger Delta with recent conflicts and civil unrest giving rise to wanton destruction and unnecessary loss of lives.

The cost of pollution from oil industry is already high and will grow exponentially if these problems are neglected. It is thus absolutely essential to reduce emissions per unit of production. This requires some investment in monitoring, new equipment and the development of new technologies.

It is in light of these realizations that the Center for Environmental Resources and Sustainable Ecosystems (CE-RASE) decided to seek ways of providing independent environmental services that translate to action at the grassroots.

Oil Spillage and its Ecological Impacts

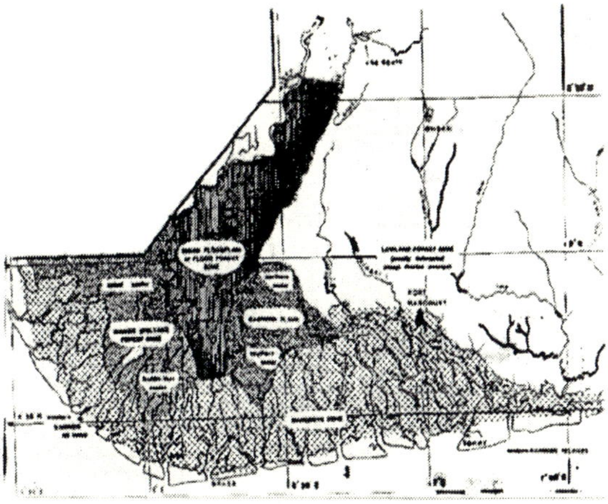

Figure 1: The Major Ecological Zones of the Niger Delta

The Niger Delta is located in the southern region of Nigeria. It is one of the world's largest wetlands, and Africa's largest delta, about 40-70,000km2 in area, the Delta is made up of five ecological zones (Fig. 1). The barrier island forests are actually tropical lowland swamp forests sometimes fringed by mangrove vegetation. These outermost coastal forest zones represent some of the last remaining pristine forest resources and centers of endemism in Africa. It is followed by massive mangrove forest zone that is the largest in Africa, and the third largest in the world. The mangrove ecological zone is followed by the lowland swamp forest zone, the most extensive freshwater water swamp forest in West and Central Africa which in turn is followed by the relatively higher terrain by the tropical rainforest zone. About sixty percent of the entire region is flooded annually. Most of the rain forests have either been lost to logging or replaced by farm bush. Most of the rain swamp forest zone and tracts of riparian forests along existing or relic river flood plans still have primary forests or natural vegetation.

These features are responsible for the variation in habitat type, species composition, abundance and distribution of associated endemic, rare and endangered species. It is these enormously rich and diverse but fragile functional ecosystems that carry the burden and inevitable consequences of Nigeria's oil industry installation activities.

Sources of Oil Spill Into the Environment

There are several sources of oil spill into the Niger Delta environment. The major oil spill sources are from crude oil mining, refining and servicing sectors. Other sources are from effluents of non-oil industrial sector and from transportation – boats, vehicles, and garages.

Oil spillage from the oil sector arises from numerous activities, which under upstream sub sector can be categorized into exploratory drilling and production operations. The down stream sector contributes especially the effluent source which has chronic waste discharges (Figure 2). There are numerous spillages in the Delta, two very spectacular ones are the Funiwa-5 well head blowout of January 1980 and the Mobil Idaho Spillage of January 1998.

Funiwa Oil Spill resulted in the introduction of massive amount of crude oil from the Funiwa No. 5 well located in the Funiwa Field about five miles off-shore Sangana into valuable coastal waters, estuarine and mangrove resources with devastating effects.

The Mobil Idaho spillage from a 12-inch pipeline rupture, which occurred on 12 January 1998, caused even more widespread damage and nuisance that covered virtually the entire Niger Delta Coastline. The exact quantity of spilled crude, may never be known as Mobil's claim of 40,000 barrels, and has been disputed at various levels.

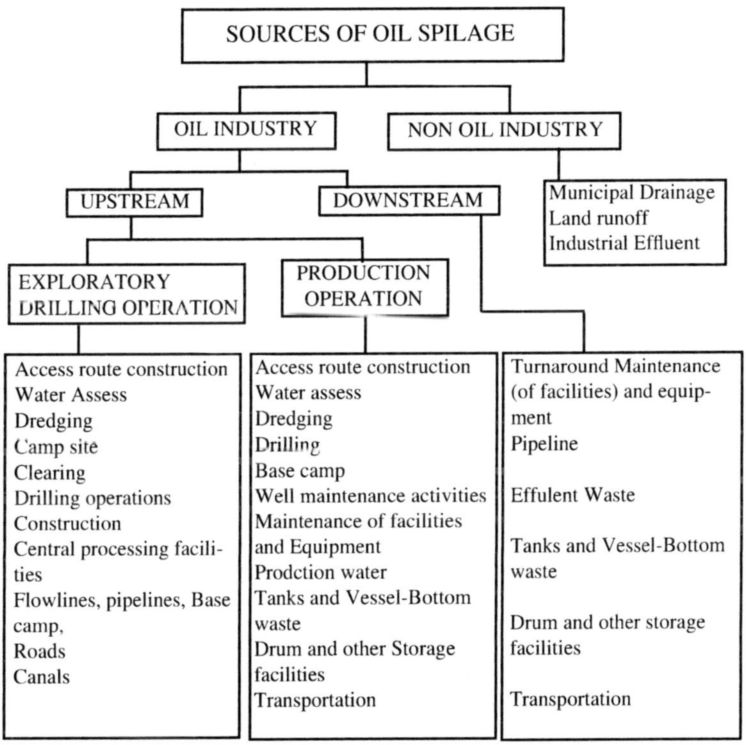

Figure 2: Sources of Oil in the Niger Delta Environment

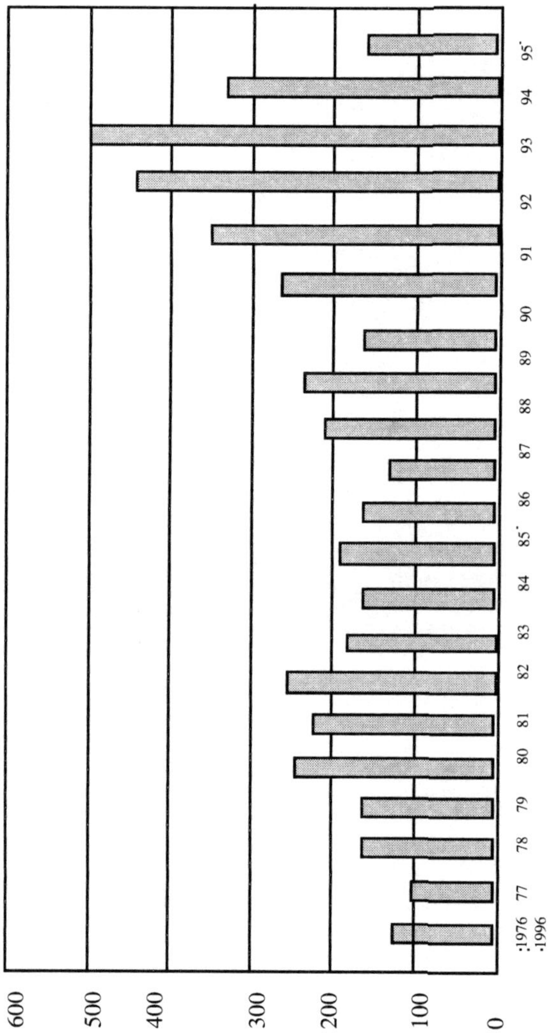

Figure 3: Frequency of Spill Recorded Per Annum

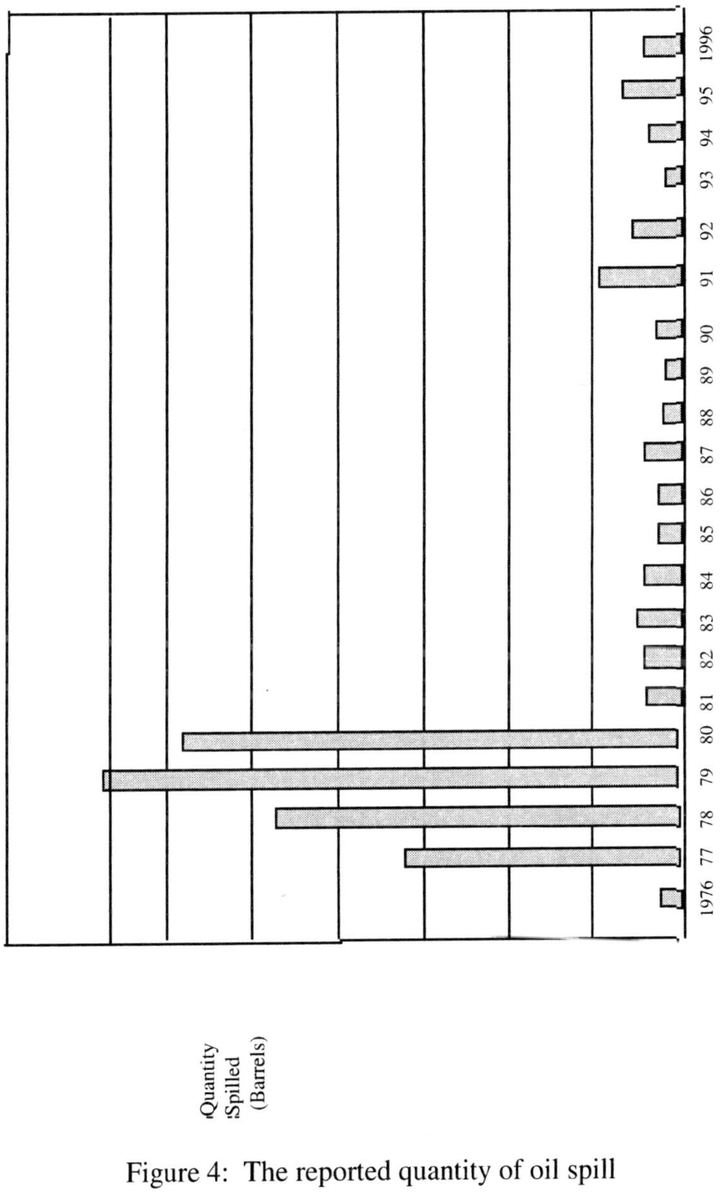

Figure 4: The reported quantity of oil spill

Available data showed that oil spills by several causes have occurred in the Niger Delta region (Figure 3). Awobajo (1981) observed steady increases in the number of oil spill incidences from 1976 to 1980. Presently, the scenario is worsening as the number of incidences has almost tripled. The quantity of reported oil spills by the operating oil companies tend to suggest that the amounts of crude oil discharge decline over the years as incidence of spillage peaks (Figure 4). However, the quantity of oil recovered from all the spills recorded showed that over 80% of the oil discharged is left in the environment (Figure 5).

Impact of Oil Spillage

The environmental effect of oil spill is a function of time, type and composition of oil, its degree of weathering and sedimentary characteristics of the receiving environment. Ecosystem type, (on – shore or inter – tidal), and the time or season of the year when spillage occurred. Figure 6 illustrates some of the processes that determine the fate of crude oil in the marine environment.

Vegetation Damage / Species Loss

Oil spillage in a humid environment generates varying ecological responses ranging from outright mortality of plants to wilting, defoliation of plant leaves, loss of reproductive cycle – fruits and species loss. The magnitude of impact depends on the physiographic nature of the area, containment and cleanup measures adopted.

In a fresh water environment (streams, flood plains and back swamps) the effects cause great damage to the plant communities (emergent, rooted, and climbers). These habitats are characterized by the presence of several rare endemic and endangered species.

Impact of oil is thus often devastating due to high retention time occasioned by limited flow (low energy of water current) caused by thick macrophytes cover and enormous herbs, shrubs and trees which make flow extremely difficult. Consequently, several of the plants are exterminated instantly, especially, the lower plant (algae, lichens, pteridophytes, byrophytes) or by gradual withering, especially the tree plants (timber).

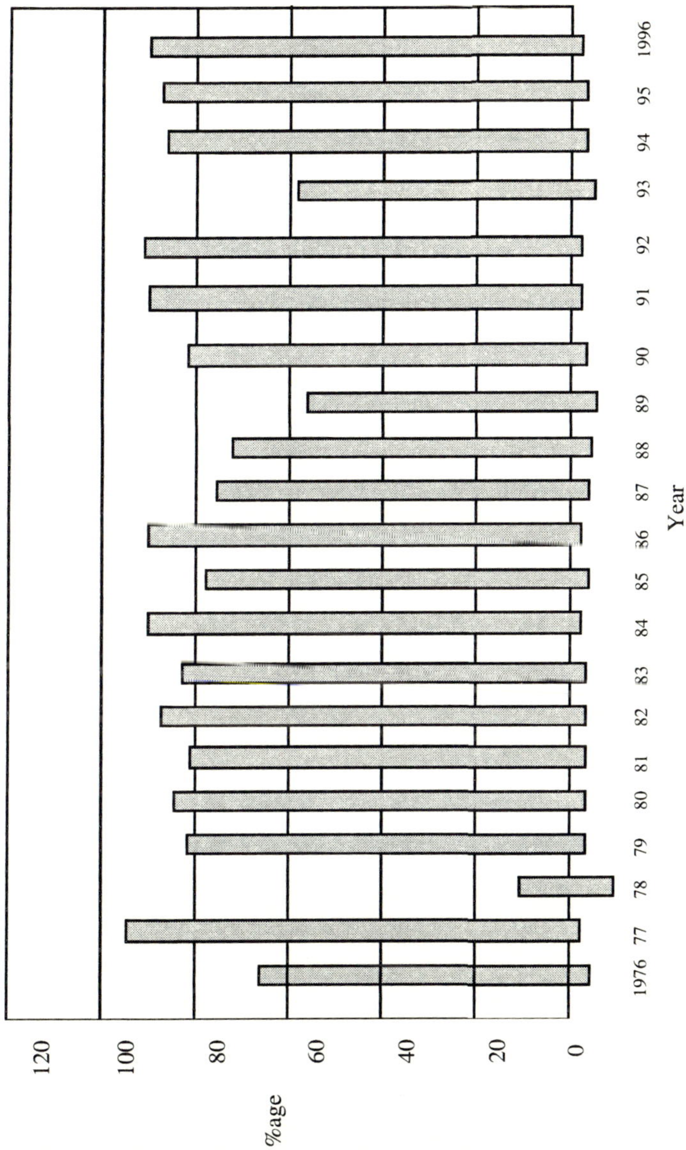

Figure 5: The Percentage of Oil Left in the Ground

The entire process results in low productivity or loss of plants species or both. The ecological implication is non quantifiable and the economic impact both at short and long term basis depends largely on the timber and non-timber forest products impacted on. In some instances where oil spill incidents occurs with fire outbreaks, more plant species are damaged (Figure 7). For example, an incident which occurred along Brass-Ogada-Rumuekpe pipeline at Odua, over 10 hectares of rich and flourishing vegetation were consumed by fire. There are a litany of examples of such spills across the Niger Delta wetlands. A similar incident occurred around Okorogba in Bayelsa State, the ignition of the spilled oil by a fishing lantern, cleared several hectares of vegetation. Often, wild life species are displaced or killed following an oil spill incidence.

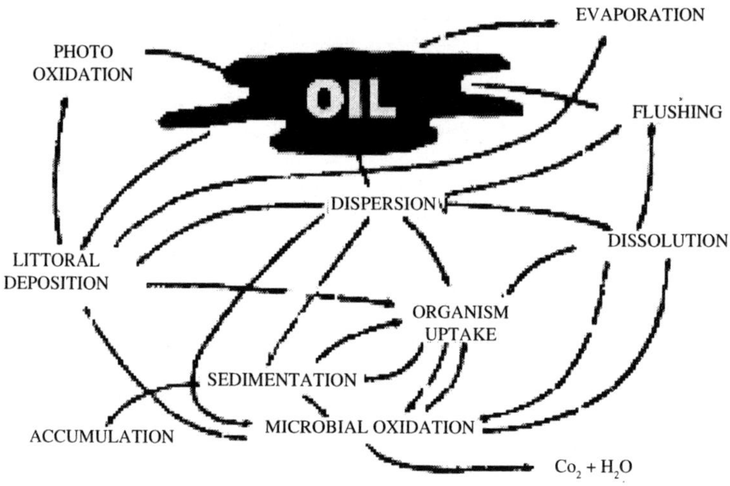

Figure 6: Fate of Crude Oil in the Environment

In dry and flooded zones of the Niger Delta, cases of disappearance of rare and endangered wild life species have been reported. In the mangrove habitat, pollution caused migration of white-throated monkey and long snout crocodile around the Ekuloma environment. The consequence of this to an environment already under intense pressure from past activities of oil operating companies in the Niger Delta is enormous. The natural role of plants in air purification such as carbon dioxide utilization and oxygen production is further reduced. The plant materials needed by the locals for various ecological and economic activities are depleted or totally obliterated by the oil spill. In Diobu and Perimabiri area of Niger Delta, for example where Raphia plant is abundant and the people depend largely on it for food and other economic needs (Raphia is used for wine, baskets, thatch for roof, and gin) such an incident will be disastrous.

Fig. 7: Vegetation Impacted By An Oil Spill

In the mangrove, plants have been impacted severely following oil spillage. In Jones Greek, for instance, oil spill resulted in the extermination of mangrove plants (Figure 8)

Figure 8: Dying Mangrove Vegetation After an Oil Spill

Similar scenarios of death of mangrove plants are seen all over the Niger Delta environment. In some cases, mangrove plants do not show early mortality depending on the severity, but the prop roots decay before eventual signs of physical death (wilting of leaves) are seen (Figure 9).

Figure 9: Decaying Prop Roots After an Oil Spill

The defoliation of mangrove plants exposes the delicate intertidal forms of intense heat, which in addition to oil stress may die by synergy. In addition, decay of the dead plant and animal materials eventually cause fouling of the environment, particularly, areas with limited capacity to flush. The decaying process may also impact color, taste and odor. This results in nutrient release to the environment (the organic nitrogen-ammonia-nitrate) with the attendant increase in the microbial load, increase biochemical oxygen demand (BOD) and depleting the dissolved oxygen in water to a level that is beyond the tolerance limit of the organisms. The water quality of the system is altered and anoxic situation may lead to fish kill in the area.

Furthermore, the prop roots of the mangrove usually have attached flora and fauna. Those include the gastropods *(Thais sp., Lithorina cingulifera, L. angulifera Nerita sp*. Others include oysters, *crassostrea gaser*), crabs-*sesarma elegans*, *S.Angloense*, and *Gomniopsi epli*). The decay of the prop roots as a result of oil in the environment may lead to direct mortality or dislodgment of the organisms from their natural environment. The eventual result is loss of resources and revenue as these organisms have high nutritive values and high economic potential. The thick mat of algae on the prop roots constitutes eggs of fish (top minnow*)* and larval forms of insects (Diptera) that are usually very vulnerable to oil spills.

Oil on Water

Apart from spills from oil industries and effluents from other industrial sources, oil from machinery and or from storage facilities during handling may leak into the immediate environment and the spread will be dependent on the quantity spilled and the current pattern. Oil film on water surfaces reduces the inter-phase with the atmosphere resulting in less oxygen dissolved in water. Low oxygen in water can induce physiological stress on organisms, which may lead to eventual death. The nuestonic forms and planktonic (phytoplankton and zooplankton) which reside close to the surface will be adversely affected. The developmental stages (Meroplankton) of oysters, shrimps, crabs and fish that are more vulnerable will also be impacted. The effect on the larval forms and early life stages will include low recruitment potential of these forms, wide generation gap and low yield (low catch potential).

The oil film also reduces the amount of light reaching the primary producers such as the phytoplankton, which in turn results in low productivity. The consequence could be low yield in fin and non-fish catch.

Socioeconomic Impact

Oil spills have caused several communities to evacuate their homestead either as a direct danger posed or consequences of the pollution problem caused by the spill. The demand for compensation for damaged resources and other properties has put individuals or communities on a collision course. The situation in most cases tore the communities apart and resulted in instability in the region.

Figure10: Abandoned Oiled Fishing Net

Impacts of Soil Erosion in Nigeria

The pathetic stories of incidents of soil erosion reflect the crisis situation in the erosion-ravaged southeastern states of Nigeria. Over 70 per cent of the land is threatened, and more than 20 per cent of ancestral and agricultural lands have been destroyed by landslides, soil and gully erosion. This loss which has continued unabated was estimated at N5.2 billion by the Federal Environmental Protection Agency, FEPA in 1992.

The Problem

In the southeastern states including some parts of the Niger Delta, the word erosion is now synonymous with danger, destruction, despair and death. The ultimate destruction created by floods and gully erosion is the occurrence of landslides. This situation is very rampant in major gully erosion areas such as: Agulu, Nanka, Obosi, Alor, Nnobi, Umuahia, Ohafia, etc.

In all these localities, this disastrous monster of indescribable dimension has rendered people homeless and destroyed farmlands worth millions of naira. The people have resigned their fate to whims of the monster. They never know where and when a gully would spring up.

There have been instances where children left at home by their parents who have subsequently been swept away along with domestic animals and properties. Death and mayhem are the aftermath of these landslides.

Some notable erosion sites are found in Agulu with the most pathetic erosion sites in the Madona gully complex (a composite of over 18 deep gullies). There are about 14 erosion sites in Obosi Town all gullies, and over 15 erosion sites in Nnewi Town all in Anambara State.

Figure 11: Solid Waste Dumping in Nigeria

Solid Waste Dumping in Nigeria

Indiscriminate dumping of refuse, traffic hold up, bad roads, and hawking, are common sights in the Nigerian environment. Hundreds of dump sites are found in every city.

Figure 12: Onitsha-Owerri Road Onitsha

Figure 13: The Popular Wetheral Road Owerri

Soil Acidification in the Niger Delta

The tropical rainforest is by nature supposed to be very fertile. One important factor affecting soil fertility in this region is acidification. The Niger Delta environment, which is notable for its enormous gas flaring activities by the oil mining industries, is likely to face a double fold impact of acidification due to acid rain. A roof watch experiment was designed to track down the rate of roof degradation. After three months in a dry season, there was no remarkable change on the new roof. Dew and rainwater analysis were also carried out as we wanted to find an answer to whether there is acid rain in the Delta traceable to gas flare. Samples tested acid but there was instability of the acid compound. Samples, that indicated acidity in the morning, turned neutral by mid-day. Similar observation has been made across the Delta by different researchers. More critical analyses are still on going, however, the relevance of this evaluation is reduced as the Nigerian Liquefied Natural Gas (NLNG) Project puts gas flaring to an end. The biodiversity of the soil seems to have been hampered, as the population of common earthworms and other small animals that inhabit the soil ecosystem has been reduced drastically.

The Concept of Sustainable Bioremediation: Towards a Universal Definition

The word sustainable refers to the maintenance or enhancement of resource productivity on a long-term basis. It connotes self-sufficiency and an implicit ideology of long-term self-restraint and self-reliance. In order words any process regarded as sustainable is expected to be self-supporting and in fact without subsidy.

Bioremediation is a term used to describe clean up techniques that are based on enhancing the microbial breakdown of contaminants including petroleum hydrocarbon compounds in nature. Bioremedial treatment includes the use of naturally occurring or artificially induced microorganisms e.g., bacteria/ fungi to breakdown contaminants in a polluted environment. It functions by generating a biologically active medium by including the indigenous bacteria and fungi population to reduce the contaminants level at optimal nutrient, oxygen and temperature conditions.

Figure 14: Degraded roofs of the Niger Delta

However, the concept of biological remediation as propounded by CE-RASE does not only cover microbial treatment, but also the direct use of other natural resources to combat degradation, in this sense, vetiveria zizanioides is a grass with extremely impressive qualities which is capable of soil stabilization.

The concept of "sustainable bioremediation" is a new catchphrase of The Center for Environmental Resources and Sustainable Ecosystems. (CE-RASE), a non-governmental organization established in 1995 to bridge gaps in environmental conservation and development. Basically, the philosophy behind the concept is the promotion of integrated environmental conservation and rural development led by community participation. The aim is to empower the people through participatory development for environmental protection using biological methods. The advantage this method has over other methods is that it involves rural people in a program to protect their lands and provide economic opportunities thereby reducing cases of civil unrest due to frustration over ecological degradation.

The concept was submitted as a CE-RASE entry for the New World Bank innovation competition under its Development Market Place Program. CE-RASE entry promoted the use of bioremediation for poverty alleviation and to combat oil spills in the Niger Delta. Some 1200 entries were submitted by various groups from over 62 countries for the competition, which featured new ideas in development. CE-RASE's award-winning concept to promote Kenaf (*Hibiscus cannbinus*) as an effective oil absorbent for mopping up spills in the Niger Delta was one of the 44 successful ideas selected at the innovation competition.

Application of Bioremediation

The Use of Kenaf for Oil Spill Response

Kenaf (*Hibiscus cannabinus)* is an annual herbaceous fast-growing plant indigenous to West Africa. It belongs to the plant family Malvaceae, notable for both economic and horticultural importance. It is related to okro (Fig. 11) and cotton plants. Traditionally, in Africa, China, India and Thailand, it is cultivated for the soft bast fiber in its stem, which resembles and can substitute for jut fibers for cordage use. In many countries, Kenaf is used to produce a range of products such as oil absorbent (Figure 16) for

Fig. 15 Kenaf Nursery in the Niger Delta

mopping up oil spills, newsprint and other pulp and paper products. Other products such as mattresses, ropes, canvas, animal bedding, packaging material, sacks, grass and flower mats, carpet backing, decorative fibers and some insulation (Taylor 1995).

In some places it is cultivated secondarily for the seeds which contain about 20% oil used for cooking, soap making, linoleum, paints and varnishes.

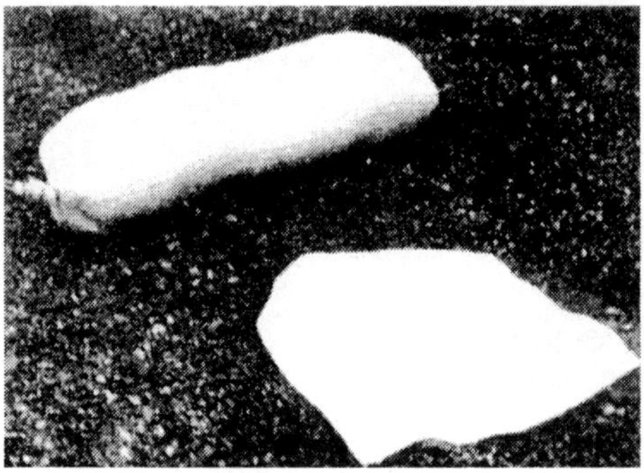

Figure 16: Oil Absorbent From Kenaf

To use Kenaf as an oil absorbent, the dried pulp is packed in special bags. It is then beaten so as to flatten the bags into sheets and subjected to more drying before final storage. In the event of an oil spill the bags are simply dropped over the spill to soak up the oil. The oil soaked bags are eventually taken to dedicated clean up grounds where they will be subjected to oil recovery process, incineration, or microbial breakdown as the case may be.

Kenaf can also be used as a forest conservation strategy. Kenaf is considered one of the most promising non-wood fibers for the pulp and paper industry. Kenaf, like many virgin wood fibers is pulped by kraft process. However, its pulping requires less chemicals, heat and time because it is not as tough as wood pulp and contains less lignin. Furthermore, its pulping is less destructive to the environment than virgin wood-based pulp production. This is due to the fact that hydrogen peroxide rather than chlorine is used for the bleaching (if need be). Although virgin wood-based paper may be cheaper, their prices do not reflect environmental costs. The destruction of ancient forests, fragmentation of wildlife habitat and the pollution of water systems are well documented. If the enormous cost of restoration is taken into account, development of Kenaf paper industry, which leaves forests intact while simultaneously reducing industrial pollution and energy consumption, begins to make both economic and ecological "cents and sense" (Taylor 1995).

The use of Vertiver

Another World Bank innovation competition entry by CE-RASE was a proposal to use vetiver grass (*Vertiveria zizanioides)* to achieve an integrated soil erosion control in Nigeria. Many grasses and trees have been tried over the years as a measure to prevent erosion but only vetiver grass has stood the test of time (World Bank 1993). Following a list of its characteristics, this truly remarkable plant is ideally suited for the vegetative system of soil and moisture conservation. It has a massive fibrous root system (Figure 17) that penetrates and binds the soil to a depth of up to 3 meters and can withstand the effects of tunneling and cracking. Vetiver is perennial, requires minimal maintenance. and grows in all types of soil regardless of fertility, PH, or salinity. Its leaves and roots have demonstrated a resistance to most diseases. It is tolerant to Al, Mn, As, Cd, Ni, Cu, Pb, Hg, Se and Zn in soils and has a high capacity of decontamination of toxic chemicals.

DENSE AND DEEP(2M) ROOT OF 10 MONTH OLD VETIVER

Figure 17: The Massive and Deep Penetrating Root of Vetiver

Grass

It is a climax plant, and therefore even when all surrounding plants have been destroyed by drought, flood, pests, diseases, fire, or other adversity, vetiver grass will remain to protect the ground from the onslaught of the next rains (World Bank 1993).

Vetiver also has numerous applications in conservation and management, which include, road highway protection, shelter belts for desertification, gully erosion and landslide control, construction, groundwater recharge, boundary demarcation and pollution control.

CE-RASE is using vetiver grass for biological remediation and soil stabilization, to control soil degradation and create wealth for communities affected by soil erosion and floods in Nigeria. This is an integrated soil conservation program, which involves the mobilization of rural communities in planting and raising vetiver grass on degraded soil or eroded land.

The Use of Oil Metabolizing Microbes

Microbial degradation of sludge is a natural process, which eliminates the bulk of oil pollutants collected after initial physical and chemical breakdown of its constituent. The occurrence of hydrocarbon – utilizing microbes with the metabolic potential to degrade polluting oils — has been reported in Nigeria by Amund and Igiri (1990). CE-RASE is therefore promoting the use of these microbial actions to reduce sludge. Various microorganisms including bacteria yeast strains, with the capabilities of utilizing petroleum hydrocarbons have been widely reported. Studies by Amund et al. (1993) have shown that hydrocarbon utilizing species like *Pseudomonas*, *Acinetobacter*, *corynebacterios*, *Bercillus* and *Micrococcus* are present as normal soil flora capable of degrading a wide variety of hydrocarbons through natural self-rehabilitating processes.

The main advantage of this remediation method includes its versatility on various types of contamination, in-situ treatment and soil rehabilitation. In the oil industry, the results of controlled laboratory experiments and field trials of actual oil spills have shown conclusively that bioremediation can enhance ecosystem restoration (Swannel et al. 1996). The successful use of this technique to clean up almost 180km of beaches following the Exxon Valdez spill of 1989 has stimulated considerable interest in its potential usefulness (Laws 1993).

Conclusion: How Can Bioremediation be Participatory

It is common knowledge that those in need fail to benefit even from "well-intentioned" aid and development programs unless they have an opportunity to participate and contribute to the planning process. Development is a learning process, not just the application of a predestined blueprint (Reid 1995). It does not start with goods, it starts with people and their education, organization and discipline. Without these three all resources remain latent and untapped (Schumacher 1973).

CE-RASE's sustainable bioremediation projects is encouraging local participation from planning to implementation. (Fig.18). The local communities will use the technology to clean up their environment and restore its productivity. CE-RASE helps people develop the competence, commitment and ultimately the self-confidence, to ensure a sustained and responsible community effort in the projects. With a participatory approach to the project, communities stand a better chance of developing their skills for a better management of their resources, which will ensure long lasting economic empowerment.

Figure18: Participatory Rural Appraisal Exercise in the Niger Delta

Both Kenaf and Vetiver projects involve nursery production. This process is already a participatory and income-generating activity because private nursery producers within the communities will emerge, from whom seedlings will eventually be sourced by both the government and private sector. This ensures and guarantees the availability of seedlings. Alongside having nurseries, community tool banks are established. This will help in lowering the cost of inputs needed for interested participants, hence helping to sustain interest. With some financial support, cottage Kenaf processing units can be developed thereby providing employment to the local people. The communities can sell Kenaf oil absorbents to oil companies to generate a steady income.

CHAPTER 21

SCIENCE AND TECHNOLOGY IN AGRICULTURAL DEVELOPMENT IN MALAWI

Ian Kumwenda and Wells Kumwenda

Introduction

Science may be defined as humanity's organized attempt, through objective study of empirical phenomena, to discover how things work as causal systems. By means of national and systematic thought, science brings together bodies of knowledge in an effort to reconstruct the world by the process of conceptualization. The purpose of science is therefore to comprehend not to invent. Technology however, denotes the whole or an organic part of knowledge that relates directly to the production or improvement of goods and services. In other words science can be schematically described as know-why while technology pertains to know-how.

Technology is traditionally used as a tool to extract and process natural resource and to relieve drudgery. But this role is changing. The changing role of S&T in development is seen in the various definitions put forward by writers and analysts. Some consider technology as consisting of machinery and equipment or techniques. In recent years as a result of the rapid advances being made in knowledge-intensive technology such as microelectronics, computers and biotechnology, technology is considered by many as knowledge required in producing goods and services. This knowledge may be embodied in equipment or machines like the computer or humans and abilities. The knowledge, on the other hand, may be implicitly contained in documents and organizational systems. Hence technology has also been defined as knowledge embodied in human beings or machines required to produce goods and services.

New knowledge is a product of both scientific and technological research and accounts for the letter R or the term Research while practical innovations are result of experimental development accounting for the letter D. The relationship between Research and Development indicates that new knowledge emanates from independent inputs to scientific research and feedback the input from experimental development while new applications emanate from independent input to experimental development and from the output of scientific research. New knowledge also feeds into the education system while new applications from experimental development feed into industry and services. A national Science and Technology system may be understood to consist of a set of organized scientific and technological resources and activities aimed at discovering, inventing, transferring and promoting the application of new knowledge with a view to achieving national goals. Internationally, a national S&T system is viewed to have four levels (Gupta 1998).

First, there is the planning and decision making level. This is the top level in a national S&T system at which planning and decision-making are undertaken on a comprehensive national scale. It is also at this level that inter-ministerial coordination is achieved. In many countries this takes concrete form in the shape of a ministry. It can also exist as a special council closely related to a very high government office. The second is the promotional, financing and coordination level. This is the level that consists of bodies that promote, finance and coordinate research and development (R&D) with various sectors of national activity. Typical among the bodies found at this level are research councils dealing with sectors such as agriculture, medicine and industry whose members are drawn to represent different spheres of science and technology, government and industry. The third is the research performing and institute level. At this level R&D work is actually executed and covers the network of research institutions and centers, and universities and their special schools or centers. The fourth is the scientific and technological service level. This level consists of institutions that provide scientific and technological services required for execution of R&D and for the production of goods and services. The key issues that are critical for development include S&T policy, institutional development, S&T manpower, financing of R&D, science and culture and international cooperation.

Science and Technology in Agricultural Development

Malawi's growth in agricultural output is increasingly becoming dependent on the development of scientific and technical capacity to develop new mechanical and biological techniques. As such growth in agricultural output, based on increases in cultivable land, is no longer possible. Farm inputs, including labor saving technologies, fertilizers and pesticides have become increasingly expensive hence inaccessible to the majority of the agriculturally dependent resource poor smallholders. Therefore, the objectives of the research program include generation of appropriate technologies for increased yields to obtain higher rates of return per unit of land; develop labor saving technologies; identification of crops and livestock for diversifying agricultural production in order to provide higher income, employment levels and increasing foreign exchange where possible; developing an institutional framework for human resource development; identification of agricultural technologies which are financially and economically viable; research into non-traditional crops and livestock which uses less land but have higher returns; identification and development of methods of disease and pest control using the most efficient and least costly means; and maintaining and updating support service system that can actively contribute to the process of technology generation and transfer.

Main Research Thrusts

The major research thrusts include:

- Development of high yielding and early maturing crop varieties that are tolerant to drought, pests and diseases, and evaluation of animal breeds suitable for various production systems.
- Development of appropriate technologies that support agriculture and natural resource diversification.
- Development of feed and evaluation of feeding technologies for increased livestock production.
- Development of improved soil fertility technologies, appropriate and improved land husbandry, soil and water conservation practices.
- Development of agro-forestry technologies that address low fertility, rapid deforestation and land degradation.
- Development of appropriate farm machinery, irrigation, storage, processing and post-harvest technologies.

Research is conducted both on-station and on-farm throughout the country. There is a major requirement that field research has to be conducted for at least three seasons before technologies can be accepted for release. Presently, there is a further requirement that the technologies are developed in partnership with extension and end-users in a participatory manner in line with the decentralization policy.

Organization of Agricultural Research

Agricultural research in Malawi has three easily identifiable streams namely, research centers that principally serve the needs of the estate sector and national and international research centers (stations) that principally provide good public research. There are two private agricultural research centers in Malawi. These are the Tea Research Foundation of Central Africa and the Agricultural Research and Extension Trust (ARET). The Tea Foundation is an inter-governmental center that undertakes research in tea and coffee for the benefit of growers in Malawi, Zimbabwe and

South Africa. The ARET undertakes research in tobacco for the benefit of local growers. Tea and tobacco have a highly organized marketing system and this provides a mechanism through which research funds are obtained. In both cases a levy is imposed on sales and the proceeds of the levy are eventually transferred to the research activities.

As for international research centers, there are four: the International Center for Agro-forestry (ICRAF), the International Center for Research in Semi-Arid Tropics (ICRISAT), the International Center for Living Aquatic Resource Management (ICLAM), and the International Institute for Tropical Agriculture (IITA). These centers conduct research and generate technologies that supplement the research efforts of local agricultural research institutions.

The Department of Agricultural Research and Technical Services (DARTS) that operates under the Ministry of Agriculture and Irrigation is the focal point for public agricultural research. The mission of the department is to conduct applied or production oriented agricultural research and provide technical services that will generate information and technologies to be utilized by the farming community and the stakeholders which should be demand driven to lead to increased agricultural productivity and proper maintenance of natural resources.

Administratively, research in DARTS is organized into five commodity groups namely: cereals, horticulture, livestock, legumes, soils and agriculture engineering. The Department of Research has three major research stations: Lunyangwa in the north, Chitedze in the center, and Bvumbwe in the south. There are also four experimental stations: Mkondezi in the north; Lifuwu in the center; and Makoka and Kasinthula in the South. Further, there are ten substations: Baka; Bolero; Meru; Mbawa; Nchenachena in the north; Bembeke and Chitala in the center; and Ngabu in the south. The 1999 Malawi Agricultural and Natural Resources Master Plan states that the research system is currently fragmented. Thus scientists and equipment are dispersed over too many research institutions resulting in inefficient use of facilities and resources, duplication of efforts and consequently considerable waste of public funds.

A weak linkage between researchers, extension staff and farmers has also been identified as contributing significantly to the low adoption rates of developed technologies besides other equally important and complex factors that affect technology uptake.

Human Resource Development and Management

Upon realizing that key to national development is human resource development, the government expanded the educational system including the establishment of the University of Malawi and overseas training of staff in most sectors of the economy using donor assistance. This is reflected in the scientific manpower of the institutions involved in research in agriculture and natural resources. Most staff are recruited from the University of Malawi after completing a BSc. Degree and sent abroad for advanced training under donor-funded projects. Currently, some researchers receive advanced training from the University of Malawi through programs supported by donors such as the Rockefeller Foundation and German Technical Cooperation (GTZ). The national research institutions have made great progress since independence in 1964, in localizing research positions. For instance, the Department of Agricultural Research and Technical Services (DARTS) had one Malawian researcher at the professional level in 1964 compared with 23 expatriates, but now most of the institutions, including DARTs, Department of Animal Health and Industry (DAHI), Agricultural Research and Extension Trust (ARET), The Rotary Foundation (TRF) and Forestry Research Institute of Malawi (FRIM) are staffed by nationals with a few expatriates in donor-funded projects. While the department of research has a good stock of researchers, retaining them in the system has been difficult due to the civil service salary structure and other conditions of service. The establishment of a semi-autonomous institution would enhance flexibility in this area.

Of the total work force of the Ministry of Agriculture and Irrigation of about 10,000 employees, approximately 2,000 are extension workers in the field, although 7,000 are involved in agricultural extension and related activities, broadly defined. Most of these employees are in the regions (Agricultural Development Divisions (ADDs), Rural Development Projects (RDPs), and Extension Planning Areas (EPAs).

Funding

Funding for research in Malawi has been unstable and inadequate. This has detrimental effects on the morale of scientists and the quality of research. In most cases funds from donors have been forthcoming but with little or no comparable local funding thus making the funding mechanism unsustainable.

As of 1997/98 public spending on research in Malawi was 0.49 percent of the government's budget and 0.13 percent of GDP, well below international norms (about 0.5 percent of GDP in low income countries, 0.85 percent of GDP in middle income countries). It is suggested in the Agricultural Research Master Plan that spending on research should be increased to 2.5 percent of GDP, which would be well above international norms. Malawi is in a position to attain the same research output consuming less than the international norm of 0.5 percent of GDP by borrowing research findings from its neighbors whose soils and meteorological conditions are similar.

During the budget cuts of 1985/96 to 1996/97, research was cut deeply (0.2 percent to 0.17 percent of GDP) while administration remained about constant (about 0.3 percent of GDP). This is because cuts fell on goods and services in research, while staffing was retained. Recent trends show a decline in government financial allocation for agricultural research. This has resulted in researchers sourcing most of financial requirements for research from donors. It is felt that donors would prove more forthcoming if the Department of Research became semi-autonomous.

Notable Achievements in Science and Technology in Developing the Agriculture

Despite the various constraints, science and technology have made some considerable contribution to agricultural development in Malawi as shown below. Research on crops has led to the development of several technologies that can be used to increase productivity. These include improved crop varieties, agronomic practices, and soil fertility improving practices, farm machinery, and irrigation technologies.

Improved Varieties

Concerted short and long-term efforts by scientists in DARTS, Bunda, and International Agricultural Research Centers have led to the development and release of a large number of crop varieties. These varieties for both field and horticultural crops are high yielding, tolerant to diseases and pests and are adaptable to local conditions. Research on maize has led to the release of several maize hybrids and composites that yield six to ten tones per hectare compared to the yields of less than one tone for local varieties. Work on legumes and oilseeds has also led to the release of a large number of groundnut, pigeon pea, soybeans and bean varieties. Work on horticultural crops has led to the release of high yielding and disease tolerant varieties of bananas, mango, tomato, cabbage, sweet potato, cassava and potato, among many others. Research by ARET and TRF has led to the release of numerous tobacco and tea varieties. The two private institutions have been able to multiply adequate basic seed of tobacco and tea and this has accelerated the transfer of new varieties more quickly to farmers. On the other hand, the public and parastatal institutions have had to rely on commercial seed companies and NGO seed programs for multiplication of basic seeds, hence slowing down the spread of some varieties.

The improved varieties have high potential yields over local varieties and the yield gap between smallholder farms and research plots ranges between 60-80 percent. Most varieties have been developed from local and introduced germplasm through normal crossing and selection programs. Where such programs do not exist, as in vegetables and fruits, varieties have been introduced and selected for adaptability to local conditions subsequent to release. A comprehensive list of crop varieties released in Malawi is contained in *A Handbook of Crop Varieties* (DARTS 1995).

Improved Agronomic Practices

Research has developed a lot of improved agronomic practices for various crops. These include recommendations for plant spacing and plant population, weed control and time of planting. Field crops plant spacing ranges from 15cm in groundnuts to 75cm in maize, while in fruits this ranges from 5m in apples to 9m in macadamia. Research has also recommended planting with first rains for all crops as late planting has been shown to reduce crop growth, yield and quality. In terms of weeding, research has recommended good weeding practices in maize and has demonstrated that with good weeding, farmers can achieve high yields at half the recommended fertilizer rates.

There has also been substantial work on cropping systems, mostly on inter-cropping where a range of crop mixtures and arrangements have been developed. Some research output in this area has been the demonstration of inter-cropping benefits from cassava and groundnuts, maize and sweet potatoes, cowpeas and beans. Research has also shown that grain-legume rotations have great benefit to soil fertility.

Soil Fertility Improving Technologies

Early research defined crop response zones in Malawi (Brown 1962). This identified areas that are generally deficient in nitrogen, phosphorous and sulphur. In response to this, research has over the years developed technologies that address issues of organic and inorganic fertilizers as well as improvements in soil organic matter. Recommendations on fertilizer types, fertilizer levels, application frequencies and methods are available for most crops. In maize, area specific fertilizer levels have been developed for all Extension Planning Areas (EPAs) in the country and applications of 90kg nitrogen has been shown to increase maize yields from one ton to about five tons. The same is true in tobacco where applications of about 98kg nitrogen and 100kg each of phosphorous and potassium have led to very high yields and improved quality of the tobacco leaf.

Irrigation

Research has developed technologies for irrigation of rice and a range of vegetables. In rice, research has recommended irrigation regimes and water management techniques. There have been some recommendations for the growing of crops such as maize, sweet potatoes and beans using residual moisture in dimbas. These recommendations include: plant populations, varieties and fertilizer level for these crops.

Pest Management

Research on control of pests and diseases has resulted in recommendations for their control in a number of crops. Several chemicals have been recommended for the control of pests and diseases, such as use of carbaryl for the control of boll worms, leaf eaters and jassids in cotton; dimethoate for the control of aphids and red spider mites in vegetables, among many others. Research has also recommended methods for the monitoring and control of migratory pests such as red locust and armyworms. There are also recommendations for the biological control of certain pests, such as the use of the parasitic wasp for the control of cassava mealy bugs. As part of an Integrated Pest Management (IPM) approach, research has established economic thresholds for the control of some pests, such as bollworms and jassids in cotton and the stinkbug and borer in macadamia.

Labor Saving Technologies

A wide range of technologies for mechanization of most farm operations including those for production, cultivation, transport, processing and irrigation have been developed. Technologies such as ox-drawn ploughs and ridges have been adopted for use by about 13% of the smallholder farmers in Malawi. Others include: ox-drawn carts and cultivators; grain dehullers and grain treating machines; small scale ram press oil expeller solar dryers, maize shellers and fruit juice making machines.

Livestock Research

Livestock research, especially in animal production is currently being conducted by DARTS and Bunda College of Agriculture while animal health research is mainly conducted by DAHI. Livestock research covers cattle, goats, sheep, pigs, poultry and more recently non-conventional livestock and wildlife. Research in these livestock commodities has mainly concentrated on improvement of breeds, folder crops and vaccines.

Some research achievements worth noting include a substantial pasture and browse program initiated in the late 1960s under which a range of recommendations was produced such as the use of Rhodes, Star, Napier and Guinea grasses, Leucaena and forage to improve livestock nutrition. Research on livestock nutrition also concentrated on improved utilization of crop residues and by products for both stall-fed beef and dairy cattle. The use of locally formulated urea/molasses/mineral blocks is now a major supplementary feeding recommendation for large ruminants.

Little research has been conducted by the private sector. However, the Malawi Industrial Research Technology Development Center has designed some low cost small-scale equipment for rural farm use while the pharmaceutical industry has conducted trials on pyrethroid, acaricides and on tick resistance to commonly used acaricides in Malawi.

All stations within the DARTS system are encouraged to hold at least a field day each year in order to popularize the technologies. Such field days attract farmers, school children, policy makers and the general pubic. In addition, scientists are encouraged to conduct targeted field days on specific technologies for particular groups.

Dissemination of Technologies (Extension)

Government Extension Service

Extension work began in colonial times as a result of estates requiring better agricultural productivity. This gave birth to the then Department of Agriculture in 1907. At that time the instructors sent out by the government to teach crop production practices used coercion. Violation by the local farmers of the recommended practices resulted in court sentences, which ranged from payment fines to imprisonment. Later the concept of the Master Farmers was incorporated into the mainstream of extension activities. These Master Farmers who were better off and innovative, received government support in terms of inputs and extension services. They followed recommended practices and therefore acted as demonstrations to other farmers.

In trying to enhance the group approach, the Block Extension System, a modified Training and Visit system (T&V), was adopted in 1981 with the aim of improving farmer coverage. The group approach then went beyond specialized groups and tried to contact a wider range of farmers, including the resource poor and women. However, it was observed that the majority of the resource poor farmers were not reached with extension messages because of its top-down approach and consequently the adoption rate did not improve. While there have been some positive results, it is evident that the present public extension system can no longer hold. A perennial problem has been that it has tended to neglect. Smallholders with less than 1.5 ha and concentrates on farmers with more land who are able to access credit. By de-linking credit and extension during the 1990s the authorities hoped to increase the coverage of smallholders. However, other people have argued that de-linking has failed because most of the extension staff have less contact with farmers than when they had four or five credit clubs to supervise.

Several reasons account for the low productivity of the extension system in Malawi. First, staff time was diverted to Starter Pack Program, which was aimed at providing farmers with basic inputs. The time and resources of extension agents were diverted to become organizers of the starter pack programs in 1998 and 1999, and other programs. In some EPAs the extension agents spent the period July-December 1999 compiling long lists of potential beneficiary households to the exclusion of their extension work. Second, there was the lack of resources for staff in terms of transport and demonstration materials. For example, spending on extension has fallen from about 3.7% of the total government budget in the early 1999s to about 2.5 percent in the late 1990s. Similarly it has fallen from about 1.0 percent of GDP in the early 1990s to around 0.1 percent of GDP in the early 1990s to about 0.7 percent in the late 1990s. Third, there is the problem of low salaries. Thus, the monitoring and evaluation system is in place but is not functioning properly.

Malawi Industrial Research and Technology Development Center (MIRTDC)

The center has a very important role to play in the development and dissemination of technologies. Once the technologies have been developed by other research institutions MIRTDC has the responsibility of demonstrating them to potential users. MIRTDC has expertise in industrial research, applied technology and technology assessment and transfer. There is opportunity for MIRTDC to link with the technical advisory service division of DEMATT to enable potential users get access to funds. MITDRC's activities include trade fairs, product development, buyer-seller meetings, and extension services to exporters and training. As far as the export market is concerned, it is in a position to facilitate co-operative development, cottage industry information base and networking. Due to poor funding and lack of experienced staff they are unable to perform their roles properly.

Non-Governmental Organizations

There are several non-governmental organizations (NGOs) operating in Malawi. Some of these NGOs are involved in the dissemination of scientific technologies such as new high yielding varieties, crops that mitigate malnutrition or drought resistant varieties. Examples of such NGOs are World Vision International, Action Aid, and Concern Universal, etc. The role of NGOs can be greatly enhanced by giving them a clear policy on what the country wants to achieve including the time period because resources are not a major constraint to this group.

Lessons

National S&T System

Since 1964 economic development planning has not paid special attention to the development of and defining the minimum national capacity in science and technology. Malawi's capability in science and technology is weak. For example by 1991, the ratio for R&D scientists/engineers per million of population stood at 42 against a world average of 1000. The recommended figure for African countries was a minimum of 200 per million of population by year 1980. This shows that Malawi has not made significant strides in improving her stock of S&T manpower. The major reason for the failure to integrate science and technology in national development planning in Malawi emanates from the existence of a weak national system for science and technology at all four levels, namely; planning and decision making, promotional, financing and coordination, research performance and scientific and technical services. The system lacks institutional mechanisms for forecasting and identifying emerging issues and problems of Malawi. There is generally a low level of national S&T awareness and this has tended to weaken the capacity of Malawi to integrate S&T in national development planning.

Institutional Structure

The institutional structure of S&T in Malawi shows the existence of all the four institutional levels for an effective national S&T system. The National Research Council of Malawi was established by Presidential Decree in 1974. This showed the recognition for S&T at the highest political level. There are sector coordination committees/councils within the National Research Council of Malawi. There are research performing institutions in key sectors of the economy such as agriculture, industry, energy and natural resources (Forestry and Fisheries). However, a number of problems exist in the S&T system in Malawi. First, There is a lack of guidance as evidenced by the lack of a clear policy and political commitment in practical terms. Another problem is the lack of integration of S&T in national development planning as shown by the failure to recognize the importance of integrating S&T considerations into the national development process. Second, the National Research Council is ineffective because it does not fund research and is consequently unable to coordinate national scientific and technological research focus. Third, the research performing institutions in Malawi are organized following the sectorization principle and are poorly coordinated despite the existence of sector research councils. The location of sector councils in the National Research Council of Malawi while the research performing institutions report to sector ministries further weakens the coordination function. The lack of collaboration between research institutions has led in certain cases to the duplication of research work thereby wasting scarce resources. Finally, there is an ineffective S&T Information Service as shown by the absence of information retrieval service and national and sector databases.

Financing of Science and Technology

The conventional budgeting procedures in Malawi do not easily permit the explicit and appropriate identification of funds allocated to S&T in the national budget. The procedures make it difficult to analyze the amount, distribution and purposes of such funding and this accordingly precludes any evaluation of their relationship to national development goals.

The financial allocation of S&T activities in Malawi is very low and is below the recommended level of 1 percent of GNP that is spelt out in the National Science and Technology Policy. There is also the problem of continuity for S&T programs especially under the cash budget system currently in force by the government.

Science Culture

A national culture for science and technology is closely related to the concept of societal utilization of science and technological research. Malawian society does not fully appreciate how it has benefited from national efforts in scientific and technological research. A large proportion of ordinary citizens of Malawi are not adequately exposed to techniques and mechanical devices at home, in school, and at work. Malawian culture, like elsewhere in Africa, is dominated by superstition, traditional ideas and beliefs. Low levels of literacy have also compounded the problem; the dominance of the hoe in agriculture, for example, reveals the low levels of innovation in Malawian society.

International Cooperation

Malawi has signed bilateral cooperation agreements in the field of science and technology with a number of countries, including the Republic of China and Malaysia. The extent to which these agreements have benefited the national S&T system is not well documented.

Recommendations

There are many factors that influence the impact of science and technology. Some of these are endogenous to the technology and transfer system itself while others lie well beyond the system. Several researchers have recorded successes and failures and some of the success has resulted in the generation of technologies, which have had a positive impact on agriculture. Some of the factors that have influenced the success of science and technology include funding, human resources, training, marketing infrastructure, research focus, collaboration, technology diffusion mechanism, government policies, consumer preference, and planning and management of research. Some of these factors are briefly discussed below. Review of past programs has revealed the factors that contributed to either successes or failures of the national research system. Lessons have been learned and a lot of knowledge gained to enable the government to develop well focused research strategies for the future.

Funding Levels and Sources

Provision of adequate funding is necessary for the success of any research program. Organizations or scientists who have both government and donor funding have recorded more successes than those solely dependent on revenue budget. Government has, therefore, to consider increasing its budget allocation to researchers as a way of increasing productivity in the research system. Otherwise scientists will continue to depend on donor-funded programs.

Human Resource Numbers and Quality

Having a critical mass of well-qualified scientists to do research is a guarantee to research success. Institutions with adequate number of staff who are also well trained have had outstanding outputs. This emphasizes the fact that research requires adequate numbers of professional staff who are well trained in their fields of specialization.

Training

Both long term and short term training courses are necessary for the advancement of science. It has been noted that outputs of researchers who were well trained were better than those who were poorly trained. Training of staff should be upgraded to MSc. and PhD. degree levels so that they become competent and confident in their discipline. In addition, short courses are necessary for upgrading the staff and updating them on the latest technologies and scientific procedures.

Government Policies

There are obviously many areas where government policy has a dominant effect on the agricultural sector. Investment in areas such as infrastructure, taxation or subsidies on inputs or outputs and land tenure has profound effects. Policies on national and household food security are also influential. In terms of natural resources development, sector linkages are obviously very important. For example, the Agriculture and Livestock Development Action Plan (ALDSAP) and the National Environmental Action Plan (NEAP) have addressed the need for inter-sector linkages in order to ensure that cross-sector problems such as deforestation, environmental and natural resources degradation are jointly addressed.

Researchers hold strong opinions on the negative effects of government policies such as the liberalization of the market and currency devaluation on the dissemination and uptake of new technologies. The removal of subsidies on fertilizer and devaluation of the Malawi currency at the time flint hybrids were being introduced meant that many farmers were unable to try out these new varieties. Clearly, as long as input costs exceed produce prices realized by farmers, there will be no adoption of the technologies. It is also important to note that government may have laudable goals and objectives in its development plans; however, it is usually fairly nebulous about the role of research in achieving these goals. Researchers, therefore, have to make the effort to show government what science and technology can offer to national economic development.

Research Focus and Target Audience

Most scientists have been able to define the problem and develop a research strategy to solve the problems. However, one problem has been a lack of focus to achieve the intended results. There is, therefore, the need to develop logical frameworks when designing research programs to enable the researchers to focus on research that will solve the problem. Targeting of the technologies has also some problems. Most researchers have usually targeted the smallholder farmer while neglecting the commercial sector. It is worth noting that the commercial farmer needs research technologies and has adequate resources to adopt technologies and commercialize them. The commercial farmer or commercial companies have the ability to significantly contribute to agricultural output, yet they are neglected in the planning process, including the development of the research agenda.

Planning and Management of Research

The planning and management of research programs and activities play a major role in the implementation and success of research programs. Poor problem identification at the farmer level tends to distort the accurate setting of research priorities. Poor funding and inadequate training of researchers tend to exacerbate the problems of management leading to the generation of poor technologies. Another weak aspect of research management has been its inability to adjust to major macro and micro economic policy changes which tend to have a significant impact on the success of research adoption and the application of scientific technologies.

Capacity of End Users

Due to poverty, illiteracy and lack of credit facilities, many end-users fail to adopt the technologies. This is an important factor to be considered when designing experiments to generate the needed technologies. Most technologies have not been adopted either because they are too complex, require money, capital or knowledge.

Research Capacity

The research system has limited capacity for post harvest research and for adding value, and thus creating fewer opportunities for agro-industries. Consequently, many researchers have not been able to conduct certain types of research due to the lack of facilities, i.e. soil or leaf analysis, or tissue culture. Some institutions have received equipment and facilities for research from donors but have not effectively utilized them. This has led to substantial importation of many products that could be processed locally. There are substantial produce losses in storage that require immediate attention in addition to premature selling of produce early in the season.

Another important problem is that of perishable horticultural produce that is further compounded by inadequate post harvest technologies. There is also serious manpower constraint in the field of veterinary science and this, apart form other factors stated elsewhere, have contributed negatively to livestock productivity. There is, therefore, the need to undertake a skills-gap assessment and correct the situation accordingly.

Collaboration

It is important to work in multidisciplinary teams and collaborate with other scientists working on similar problems. Collaboration is currently weak and scientists tend to work as individuals, hence failure to produce a total package of information needed by the end users.

Consumer Preference

Researchers whose technologies take into account consumer preference usually succeed. Hence, scientists should implement demand driven research in order for them to achieve success because such technologies are readily adopted. The maize program in the mid-70s disregarded consumer preferences. Research priority was given to hybrid dent varieties, whereas Malawians prefer hard seeded, flint varieties that can be pounded and are resistant to storage pests. The dent hybrids, despite being high yielding, probably addressed the needs of only about 10% of maize growers, mostly medium to large estate owners, who have access to credit for essential inputs. Another example of disregard of consumer preference was that of rice in which the high yielding, unscented variety failed to replace the low yielding, scented varieties preferred by the consumers. In capture fisheries research, the large fishermen were favored at the expense of small fishermen, yet the latter catches most of the fish consumed in Malawi. In forestry, consumers prefer hard wood for timber and firewood while researchers recommended the fast growing soft-wooded species. These are clear examples of the need for researchers to consider consumer preference when designing research programs.

Climate

Agriculture is dependent on the natural resources and the environment. The impact of technologies will depend on the existence of a favorable climate such as rainfall, temperature and soil fertility, and soil structure. Years of adverse climatic conditions have had a negative impact on the adoption of certain technologies. However, these factors are beyond the control of the researcher and end users of technologies.

Extension Mechanisms, Uptake and Impact of Technologies

The establishment of clear mechanisms of technology transfer is essential. The number of technologies transferred and adopted by the end users also measures the productivity of the research system. Where technology transfer procedures were nonexistent or weak, most technologies remained on the shelf; hence, limiting their impact in the agricultural and natural resources sector. Research institutions with strong and clear mechanisms such as ARET and TRF have been able to transfer their technologies, which have had an impact. There is a general perception that there is inadequate dissemination of useful technology and that many results remain on the shelves of experimental stations in unprocessed and unpublished form. On the other hand, the non-scientific community often presents research findings in technical language that is difficult to understand. The responsibility for removing these barriers of communication to aid technology transfer lies with both the research and extension staff.

The ultimate measure of success of any agricultural technology is its uptake by farmers and impact on production. However, a general concern of many research projects is the perceived lack of update and impact of the technologies. There is thus an increasing emphasis, especially by donors, on studies dealing with farmers' needs assessment in the hope of identifying potential technologies that will diffuse quickly and have a rapid impact at the farm level.

One of the problems in measuring uptake and impact of research in Malawi is the lack of information. Maize is one of the few commodities for which substantial information is available. This is because of its dominance in smallholder agriculture so that its uptake and impact have been studied extensively. For other commodities, only aggregated information is available. The liberalization of hybrid maize production and marketing in 1993-94 and the removal of fertilizer subsidies between 1992-1994 have had major impacts on maize production. Research has shown that the present hybrids can be widely accepted by farmers if they can afford to purchase the seed; hence, the tendency to recycle seed from F1 hybrids. Since fertilizer is beyond the reach of most resource-poor farmers, the yield advantages of hybrids, though positive, are small and likely to decrease with decreasing soil fertility.

Staff Conditions

Conditions of service in the Civil Service and the University of Malawi have led to frustration of the scientists leading to low working morale. Attempts to boost morale have not been successful. Other possible measures to address the problem may include restructuring of the research system. If institutions became semi-autonomous, they would probably be more effective and have better incentives for the research staff. The assumption is that government would continue funding research while donors would increase their allocation of funds to research.

CHAPTER 22

(MIS)UNDERSTANDING CHANGE IN AGRO-ENVIRONMENTAL TECHNOLOGY IN AFRICA: CHARTING AND REFUTING THE MYTH OF POPULATION GROWTH INDUCED BREAKDOWN

Aaron deGrassi

Introduction

This chapter explicitly counters conventional policy wisdom on the locus of change in agro-environmental technology in Africa. I show which assumptions are unfounded, under what historical conditions such assumptions were made, with what affect, and what is a more precise understanding of agro-environmental change. Academics, policymakers and the public at large often assume that although previously stable shifting cultivation systems functioned for primeval societies, such systems have recently broken down because increasing population has reduced the amount of land, which farmers can use to fallow (set aside to allow replenishment). The assumed result is the apparently visible spiral of regional underdevelopment, reduced incomes (and hence local and national level under-nutrition), and environmental degradation. Such notions rose in colonial times to justify colonization and concerns for settler welfare, administration revenues, imperial environmental heritage, and social stability. These notions were fortified in the 1970s when the paradigm and practice of farming systems research arose, building off previous socially and agronomically synchronic structural-functionalist and unilinear, teleological evolutionist thought, and was institutionalized throughout the world, having initially been spawned by critiques of the original Green Revolutions in Mexico and Asia. Finally, life-sciences corporations, academics and policy makers have revived the myth in the late 1980s and 1990s with cruder varieties of development populism, co-opted environmentalisms, and pro-biotechnology public relations campaigns.

In contrast to this dominant meta-narrative of population-induced breakdown in shifting cultivation, *social relations* at multiple levels are key in understanding the patterns and changes of the socio-temporal production of agriculture and environments. Among the factors that combine in historically and geographically specific conjunctures include: gender relations, labor relations, land tenure and distribution, direct and indirect state policies (colonial and post-colonial), agricultural development projects, armed conflict (local, national, international), foreign aid, national parks, subterranean resources (oils, diamonds, etc.), and commercial timber production. The incorrect emphasis on population as the only motor of change—and consequent occlusion of history, politics, power and agency—underlies top-down strategies, inappropriate prescriptions, and ill-fated projects promoting green revolution technologies.

The essential claims in "conventional development studies and policy wisdom" on change in African agro-environmental technology are as follows. African farmers have long practiced "shifting cultivation," wherein they have cultivated plots for a few years until exhaustion of fragile tropical soil, and then moved on ("shifted") to cultivate a new plot cut from wild areas, leaving the old plot alone for ten or so years to naturally regenerate its quality (fallow).[1] As populations increased, there was less land for farmers to move on to, and hence they had to either come back to their old plots before fields were replenished or cultivate marginal forest and dry areas. The apparent result of population growth-induced breakdown in shifting cultivation is poverty, hunger, and environmental degradation.[2] This dominant narrative of African agro-environmental change began in the colonial period, took hold in the late 1970s and early 1980s, and has been revived in the late 1980s and 1990s. In the last section of the chapter, I very briefly present a counter-claim that agro-environmental technology actually changes according to multiple social processes occurring at multiple historical and geographical scales. These factors, which often take specific forms and combine in specific ways, include: gender relations, labor relations, land tenure (and property rights regimes more generally), land distribution, direct and indirect state policies (colonial and post-colonial), agricultural development projects, conflict (local, national, international), foreign aid, national parks, underground resources (oils, diamonds, etc.), and commercial timber production.

Such policy narratives are in some cases important material forces in the lives of people in the Third World. The continuing tacit acceptance and occasional explicit espousal of the conventional wisdom has often helped to remove important issues of politics and power from discussions of agro-environmental change. The narrative leads to a distortion of intellectual, financial, and social resources towards three narrow development programs: first, developing and extending new, more productive agricultural technology; second, delimiting nature reserves; and third, restrictive population control measures. The African version of the myth has been conclusively—but always implicitly—disproved in case studies by academics, and has been reworked by various people and institutions for different, frequently political aims, but it has not to date been explicitly charted and refuted.[3]

The Colonial Era (1800 to 1960): The Myth Begins

This section briefly describes colonial interpretations of African agro-environmental technological change, and how colonial powers became so concerned with Africans' techniques of production and environmental management. A mix of ignorance, selective and partial description, racism, justification for control, and Victorian evolutionism shaped colonial understandings and representations of African agro-environmental technology. As a result of these influences, colonial ideologies came to hold three basic tenets: first, African farmers were *not* rational;[4] second, farming systems were bounded, coherent, uniform and homogenous systems (Niemeijer 1996); and third, farming systems were more or less unchanged in structure until they were broken down by population increases.

These colonial understandings of African agro-environmental change took such a hold because of imperial justification for colonization, as well as concerns over protecting settler interests, preserving the natural heritage of the empire, ensuring a steady flow of revenues, and maintaining sovereignty by regulating resource extraction so as to prevent politically explosive social instability. Administrators believed that beneficial colonial health measures and protection from internecine warfare had allowed African populations to expand, thereby inadvertently causing widespread disruption in their delicate agro-environmental systems—which were viewed variously as innately destructive and irrational, or as Rousseauean examples of primitive harmony with nature. Notions that African farmers needed guidance and trusteeship (occasionally supplemented with forced compulsion) provided European powers with justification for colonization. The purported destruction by shifting cultivators made them—rather than timber companies—easy targets, as increasing imperial control took the guise of environmental protection. The economic and cultural sanctity of European settlers was also at stake. Certain factions of settlers, often new European immigrants attempting to solidify their own identity, espoused concerns that shifting cultivators were threatening to destroy scenic wildlife, cause deforestation, soil erosion, and spread crop and livestock diseases.

Shifting cultivation also aroused concern because it was believed to threaten state revenues gained through the sale of timber, settler export crops, and game—revenue necessary for funding government in the colonies, given that there was some contention regarding the profitability of the colonial enterprise.[5] In addition to justifying domination and ensuring sufficient resources for settlers and administrative costs, colonial powers were also concerned with African farming because of broader imperial environmental concerns. As Grove (1998) has documented, with the changing ideas of the Enlightenment and Renaissance, European imperial powers' concerns shifted from the amount of timber on tropical island available for naval batteries, to instead see these islands as "Edenic" tropical paradises needing protection from the "destructive races." Natural resources were seen as the empire's heritage.[6] From the islands, conservation notions were transferred to India, and then Africa. It was in India and Africa that concerns about deforestation, drought, famine, and social breakdown became linked through the work of an international circuit of professional scientists and planners in the rising disciplines of ecology and forestry. The understanding of agro-environmental change grew increasingly salient as colonial powers became more concerned about resource scarcity in France, England, and the United States, and as they witnessed heightened disputes, regulations and infractions around forest resources in the colonies. The transnational networks of colonial scientists and administrators also believed that the US dustbowl foreshadowed the future of unchecked degradation in then drought-ridden southern African (Anderson 1984).

Colonial powers also believed that the resource degradation brought on by various mixtures of population growth, irrational techniques, commercialization, and colonial demands could threaten the fabric of African society and lead to dangerous rebellion (Isaacman 1993; Anderson and Grove 1987).

In order to protect imperial heritage from the growing legions of African shifting cultivators, colonial guardians felt compelled not only to use the "trickle-down" approach of extending technology to large wealthy "progressive" farmers,[7] but also to exercise brute coercion through fines and imprisonment of small agriculturalists deemed too irrational, ignorant, or imprudent to adopt conservation measures on their own initiative. Africans were forced to work on state- or chief-owned plantations, change their crop compositions, contribute to emergency food reserve silos, decrease their cattle holdings, and practice laborious soil conservation measures, such as constructing terraces and ridges (Anderson and Grove 1987). These measures ultimately provoked considerable resentment and fed into anti-colonial independence struggles (Grove 1998: 177-223; Salih 1999).

Colonialists thus generated the myth of how changes occur in African agro-environmental technology. For instance, E.P. Stebbing, a highly influential colonial forester from India who toured Africa, gave an account in 1934 that is strikingly similar to current narratives:

> In West Africa the process [of the advancing Sahara] owes its commencement to the system of farming the bush or degraded type of forest, which covers much of the countryside, this system being a form of shifting cultivation. With an increasing population the same areas are refarmed at shorter intervals, with a consequent more rapid deterioration of the soil constituents, until a stage is reached when the soil is no longer sufficiently productive for agriculture. (Stebbing 1938:12, quoted in Swift (1996: 74).

Stebbing called for an international effort to stem the supposed agriculture-induced desertification, which he believed was enlarging the Sahara and changing the climate of the entire region. An investigative commission was soon sent, but during a particularly rainy year, and hence their conclusions were much milder. Concerns about over-population remained less pronounced until they were revived in the 1970s in the midst of severe droughts, famines, and Cold War politics (Swift 1996: 75-77).

Fortifying The Myth: Farming Systems Studies in the 1970s and 1980s

A rising interest in the 1970s in the *rationality* of small farming occasioned a number of "farming systems" studies that repeated the dominant myth of agro-environmental change in Africa and fortified it by meshing it with underlying currents of ahistorical "systems theory"-based structural-functionalism and teleological, unilinear evolutionism. The myth has been disseminated worldwide as farming systems studies have risen in prominence amongst academic circles, national governments, NGOs, and international donors.

The burgeoning interest in farming systems stemmed from controversies over new sets of agricultural technologies, known as the "Green Revolution." New technologies—such as high-yielding plant varieties, fertilizers, pesticides and herbicides, and irrigation—had given dramatically higher yields (which were hoped would alleviate political and environmental threats of growing population pressure on resources), but often increased yields only when used together in favorable growing areas. Other problems included smaller, more marginal farmers' inability to adopt the technologies, implications for human health and the environment, and rates of poverty and hunger that had not declined despite increasing national food supplies. Through a number of innovative studies in various parts of the world (some drawing on the farm management literature of the early Twentieth Century USA), there arose a group of researchers emphasizing that farms had multiple, complex aspects that were not being addressed by mainstream technological research and development (Collinson 2000; Norman and Matlon 2000; also, perhaps see Oasa 1987).

In their theories and practical efforts, experts began paying more attention to how farmers adopted technology that matched their specific needs and resources. A crucial recognition was that poorer, marginal farmers were rational—not ignorant or conservative—in rejecting new technology, since much of it was not well suited to their bio-physical and socio-economic constraints.[8] The fundamental argument is as follows:

- The specific form of technology adopted is that which yields the greatest benefits per unit cost.
- The technology most likely to yield the greatest benefit is the one alleviating the most limiting factor to production (that is, if factor prices reflect factor scarcities).
- Factors of production—either land (with potential variation in water and soil fertility), labor or capital—are limiting in accordance with their relative proportions and qualities.

Thus, technology would only be adopted according to specific factor proportions. In an era where some development practitioners bluntly believed that any poor person would benefit from any sort of Western technology, these studies were important because they suggested that technology would not be adopted or have its intended affect if it was imported without change from a developed country into a drastically different Third World context.

A key work stimulating thinking and policy on this issue was Ester Boserup's 1965 book, *The Conditions of Economic Growth*, which considered the effects of one variable, population density. Boserup emphasized that the structure of farming systems varied according to the amount of people per hectare of land. Higher population densities meant that land was a scarcer commodity, and farmers relieved this constraint by using technology that gave relatively high output per unit of land. In fact, she claimed, growing or shrinking population densities caused farmers to change their cultivation practices accordingly.[9]

Soon thereafter, Hayami and Ruttan (1971) presented a more elaborate and explicit theory—termed "Induced Innovation"—to explain patterns of technology adoption.[10] They argued that mechanization is more prevalent in the U.S. because, although mechanization is capital intensive, it alleviates the limiting factor of scarce and expensive labor and takes advantage of the abundant and cheap factor: land. Conversely, Japan, with substantially more people and less land, has adopted technologies that are more labor-intensive and "land-saving," such as detailed irrigated paddy rice. In other words, the form of new technologies is such that they will substitute for scarce, expensive factors and take advantage of cheap, abundant ones. Hence, where factor endowments differ, then there will consequently be different paths of technological development. "Inappropriate" technology is that which is not called for by the factor endowments; technologies bare the "foot print" of the societies that create them, and therefore may not be applicable in differing societies. These issues were further explored in collaboration with Hans Binswanger (Binswanger and Ruttan 1978; Koppel 1995).[11]

A third key work was Hans Ruthenberg's classic pedagogic text, *Farming Systems in the Tropics*, published in 1971, 1976, and again in 1980. Ruthenberg added ecology as another characteristic to be considered in farming systems analyses of patterns of technology adoption. For instance, farmers with abundant land might burn brush not only because it reduces labor needed for clearing, but also because the ashes increase soil pH and hence available potassium.

In accordance with these growing sentiments towards technology appropriate to small farmers, there developed a whole new methodology for doing agricultural technology research and extension, termed Farming Systems Research and Extension (FSR/E).[12] FSR/E's novel methodology was to forge closer links between researchers, extensionists, and farmers so that researchers could better tailor technology to farmers' complex bio-physical and socio-economic conditions. In short, the fundamental premises of FSR are (a) farmers are rational and (b) they face complex systems of constraints, which implies (c) they will adopt technology that is suited to them, and so (d) technology will have to be based on in-depth analyses and actual on-farm testing (Merril-Sands 1986).[13] Initial farming systems proponents emphasized adapting technology for both the "downstream" factor of inappropriate R&D (such as corn husks that were too short) and the "upstream" factor of political and socio-economic conditions (such as poor prices for crops or civil war).

However, the range of explanatory factors that farming systems studies actually came to consider was quite restricted, mainly limited to the few variables initially identified by Boserup, Hayami and Ruttan, and Ruthenberg: population density, market factor prices, and agro-climate. Part of the reason that only phenomenon within the circumscribed "farm-unit" were considered was because these analyses of farmers' conditions came only as reluctant responses to appease critics; their *raison d'être* was to show a concerned public and donors that scientists were indeed trying to "aim agricultural research at the needs of farmers" (Winkleman and Moscardi 1982; Oasa 1985). Early on, several researchers recognized that, in excluding consideration of larger "upstream" factors, Farming Systems (FS) studies were focusing only on small adjustments and "technical fixes;" broader constraints were treated as fixed parameters and thereby eliminated from consideration. However, changes in the scope of analysis were blocked, partly by institutional politics. Center staff—mainly breeders and agronomists—were often indifferent, and sometimes hostile, to the idea of more social science because they viewed it as making their own work just too complicated to be effective.[14] Study of the influence of policy was largely partitioned off to one isolated center based in Washington DC, the International Food Policy Research Institute. Much of what social science was allowed was based in fundamentally limited classic concepts (see below) and even then was ignored or used selectively.[15] An economist, Donald Winkleman, was hired at one influential center in order to deter, rather than investigate, critical investigations of the broader social, economic and environmental implications of, and reasons for adopting, green revolution technology (Winkleman went on to become director of the center and then chairman of Nineteen-Center Consultative Group on International Agricultural Research, CGIAR) (CIMMYT 1992). Consequently, much farming systems research remained focused on technical germplasm issues, albeit supplemented with attention to the complex micro-dynamics of farming systems, but little analyses of the wide array of broader socio-economic and political processes and structures shaping land-use (Norman and Baker 1986: 37, 38, 54).[16]

Many farming systems analyses were also more profoundly constrained by the larger and older underlying conceptions of "systems theory"-based structural-functionalism and evolutionism. The limitations inherent in these concepts constrained FS analysis and thereby caused it to perpetuate the dominant myth of African agro-environmental change.

In the systems thinking of the 1960s and 1970s, also sometimes known as "holism," linked networks worked through elaborate feedback mechanisms to maintain a balanced equilibrium of the system as a "coherent whole."[17] Such thinking prevailed in academic and popular writing and in development policy as well.[18] While applications of systems theory to agriculture did illustrate the importance of intra-farming system connections as well as the other components of farming besides (mono) crops, it suffered from a focus on bounded, coherent, and generally static and parsimonious farm-level systems, thus giving a misleading caricature of the settings of rural life. The tendency towards drawing up a structure of groups, of classifying farming systems," was incredibly reductionist in its analysis of (1) rural differentiation and (2) broader connections. Connections to politics and economics beyond, and heterogeneity within, the farming systems were not emphasized, as farming systems were viewed largely as homogenous self-contained units (Little 1984). Systems theory's emphasis on bounded and coherent systems in an equilibrating stasis led to categories so general that they obscured more than they revealed.

As Hart (1986: 8) notes based on her examination of similar theories applied in rural Java, "the problem with all of these theories is that they omit critically important variables." The reading out of important, but numerous and historical and complex, variables (such as labor mobility, changing class relations, and processes of state consolidation, etc.) led to a narrow focus on the influence of the affects of factors "endogenous" to farming systems, such as population density, factor markets and agro-climate (see also Biggs 1995).

What resulted from such narrow, synchronic analysis was a functional understanding. Functionalism is the notion that existence is explained not by historical development, but by contemporary function. The functional explanation of shifting cultivation is that it exists because it functions to allow delicate tropical soils a period in which nature can restore their structures and nutrient content. Functionalism is, however, built upon three fundamental limitations. First, if one explains form from need rather than historical processes, one encounters problems of *teleology*; that is, of establishing a logical sequence of causality in which cause precedes effect.[19] Second, one often cannot explain *temporal diversity*: why forms change when there does not appear to be any corresponding change in form's function, or, conversely, why forms stay the same when it appears that a previous function no longer exists or there has arisen a need for a new function. Third, one often cannot explain *spatial diversity*: why we may observe different forms that appear to fulfill the same function, or, conversely, why there are similar forms in differing contexts.[20]

Having omitted historically informed understandings of

transformation,[21] synchronic functionalist analysts proceeded to derive a framework with which to explain the apparent lack of significant change in agro-environmental technology (slow, gradual changes are conceded) (Norman and Matlon 2000: 22-23). In these functional frameworks, peasant risk explains the apparent absence of change: poor farmers do not want to take risks because they are so close to the margin of life or death that a chance failure would be catastrophic. Since it is risky to try out new options (being literally dangerous to experiment), farmers are unable to develop new, more productive technologies: "The African farmer is interested first and foremost in yield stability. He prefers a safe but modest yield to an erratic high one. The chief threats to stable yields come from diseases and pests, which flourish in conditions of higher temperature and high humidity" (IITA 1977: 7).[22] The widely accepted wisdom is thus that, for poorer farmers, risks pose greater dangers, and hence returns on a risky investment must be greater in order to try it out—in short, the poorer, the more agronomically conservative.

A historical structural-functionalist account of peasant agriculture was even more appealing because it seemed to explain not only the stagnation of farming systems, but also parallel conservative socio-cultural traditions: the notion of risk at the farm level was superimposed with conceptions of traditional *socio-cultural systems*. Farmers who face both risk and poverty will develop cultural attitudes and social structures passed on from one generation to the next, that change (cultural, technological, social, etc.) is potentially risky and ought to be avoided. Conservative social systems automatically arise that function to reduce risk. A prime example is the redistributive mechanism, whereby a leading figure (a chief or feudal lord) in a community would accrue a small portion of each farmer's production and store it, so that should a mishap occur, this surplus would be available to bail the victim out. The head-person was restrained from personally benefiting from this exercise through social sanctions (such as gossip), which necessitated that they redistribute excess surplus, often done in elaborate feasts. Further, external relations were viewed as a threat to the head-person's authority, and to the stability of the community. What evolves, then, in functionalist accounts, is a bounded community with farmers using old, rudimentary technology, and a coherent, harmonious social system which has evolved and continues to exist only in order to reduce risk (Scott 1972; Migdal 1974).[23]

As farming systems analyses were premised upon the general inaccuracies inherent in ahistorical systems theory-based structural functionalism, they in turn expressed severe inadequacies in attempting to explain empirical reality. Because they ignored history and geography, farming systems studies could not therefore accurately account for diversity or change through time or space, and they consequently came to rely on static, stereotypical representations, abstracted from any larger context or internal differentiation.[24] For example, the farmer in the quote from IITA above is a bland, stereotypical, subsistence-oriented male. The ahistoricsm was evident in descriptions of farmers as "endowed" with sets of resources, refraining from actually examining the historical process of endowment. Likewise, there was little attention to internal differentiation (at regional, local or household levels) and its influence upon farming systems:

> The premise of the FSR approach is that farming households *are endowed* with different characteristics ... One of the first activities in FSR is to identify the groups of households within a target region that are "relatively *homogeneous*" in their characteristics, on the assumption that they will respond to new technology in a similar way (McMillan 1987: 295, emphasis added).[25]

The untenable assumption of homogenous farming systems has met sharp criticism from scholars of women in particular (Moock 1986; Poats et al. 1988; Feldstein and Poats 1989). Nevertheless, farmers were depicted as uniformly poor (a shared poverty), and as being so since time immemorial (aboriginal or primordial poverty), primarily because of "the technical element—climate and soil" (Norman and Baker 1986: 47).[26]

In addition to ahistorical systems theory-based functionalism, FS studies are also pervaded by evolutionism. At the turn of the twentieth century, Victorian ideology lent to the concept of evolution a distinctly unilinear and teleological overtone, in direct contrast to Darwin's original conception of evolution as a fundamentally contingent process, one not pre-determined, guaranteed or singular, but rather shaped by the convergence of multiple selective forces upon populations with diverse, heritable traits.[27] It is the unilinear, teleological conception of evolution that classic cultural anthropology picked up and made popular, through such cross-sectional "arm-chair" anthropological works as *The Golden Bough* (Frazer 1890), and it is this conception which farming systems studies latched on to.[28]

Farming systems studies suggest, in their teleological, unilear evolutionist aspect, that there are different stages of agriculture that represent different rungs on a single ladder of intensification. As population density increases (and also, perhaps, market access and favorable ecology), farmers switch from forest fallow, to extensive shifting cultivation, to bush-fallow, and then to increasingly sedentary and intensive methods. In fact, one IARC Annual Report even included a narrative of this sort with fill-in-the-blanks for a stereotypical Country X. Because such unilinear, teleological evolutionist frameworks ignore the influences of multiple variables and the ways that specific conjunctures of historical processes give rise to patterns of production, it consequently appears as if, for each change in the value of the few determining variables, there is *one* corresponding optimally functional farming system structure. Looking over the range of a variable's values (e.g. over population densities), then, there is a discernable single path or set of stages, a "ladder" of uniformly applicable discrete classifications or "rungs," through which all systems inexorably pass. All that remains after describing these stages, these ideal-type molds, is to take the collection of cross-sectional case examples—each of which has been measured according to a few values of a few variables—and plop them into their appropriate category or "stage" in the unilinear typology.[29]

Such farming systems studies that were flawed in their foundation in ahistorical systems theory-based structural functionalism and evolutionism nonetheless appeared particularly accurate and apposite when applied to Africa because they meshed well with prevailing misconceptions of African societies and ecologies. Colonial understandings of destructive shifting cultivators colored analytical lens (in fact, many analysts began their farming systems studies during or shortly after the transition from colonial rule). Also, Boserup and Ruthenberg's original studies had also been based heavily on a cross-sectional citation of secondary evidence in Africa.

Numerous studies began analyzing Africa in terms of the evolution of functional farming systems, influenced directly or indirectly by Boserup, Hayami and Ruttan, and Ruthenberg (Boserup 1999; Bryceson 1995). The studies reached their apex in the exemplary attempts by key figures in the original farming systems movement to explicitly apply their elaborate analytical frameworks in explaining patterns of technology (non) adoption in Africa. Thus Pingali, Biot and Binswanger's (1987) *Mechanization and the Evolution of Farming Systems in East Africa* "owes important intellectual debts to Ester Boserup and to Hans Ruthenberg" (vii).[30] Pingali et al. began their research "by narrowly focusing on the mechanization question," and specifically asking why Africa as a land-abundant continent has not widely incorporated the use of plows or tractors. They correlated mechanization in fifty cross-sectional examples with the two variables of population density and access to markets, and consequently "quickly ventured into carefully examining and testing grander evolutionary themes"(vii).[31] Similarly, McIntire, Bourzat and Pingali's (1992) study examined *Crop-Livestock Interaction in Sub-Saharan Africa* by considering correlations with variables of agro-climate and population density across thirty three cross-sectional examples.[32] Unfortunately, these African applications of farming systems studies have not been critiqued as thoroughly and explicitly that prior applications in Asia had.[33]

These applications of farming systems frameworks to Africa seemed so apposite because they meshed with sedimented ideologies of systems thinking, functionalism and evolutionism that had a peculiarly African character. The ideological bedrocks of ahistorical structural-functionalism and evolutionism upon which FS studies flourished arose from specific colonial discourses on African tradition and "tribal" society. Farming systems studies complemented and built upon these underlying currents and fortified a myth that was particularly *African* in two ways: socially and ecologically.

Firstly, FS's more general narrative of agro-environmental change appeared to be especially accurate when applied to Africa because there one could clearly identify the exact social systems that had evolved with a functional structure. Drawing largely off of a largely synchronic British structural functionalist anthropology of Africa (see footnote 23), there had been a focus on the bounded, coherent, and static nature of African social units at concentric levels of the "tribe", the village, and the household, corresponding to analyses of political systems, communities, and kinship respectively. People were seen to remain rooted in their place: there was an emphasis on exclusive "tribal identity" and sharp ethnic and territorial boundaries, "culture areas" as Kroeber (1931) and Herskovits (1930) called them, with typical "tribes" men remaining more or less in their area over their lifetime, within their village for much of their life, and in the household area for most day-to-day activities (Driver 1962; Malki 1997; e.g., Mensah 2000: 16). Political and economic forces were largely those within these bounded arenas. These units of "tribes", villages, and households were perceived (at least implicitly) as a largely uniform, unstratified set of homogenous households (Guyer 1981; Hyden 1983). Britain brought in its functionalist anthropologists to help codify and classify "tribes" and their customs.[34] Cultural customs and traditions were seen to change very slowly because they served an underlying function, and hence societies appeared static, existing today much as they had hundreds of years ago. Africa was hence seen to be a place without history, a "Dark Continent," partly because of Europeans" ignorance, partly racism and Victorian notions of superiority, partly because propounding the need for trusteeship over "primitive," "child-like" people served as a justification for colonial control for political and economic gain. Their farming systems had evolved slowly over the millennia by trial and error as natural processes of adaptation to the environments. Each "tribe" had a chiefly leader that demanded and enforced loyalty in exchange for redistributing wealth to cover nature's vagaries. These primitive societies and polities occupied a lower rung on the ladder of civilization (with Europe, and specifically England, at the top) (Comaroff 1984; Isaacman 1993; Moore 1994; Gupta and Ferguson 1997b; 1997c).

However, although we are accustomed to images of isolated "tribes", unchanged throughout the centuries, stubbornly holding onto traditional shared systems of belief and action, in contrast, African societies and traditions were deeply infused with multi-leveled power, inequality, politics, and above all, change and contestation (e.g., Wilmsen 1989). Mamdani (1996) has perhaps given the most elaborate exposition of how tradition was perceived, constructed and contested in Africa under terrain of unequal power inherent in indirect colonial rule. In order to balance the objectives of minimal administrative costs and maximum resource extraction and still maintain socio-political control, colonial powers sought to rule through what they believed were pre-existing socio-political structures. Administrators and anthropologists alike perceived the unit or "container" of traditional social organization to be "the tribe," which was "rooted" in a communally owned, bounded geographic area and headed by a chief (Mamdani 1996: 49-55, 107-170; Ranger 1983; Malki 1997; Vail 1988; Bryceson 1995; Bassett 1993), partly because indirect rule through the stable, functional socio-political systems therefore allowed "colonialism on the cheap," or, as Berry (1993), puts it, "hegemony on a shoestring": "By this dual process, part salvage and part sculpting, they crystallized a range of usually district-level Native Authorities, each armed with a whip and protected by the halo of custom" (Mamdani 1996: 49). In so doing, however, colonial administrators began making claims upon resources and meanings, thereby intensifying struggles and prompting increased investment in the social means of retaining access to land, trade and labor resources: "Administrators sought information on traditional social structures and identities in order to know how to apply customary rules in governing colonial peoples, while African colonial subjects *renegotiated* rules and social identities in order to cope with or take advantage of colonial rule and commercialization" (Berry 1993: 34). What colonial powers and popular anthropologists produced in the colonial period was thus not only a set of ideas about agriculture and the environment, but more importantly, a corresponding set of ideas about socio-political organization in Africa that found new light in the 1970s with the rise of the emerging Farming Systems Research paradigm (Vail 1988).

Farming systems studies, in addition to building off of *social* theory that had particular connections to Africa, also was shaped by, and reinforced, an *ecological* narrative specific to Africa. The specifically African ecological portion of the myth—that Africans practice (d) shifting cultivation—was fortified through work at another CGIAR research center, this one in Ibadan, Nigeria: the International Institute for Tropical Agriculture (IITA). African agriculture was seen to be largely comprised of shifting cultivation, in contrast to the high-density intensive Asian agriculture and the unequal Latifundio structure in Latin America. It was believed that as colonialism brought health improvements and pacification of warring "tribes," population had increased and was causing a breakdown in shifting cultivation, an abnormally rapid transition in stages of farming systems to which African farmers could not react fast enough.[35] Thus, while the foundational farming systems works recounted above do collectively emphasize population density, markets, and agro-climate, in effect the attention has come to focus exclusively on increasing population growth alone as the primary cause of breakdown.

Consequently, a broader form of farming systems research for Africa was embodied in IITA's "dual mandate," which consisted of increasing food production on the one hand, and maintaining a healthy natural resource base on the other—essentially, making the transition from what many perceived to be a now destructive system of shifting cultivation:[36]

> In writing a dual mandate for IITA—one that embraced the genetic improvement of major commodities as well as a search for viable alternatives to shifting cultivation—the Institute's founders showed a clear awareness that improving agriculture in the humid and subhumid tropics of Africa would require a somewhat different strategy from the one that had yielded such dramatic results elsewhere … What Asian agriculture had then, to a large

> extent, was a basically sound engine that badly needed a more potent kind of fuel ... This sound engine was precisely what the humid and subhumid tropics of Africa lacked for the most part. [The difference is one of] making fundamental changes in farming systems, as opposed to merely injecting new technological components into them (IITA 1992: 25-26).

In contrast to the new varieties used in Asia and Latin America, Africa would need a complete overhaul. IITA's major approach was to find an environmentally sound way to sedentarize Africans—to get them to stop having to move around and cultivate new land.[37] Through IITA, FS studies took on a specifically African ecological aspect, in addition to the specific socially functional, evolutionary and systems ideas.

The result is a fortified conventional narrative of change in African agro-environmental technology: Uniform, internally harmonious and externally bounded African "tribes" previously existed in a static ecological balance with nature using the only slowly changing rudimentary practices of shifting cultivation, and being governed by social norms to maintain this balance. Population growth was brought about by Western advances in health and colonial protection from internecine conflict and has led unfortunately to environmental, agricultural, and social crises spiraling out of control (see diagram, page 494).

This notion of agro-environmental change was disseminated the world over by donors, governments and researchers as FS orientations became institutionalized in academic expertise, policy practice, and public common sense around the globe:

> FSR quickly progressed from a rough amalgam of ideas regarding technology development to become dominant concept in the 1980s in both the literature of agricultural development and the organization of agricultural research and extension systems (Tripp et al. 1990: 384).

The FSR methodology to closely investigate why small farmers had not benefited from agricultural technology became increasingly utilized in Africa after the devastating famines of the 1970s and 1980s. Social scientists (comprising anthropologists, geographers, and sociologists) based in classic structural-functionalism and evolutionism were hired *en masse* to conduct studies and evaluations and disseminate the new FS practices and technologies to development projects, with even those interpretations invoked highly selectively and in a reductionist and simplified manner by economists and natural scientists.[38] IARC staff in the Economics Department carried out many of the initial FSR Projects in the 1970s with African governments agencies (mainly in Kenya, Zambia, Tanzania, Zimbabwe, and Ghana) (Collinson 1982).[39] The centers trained the staff of national governments, while key farming systems progenitors themselves went on to influential positions in powerful international organizations.[40]

FS studies also proliferated throughout books, journals, and conferences. The CGIAR Secretariat published a key report on FSR in 1978 (TAC 1978), while Westview Press published a series of half a dozen texts on farming systems and agricultural research. In the early 1980s, a USAID-funded FS program at Michigan State University and the Economics program at CIMMYT put out a number of key farming systems papers.[41] There have been annual conferences every year since 1981, mainly in Kansas, Arkansas and Michigan. USAID began immense collaborative research and degree programs with several universities, including Florida, at Gainesville, and Michigan State. Around the world, USAID funded seventy six farming systems projects between 1978 and 1988, forty five of which were in Africa (Norman and Matlon 2000: 18). Many journals began featuring pieces on FS, and some devoted part, or all, of their issues to FS studies—including, *Human Organization* and *Agricultural Administration* (later *Agricultural Systems*). From 1986 to 1991, *Experimental Agriculture*, a journal used by many IARC staff, featured an FS series of thirty four articles, and eventually a separate *Journal for Farming Systems Research-Extension* was formed. There have been a number of accounts of "the institutionalization" of FS in agencies around the world.[42]

The dominant explanation of change in African environments and agriculture—that population growth is the primary force driving the breakdown in previously functional shifting cultivation systems with consequent poverty and degradation—has been revived by at least three factions: crude-populists, pseudo-environmental neo-Green Revolutionists, and proponents of biotechnology. These movements have added little to or changed the basic postulate and are here thus treated more in how they recapitulate the basic themes developed and fortified in the preceding eras.

Cruder varieties of populism emphasize that farmers are well stocked with a cache of indigenous knowledge that they have built up over the millennia, and hence do not need any new technology.[43] The current emphasis on indigenous institutions and knowledge as opposed to foreign impositions perhaps comes as a response to the denigration and occlusion of African capacities during colonialism as well as in the post-war heyday of modernization theory (Amanor 1999: chapter 1; Kizza 1999).

The crucial assumptions by populists, progressives and conservatives alike are that Africans once lived in harmony with nature (the primordial equilibrium of shifting cultivation) and that social institutions are essentially static stores of this creative knowledge. Given the insights by Berry and Mamdani above, the origin, practice and longevity of pure, bounded, indigenous techniques are questionable (see also Bebbington 1996; Fairhead and Leach 1996). Populists overlook the colonial and post-colonial politics of custom, indirect rule, and social institutions, and access to resources (Ribot 1999), recapitulating prevailing conceptions of indigenous people as evolutionary adaptations to their environments.[44]

The Vicious Cycle of Population Growth -Induced Poverty and Environmental Degradation

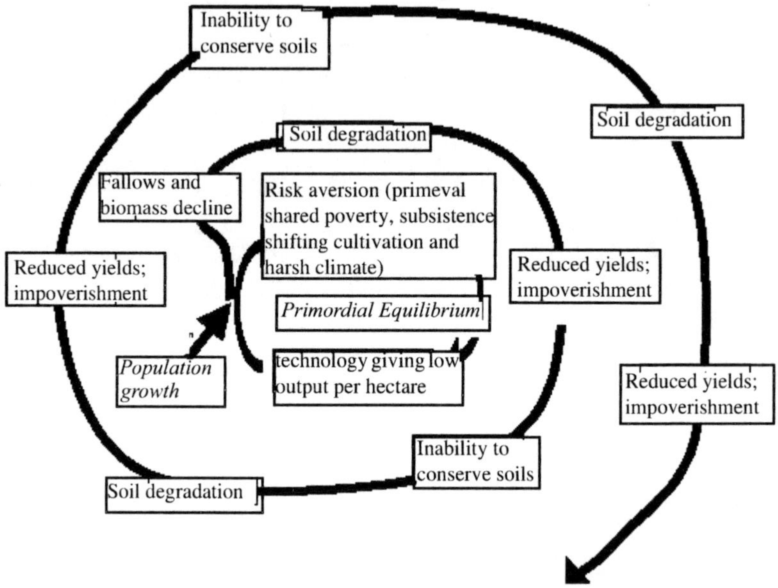

Reviving The Myth in The Late 1980s and 1990s: Crude Neo-Populism, Pseudo-Environmental Neo-Green Revolutionists, and Biotechnology Public Relations

The dominant myth is also being reproduced, ironically, in a recent co-optation of environmentalism. In an odd twist of history, Green Revolutionists have co-opted their environmentalist critics by now suggesting that high productivity Green Revolution technology not only solves population-growth induced scarcities in food supplies, but also that, by allowing more food to be produced from less land, the technologies are essential to preserving nature. As development institutions have been pressed to adopt environmentalism, at least rhetorically, [45] the CGIAR's focuses shifted.[46] However, the same arguments used in the Green Revolution continue, albeit now with "sustainability" added in to form *The Doubly Green Revolution* (Conway 1998): clean, high-yielding technology will save land by allowing farmers to intensify on their existing plots. Borlaug, the veritable father of the Green Revolution, has become the unofficial representative, and is summarized well here by Easterbrook (1997: 76, 78):

> By producing more food from less land, Borlaug argues, high-yielding farming will preserve Africa's wild habitats, which are now being depleted by slash-and-burn subsistence agriculture ... In developing nations where population growth is surging, high-yield agriculture holds back rampant deforestation of wild areas.

Green Revolution technologies are now said to be not only consistent with, but also essential to, environmental protection. Usually statistics from India are cited as proof of how hi-tech agriculture can save land in Africa.[47] Through these attempts to re-portray Green Revolution technology as environmentally friendly, analysts implicitly revive the dominant narrative of African agro-environmental change. That is, without new high-yielding technology, population growth leads the ever-increasing droves of shifting cultivators to bring their ruthless practice into fragile forests and arid zones.[48]

In addition to crude populists and pseudo-environmentalists, the biotechnology industry and its apologists have used images of African farming systems in crisis, this time in order to assuage public apprehension over new agricultural biotechnologies. In response to consumer concern in Europe, transnational agro-chemical companies have recently joined together in a $50 million advertising campaign telling us that biotechnology will bring an automatic twenty five per cent gain to help beleaguered African women break out of the vicious cycle of low yields, poverty, hunger and degradation (Barboza 2000). Biotech and friends accuse anyone voicing concerns about the environmental, economic, and social cost of such technologies as spoiled, romantic and paranoid Luddites, insensitive to the benefits biotech can bring to hungry shifting cultivators, and ignorant to the protection it can provide for the nature cultivators are endangering.

The biotech industry and research organizations flagrantly play the "Africa card," using decontextualized images and stories of stereotypical African farmers, in publicity reports, radio and TV commercials, newspaper and magazine ads, radio talk shows, and academic and policy conferences and papers.[49] The myth is given public legitimacy by expert testimony from such figures as Per Pinstrup-Anderson, the head of the Washington, D.C. based International Food Policy Research Institute.[50] Likewise, Gordon Conway, head of the Rockefeller Foundation, uses a similarly quaint story about a stereotypical poor woman in Kenya (Conway 2000). Biotech companies have even financed genuine, authentic Africans as spokespersons for the continent in conferences, prestigious journals, and newspapers around the world, imploring the West to let them have access to the lifesend biotechnology—most notable is Florence Wambugu, a Kenyan sweet potato scientist whose work has been funded by Monsanto and other industry groups.[51] Each pro-biotech statement begins by restating the conventional narrative that growing numbers of shifting cultivators have run out of land to fallow, resulting in degradation and poverty, and hence the need for productive biotechnologies. In sum, through crude populism, pseudo-environmentalist Green Revolution rhetoric, and biotechnology public relations, the dominant narrative of African agro-environmental change has been revived with vigor in the late 1980s and throughout the 1990s.

Why the Myth is Wrong: Political Ecology as an Understanding of Social Relations Spanning Multiple Historical and Geographical Scales

History, power, agency, and social relations shape processes of change in agro-environmental technology. Demography, formal markets, and ecology do not drive agricultural production and environmental degradation/preservation alone—they are fundamentally social processes.[52] Two examples: from 1960 to 1984, more than three and a half million black South Africans were removed from their lands; federal timber concessions to large export firms in Ghana have shaped land-use.[53] A single explanation or a neat typology for the continent is both impossible and inaccurate, so what follows next is a sort of laundry list of the factors that can influence agro-environmental change—the particular combination and forms of the factors will vary according with each historical and geographical context. The factors can be roughly grouped into fourteen categories:

- armed conflict
- economic policy
- gender relations
- labor relations
- inter-generational relations
- land distribution and tenure
- foreign aid
- mineral resource extraction
- timber harvesting and production
- rural development projects
- migration
- indirect and direct state mandates
- settlement schemes
- national parks

To be clear, I am *not* suggesting that population has not grown or is not a strong influence. Rather, population, or any simplistic set of "factor markets," are not the only influences. The dramatic increases in population have come largely in specific patterns during the twentieth century, a century of tremendously dynamic change—demographic change and agro-climactic variability exist in a context of multi-layered social and physical forces operating at multiple spatial and temporal scales. Of course at a very broad level, the issue of whether increasing rural population and markets require some sort of intensification seems so clear as to be self-evident. The difficulty, however, lies in the numerous, heterogeneous patterns, processes and forces between say two poles (not necessarily one leading to the other) of completely empty land and highly urbanized land. Different farmers cultivate and fallow different parts of their land in different ways according to the specific formation of their historical and geographical contexts.

The complex configurations of multi-leveled social forces make the actual pattern of agro-environmental practices much more heterogeneous than suggested by the conventional narrative of now dysfunctional shifting cultivation (Silberfein 1998). The myth homogenizes production systems and population distributions—whose tremendous variability contradicts discrete categories of production or urban/rural—by reading macro-processes of change (i.e. degradation and poverty) merely as the aggregate result of choices that are individually rational at the micro-level but cumulatively destructive (continued shifting cultivation under over-population).[54]

Conclusion: Social Relations, Power, and Method

Agro-environmental change in Africa has been a dynamic, diverse, flexible, and social phenomenon, but through a variety of forces has come to be widely seen as a stasis being broken down by population growth alone. There are four general methodologies used to legitimate the conventional narrative: (1) aggregated statistics, used at broad, national, or continental levels—summed, tabulated, and manipulated with little or no caution despite logistical and political difficulties in their collection; (2) functionalism and evolutionism depend on emphasizing *cross-sectional* study, rather than longitudinal study (a practice which may reflect the come and go nature of development experts and academic interests); (3) an almost exclusive reliance on models and quantitative equations to understand processes (either environmental or social)—a practice which probably stems from the economistic training of many analysts; and (4) deducing the past from the comparisons of snap-shot statistics, surveys and models, rather than examining the historical record (through written, material, photographic, and/or visual sources). The conventional myth can give rise to global crisis scenarios that are quite at odds with the local realities on the ground, and lead to ill-fated prescriptions for green revolution projects (Borlaug and Dowswell 1995).[55]

Despite mislabeling of the dominant myth as historical by its proponents, if politics is understood as essential to historical processes, then in excluding politics from its consideration, the dominant myth essentially denies Africans' history. Boserup

(1983: 3), for instance, suggests that "human history can be viewed as a long series of technological changes." Such unilinear, technocratic, functionalist version of history is not history at all (cf. Berry 1997), it is teleological evolution wherein population and biology determine everything, and people react mechanistically with no agency, no memory, no foresight, and no relation to one another. In fact, the exclusion of agro-environmental politics is itself one way of making a claim upon resources (see footnote 3). In reading out politics, analysts deny Africans their history, and ultimately, their humanity.

A better understanding would see change in techniques and tools of agricultural production and environmental management in terms of trajectories or transformations (Batterbury et al. 1997), for the same reasons as Hart suggests in arguing for using the concepts of multiple trajectories of capitalism and modernity (1996; forthcoming). Attention to "everyday politics and the exercise of power" allows one to see the "multiple, nonlinear, and divergent *trajectories*" of development (Hart 1997: 56). The key point is that "social processes are the product of interaction at multiple societal levels, they cannot simply be read off or deduced from organizational forms or relational structures" (Hart 1997: 63). So, "Instead of asking *what* are the rules or relational structures that produce particular outcomes," "the focus of the processual approach taking shape in agrarian literature is instead on *how* negotiation and contestation take place within and across multiple social arenas" (Hart 1997: 63, 56). This approach provides a profound critique of much of the literature on agro-environmental technology change in Africa, and points to the more general limitations of ahistoricism, systems theories, structural-functionalism, teleology, and unilinear evolutionism (Batterbury and Bebbington 1999).

Endnotes

1 I here use the term "shifting cultivation," sometimes popularly known by the misnomer "slash and burn" agriculture, to describe systems at a range of cropping frequencies, what Ruthenberg (1980) calls "R values," such as bush-fallow, ley farming, etc. since the

variation identified in systems is usually in frequency and not qualitative type.

2 In the course of a research project, I have reviewed over 1,500 technical agronomy and policy reports that repeat this same assertion, and do so without citing any hard evidence, but because those are the shared assumptions and lens through which they have been trained to perceive anything. See especially Cleaver and Schreiber (e.g. 1994); IITA (1977); Harrison (1996). I conceive of technology as not just a material item, but as involving the spatial and temporal organization of production; it is "the sum of knowledge of the means and methods of producing goods and services" (Bannock et al. 1992: 419). For example, technology is not only machetes and tractors, but also seasonal work rhythms and crop arrangements.

3 For rural Java, Pincus' (1996) empirical work refutes the "agrarian myth," building off of Hart (1986). Also see Aston (1985). On reworking see Collinson (2000); deGrassi (2000a, chap 4). On politicized re-workings in Asia, see Hart (1986); Brass (1995); Gupta (1998); Pincus (1996). This paper is an outline of a framework synthesizing conclusions gathered from empirical data that are too lengthy to be presented directly here.

4 with decisions being governed rather by conservative social traditions, emotion, physical predilections towards inaction and laziness, and/or mental incapacity for abstract logical reasoning, including memory and foresight (e.g., Collinson 1981: 438; Neumann 1998: 108).

5 In much of Francophone West Africa colonial rulers used ideas about scientific forestry to divide use-rights, allocating rights to non-commercial trees to rural subjects, while valuable commercial species rights were given to urban-based citizens (Ribot 1999).

6 See also Neuman's (1998) chapter on the "Society for the Preservation of the Fauna of the Empire" (SPFE).

7 Technical change would come about by targeting those "progressive" farmers (usually wealthier, more educated, male, etc.) who had broken out of the shackles of rural traditions and had come to embrace modern science and technology.

8 See Schultz (1964); Norman (2000: 294).

9 "Shrinking supplies of land and other natural resources would provide motivation to invent better means of utilizing scarce resources or to discover substitutes for them" (Boserup 1983: 5).

10 Ruttan had been influenced, in his position as an economist with IRRI, by early Green Revolution dilemmas. To clarify, for the

purposes of this essay, I consider II as a theory whose general orientation is subsumed under the general category of "Farming Systems" (FS) studies. FSR/E is a *methodology* spawned by, and reinforcing, FS studies.

11 Obviously much of this thinking was affected by the rising ideas of "appropriate technology" (e.g., Schumacher 1973) and participation. See especially Hayami and Ruttan (1972: 174-5).

12 Particularly important was the Economics Department at the Mexico-based International Corn and Wheat Improvement Center (CIMMYT), formally started in 1969.

13 In this approach, researchers and extensionists used five key FSR/E procedures: diagnosis through surveys and tabulation, planning, on-farm testing, assessment of new options, and extension of the chosen package (Tripp et al. 1990: 387).

14 Merrill-Sands et al. (1992).

15 See McCorkle (2000); Fine's (2001) more general argument about disciplinary selectivity is applicable here.

16 I have elsewhere explored in more detail how FSR came to be limited and diffused (deGrassi 2000a). See also Norman and Gilbert (1982: 20); CIMMYT (1992. 47); Berry (1986); Little (1985); Jennings (1988).

17 See von Bertalanffy (1968); Olivier de Sardan (2000); van Eijk (2000); McCorkle (2000); Schiere et al. (1999).

18 Ecological anthropologists emphasized, for instance, what they thought were isolated societies in static equilibrium with their environment (Vayda 1969). Popular influences included early fumbling with computer programming and such popular texts as James Lovelock's (1979) book about the earth as a single, complex living whole, *Gaia*, a superorganism with various balanced channels of positive and negative feedback. The systems view was also prevalent in the numerous *Integrated Rural Development Projects*, which focused on developing all aspects of rural life at once since all aspects life were seen to be interconnected.

19 See Gupta's (1998) use of non-teleological strategy in his analysis of transformation in Indian agriculture.

20 On inability to explain diversity, see Hart (1986: 7, 8, 192, 194).

21 What Ruttan and Hayami recognize, in their own words, as "the lack of a theory of action" (1995: 31).

22 IITA hence advertised its work as increasing productivity by developing and introducing varieties that reduced risk because they were less susceptible to the vagaries of disease and pests.

23 Resulting from early functionalist French sociology (Durkheim),

British Anthropology (A.R. Radcliffe-Brown and his British cohorts, Meyer Fortes, E.E. Evans-Pritchard, C.D. Forde, and others), and post-WWII U.S. sociology (Talcott Parsons). Additionally, in the 1970s, interest grew about the causes of rural-based revolutions in what were thought to be stable, functional societies (Wolf 1969; Migdal 1974; Paige 1975). A debate ensued over whether farmers "averted risk" and followed moral sanctions (Scott 1976), or whether they individually took risks in order to take advantage of market opportunities (Popkin 1979).

24 Skocpol comments by ignoring broader social and political forces and structures, the risk debate has dwelled on "ahistorical conceptions of the nature of "the peasant" as a supposedly general human type" (Skocpol 1982: 214, 229). Indeed, a problem with theories about ideal-types, such as "*the* African farmer," and static, bounded systems is that it denies people a geography and history (for a potent critique, see Wolf 1982; cf. Asad 1987). This is not to say that risk is some fiction made up by academic and policy discourse. Cooper (1993), Bryceson and Jamal (1997), and Bryceson et al. (2000) help put conceptualizations of peasantries in historical context. In fact, the British functional anthropologists neglected to factor into their studies of the structural minutiae of African kinship and political systems the massive disruption caused by two-hundred years of global slave trade and half a century of colonial "pacification" and rule.

25 On homogenous sub-groups, see also Collinson (Collinson 1980: 382); Berry (1993: 6); Hildebrand (1977: 8); Norman and Baker (1986: 46).

26 See Ferguson (1990) on "aboriginal" poverty; Norman and Baker (1986: 47) on "shared" poverty; Geertz (1963) on shared poverty in Indonesia, with a critique by Hart (1986).

27 See Greene (1959); Gould (1992; 1994; 1997).

28 See Erickson (1998); Gupta (1998); Ferguson (1997); Fabian (1983); McCorkle (2000: 303). There is, however, a long history of critiques of evolutionism, from anthropologists and those in development studies, who emphasized the importance of historical, empirical investigation in contrast to theoretical inferences from scattered case studies and shaky statistics, including Franz Boas in anthropology and Gershenkron and Albert Hirschman in development studies Evolutionism reappeared in the 1950s debates between the neo-evolutionisms of Leslie White and Julian Steward.

29 As Boserup (1983: 6) notes, "The method used is *cross-country* comparison of recent data; the main purpose of the exercise is to

throw light on *historical* interrelations between population size and technology." Also, while Binswanger and Pingali (1987: 287, and elsewhere) do make the qualification that farming "systems are not mutually exclusive. It is quite possible for two or more of the systems to exist concurrently," in practice they treat their case material largely as occupying one of six exclusive, discrete stages.

30 This book builds off, and contributed to a study and series of articles by Binswanger and Pingali (1986, 1987, 1988), Pingali and Binswanger (1984, 1987, 1988), Pingali, Bigot and Binswanger (1987), as well as Ruttan and Thirtle (1989).

31 There is often token treatment of other factors (a few paragraphs or pages), perhaps to appease critics. For example, Pingali et al. (1987) do mention transport, immigration, urban connections, land laws, and history (meaning slave trade and "tribal warfare"). However, these are only noted briefly (5 pages), treated only as secondary influences affecting the primary locus of change (area per capita), and are largely ignored throughout the rest of the analysis.

32 Other significant recent works on induced intensification include Tiffen et al. (1994), Bourn (1994), and Turner and Hyden (1993) among others. A series of, I believe, fatal critiques to Bourn's piece have been compiled in ODI (1994). Other contrasting longitudinal studies include Scoones (1994); Vaughn and Moore (1994); Fairhead and Leach (1996; 1998); Watts (1983b); Guyer (1997); McCann (1995).

33 Although see Hutton and Robin; Lele and Stone (1989); Berry (1993). On Asia, see Hart (1986); Hart et al. (1989); Koppel (1995); Koppel and Oasa (1987); Pincus (1996). There is a broad range of works on Africa, from Bernstein, Cooper, Davidson, etc., whose arguments, although elegant and I believe more accurate, often only *implicitly* address and critique the dominant myth of agro-environmental change as it has developed and been applied to Africa.

34 Many of the ideas stemming from colonial rule derived from, and in turn contributed to, the discipline of anthropology. Anthropologists were often complicit in the colonial enterprise, at least discursively, although there were progressive anthropologists working against, or at least to dampen, colonial forces, albeit in specific idioms (Asad 1973; Moore 1994; for Zambia see Ferguson 1999: 24-37; Crehan 1997: 53-75).

35 See Harrison (1996, esp. pp. 41-42) for a particularly exemplary description. The actual causes, nature and implication of demographic change are too complex to address here.

[36] IITA began explicitly as a reformist program, with Haldore Hanson arriving in 1971 straight from his position as the embroiled director of CIMMYT ready to make the Institute's mandate and scope broader—more regions, more crops, more R&D techniques.

[37] One (largely unsuccessful) method developed was alley cropping, where in farmers would plant their crops in rows between rows of leguminous shrubs—shrubs that would provide nitrogen to the soil, and would provide biomass for mulch, fodder for livestock, and for burning as household woodfuel. Alley cropping encountered unexpected difficulties and complexities. See McGuiness (1993); Carter (1995); Degrande and Duguma (2000).

[38] DeWalt (1985); Tripp (1985); Olivier de Sardan (2000); McCorkle (2000); cf. Escobar (1991); cf Little and Painter (1995); cf. Fine (2001).

[39] Much of the formative FSR work was done by Robert Tripp (an anthropologist), Derek Byerlee (an economist), and Mike Collinson (also an economist) of the Economics Department at CIMMYT.

[40] In addition to Winkleman chairing the CGIAR, Byerlee went on to become director of the Economics Program, and is now at the World Bank, while Binswanger is head of Agricultural Development in Africa at the World Bank.

[41] The MSU papers include Gilbert et al. (1980); Eicher and Baker (1982); Crawford (1982); Collinson (1982). Important CIMMYT works are Byerlee and Collinson (1980); Byerlee et al (1982); Byerlee and Tripp (1988).

[42] Accounts of the development of FSR include Collinson (2000); Tripp (1991); Moock (1986); Singh et al. (1986); McMillan (1990). FAO continues to publish its series on farming systems (now totaling more than 15 works). FSR practitioners would likely point out, in addition to the picture I have painted here, the diversity of experiences of FSR, and the difficulties of actually implementing it successfully, and meeting the intended objectives. I wholeheartedly agree, but the overall effects of the FSR were to disseminate this paradigm; as bits and elements of these often times detailed descriptions are selectively invoked, and the cumulative effect is to reproduce the myth.

[43] Key populist figures, whose work has commonly been mimicked, appropriated, developed and distorted in many ways, include Robert Chambers and Paul Richards, with the latter authoring the landmark 1985 book *Indigenous Agricultural Revolution*. See Kitching (1982); Richards (1983; 1985); Watts (1983a); Chambers et al. (1989); Scoones and Thompson (1994); Ayittey (1998).

44 Populist views differ from conventional accounts only in laying the blame for the recent breakdown, which populists put not so much on population growth as exploitative capitalist relations and poorly conceived development projects.

45 Rachel Carson's (1965) book *Silent Spring* on agro-chemical pollution made the wider public aware of environmental questions. Other more ecological books followed, including Ehrlich's (1968) *The Population Bomb*, as well as the gloomy quantitative modeling of Meadows et al's (1972) *The Limits to Growth* and Newsom (1995); Adams (1995). By the 1970s environmental concerns had gained increasing visibility in discussions of population, economy and resources. Concerns over resource scarcity increased in the 1970s with the UN Stockholm conference on the environment, newly formed UN Environmental Program, Sahelian famines and publicity around desertification. For many, OPEC's oil price shocks and the inflation of the 1970s seemed to indicate a rise in prices resulting from population-induced natural resource scarcity. Similar shifts affected the World Bank (Wade 1997).

46 Embodied in a key report (TAC 1988) and a significant book by Conway and Barbiers (1990). Both books focused on the environmental consequences of the overuse of Green Revolution technologies of pesticides and fertilizers on agricultural land, water sources, and directly on human health.

47 "...if LISA [Low Input Sustainable Agriculture] food-production rather than the Green Revolution food-production model had been pursued in South Asia since the 1960s, 44 million hectares that are now under forest would instead have been plowed and cropped. This suggests that biodiversity has been enhanced in Asia by the yield-increasing and land-saving Green Revolution model" (Reardon 1998: 449). The piece by Reardon from which this quote is taken is explicitly and importantly premised upon the work of Binswanger and Pingali.

48 This environmentalist co-optation has been legitimated through shaky studies of soil properties, scare tactics (Weight and Kelley (1999), among others, even revert to comparisons with the Dustbowl), and bulldogging by fertilizer industry representatives and sympathizers in the major international development institutions (Scoones and Toulmin 1999; Keeley and Scoones 2000; Henao and Baanante 1999).

49 In addition to the current conference, see also two recent ones at Harvard, The International Conference on Biotechnology in the Global Economy: Science and the Precautionary Principle, 22-23

September 2000; Raising Agricultural Productivity in the Tropics: Biophysical Challenges for Technology and Policy, 16-17 October 2000.

50 Per Pinstrup-Anderson (1999) writes, "It must be hard for an African farmer to understand the debate currently raging in Europe about the use of modern biotechnological methods in agricultural research. Health and environmental risks may dominate discussions up north, but she wonders why they don't discuss the enormous possibilities that biotechnology offer … the dreams of the poor Africa farmer may not materialize [because] some European countries oppose the use of modern biotechnology for agriculture and food." See also the National Public Radio debate transcript at *http://www.sciencefriday.com/pages/2000/Apr/hour1_041400.html.*

51 Wambugu (1996; 1999); Conway and Sechler (2000); Adamu (2000); Gressel (1996); Pretty (1998).

52 "In production, men not only act on nature but also on one another. They produce only by co-operating in a certain way and mutually exchanging their activities. In order to produce, they enter into definite connections and relations with one another and only within these social connections and relations does their action on nature, does production, take place" (Marx 1847).

53 For citations and more empirical examples for each of these factors see deGrassi (2000b).

54 This is an approach known as atomism, first countered by Durkheim. It could also be considered a case of the ecological fallacy: "using population or group-level data to draw conclusions about individuals" (Murray and Taylor 2000: 1739, citing Blalock 1982). Such an approach parallels and draws from neo-classical economics and more recent "New Political Economy" or neo-instrumentalists: see Bates (1981); cf. Stein and Wilson (1993); Berry (1984; 1993, esp p. 14).

55 Fairhead and Leach (1996; 1998); Gray (1999); Leach and Mearns (1996); Richards (1996); van Horen and Eberhard (1995); Turner (1999).

CHAPTER 23

FOOD SECURITY, PRODUCTION MODELLING AND EARLY WARNING SYSTEM IN ZIMBABWE: APPLICATION OF REMOTE SENSED PERCEPTION

Denis Rugege

Introduction

Agriculture in Zimbabwe is comprised of two main sectors, viz small-scale mainly communal farming and large-scale commercial farming. Figure 1 shows the land distribution between the small-scale, mainly communal sector and the large-scale commercial sector. In 1992 (Rukuni and Eicher 1994; Whingwiri et al. 1992), the small-scale sector was estimated to include 1 million households utilizing 16 million hectares (42% of pre-independence communally owned land); 52,000 households utilizing 3.3 million hectares under post-independence resettlement schemes and 8,650 households were found on 1.2 million hectares of privately owned land. In contrast, only 4,500 commercial farms held 11 million hectares of prime agricultural land under private ownership.

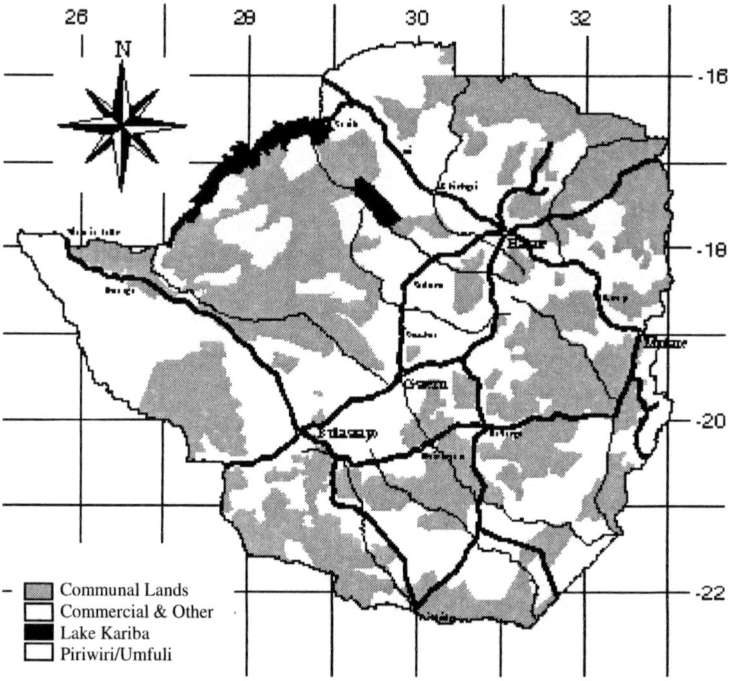

Figure 1. Map of Zimbabwe showing the location of communal lands and the areas of Piriwiri and Umfuli that were surveyed for the study of yield constraints of maize-based land use systems.

Developed under the previous Rhodesian regime, the largely whites-owned large-scale farming sector is/was relatively successful by international standards with adequately invested modern technology. Small-scale production is also hampered by a backlog of institutional and infrastructural development from colonial neglect evidenced by the exclusion of the main road network and urban areas from the communal lands.

Agro-ecology

Zimbabwe was first delineated into five agro-ecological zones (known locally as Natural Regions I to V) by Vincent and Thomas (1960; described in Anderson 1993). Whingwiri et al. (1992) and Rukuni and Eicher (1994) also used "effective rainfall" as the main differentiating criterion. Later, modifications of the zone boundaries were based on the so-called rainy pentad defined as "the centre one of the three pentads (consecutive five days of rain) that together receive more than 40mm of rain and not more than one of the three having less than 8mm". Over 90% of communal farming areas are located in Regions III, IV and V where cropping is done under severe biophysical constraints; see Table 1. Various researchers have suggested that rainfall deficits of 13 to 19 years periodic cycles in Southern Africa are related with the El Nino effect (Scoones et al. 1996; Eastman 1996) rendering drought a regular occurrence.

Table 1. Distribution of communal lands including resettlement areas by Natural Regions (after Rukuni and Eicher, 1994; Whingwiri et al. 1992)

Natural Region	Mean Annual Rainfall (mm)	% Total	% Communal Land	Production Reliability	Farming Practice
I	>1000	1.8	0.8	Reliable	Tea; coffee
II	750-1000	15.0	7.8	Reliable	Crop & animal production
III	650-800	18.6	17.2	Unreliable	Crop & animal production
IV	450-650	37.8	44.9	Unreliable	Crop & animal production with high drought risk
V	<450	26.7	29.3	Erratic & very low	Crop & animal production with high drought risk

Maize-based land use systems in communal lands

Maize plays a dominant role in the food economy of Zimbabwe, accounting for 88% of the national coarse grain production. The crop provides about half of the calorie intake of the average Zimbabwean (Rukuni and Eicher 1994). Maize is the main crop grown in the communal lands in all Natural Regions except Natural Region-V. The maize crop is usually intercropped with smaller proportions of sorghum, pumpkin and sunflower. The fields are typically small (0.2-0.5 ha) and irregularly shaped. Communal farmers grow maize for subsistence as the staple food, but also for cash when there is surplus. Early maturing hybrid seeds suitable for areas with little and erratic rainfall have been adopted. Farm operations in communal lands largely rely on livestock for power. Cattle and donkeys provide draught power for ploughing, planting, weeding and for transportation of inputs to the fields and of harvested products to the homestead and to the markets. Manure from the animals is a valued supplement to soil fertility.

About two-thirds of Zimbabwe is covered with sandy soils that are derived mainly from granites. Like most of the heavier textured soils, these sandy soils are inherently deficient in nitrogen, phosphorous and sulphur (and in many cases in potassium as well). The sandy soils have low buffering capacity, low nutrient reserves, and are relatively acidic. Compaction, capping and low water holding capacity are common. Management inputs for fertility maintenance and for crop protection against pests and diseases are low, largely imposed by lack of capital or credit facilities.

Food Security Early Warning Systems

Zimbabwe has had to import maize in increasing amounts since the early 1980s, with imports reaching a crisis level during the 1991/92 drought when lack of timely action cost the country an estimated US$ 340 million or 11% of the (then) gross domestic product.

Events and actions during this period provide an opportunity to evaluate the effectiveness of the early warning system. Rukuni and Eicher (1994) argue that the devastating maize shortage of 1992 reflected failure on the part of senior decisionmakers to initiate import procedures once impending shortages were anticipated and reported by the Grain Marketing Board and the National Early Warning Unit. This failure to act resulted in a fourfold increase of the cost of maize-meal as the government embarked on emergency and naturally much more expensive means of processing and distribution.

Questions then arose as to why the authorities failed to heed the warning. Walker (1989) compiled the following attributes that contribute to an information system's value and by which its usefulness can be judged:

- Clarity: The information must be unambiguous i.e., quantitative when describing quantities.
- Accuracy: The information might be quantitative and clear but of no value if inaccurate.
- Significance: The actual implications of the information must be clearly put across to the intended recipient.
- Timeliness: Crop yield predictions tend to gain accuracy closer to crop maturity but lose in timeliness.
- Adequacy: An information system should aim at providing all the facts necessary to make a decision. The fewer the facts it provides, the less adequate it is.
- Validity: Validity of information is illustrated by the example that economic surveys may gather high-grade data on the price and availability of food in the market but if most of the poor and those likely to suffer operate outside the market, the information is of little use.

Any combination of deficiencies in the above mentioned attributes could have impacted on the decision making process that led to the food security crisis in Zimbabwe in 1992. Surely, the decision-makers would have known that the accuracy of crop yield predictions presented by NEWU was not established. Although there is no guarantee that decision-makers would be prompted to take appropriate action, an improved early warning system would provide information, within its affordability, that combines maximum clarity, accuracy, significance, adequacy and validity. This case study considers the accuracy attribute in the early warning system employed by NEWU.

Monitoring and accurate forecasting of maize production and yield put Zimbabwe in a better position to timely procure needed supplies at least cost when production falls short and to profitably dispose of surpluses in times of plenty. An effective early warning system is clearly of great value for identifying emerging trends, while a mechanism to translate this information into timely action should be in place. The accuracy with which crop yield estimates are carried out is vital to the management and planning of a country's food security system. This is becoming even more important now that the maize industry of the region is increasingly confronted with free market mechanisms (Brink 1996).

Food Security Early Warning in Zimbabwe is the responsibility of the National Early Warning Unit (NEWU) of the Department of Agricultural Technical and Extension Services (AGRITEX) under the auspices of the Southern Africa Development Community (SADC) Regional Early Warning Unit (REWU). REWU is composed of National Early Warning Units (NEWUs) from each of the SADC member-countries. In Zimbabwe, the National Early Warning Unit is composed of experts from the Department of Agricultural Technical and Extension Services (AGRITEX), the Department of Meteorological Services and the Central Statistics Office. NEWU provides early warning information to government and other decision-making agencies on expected grain and other agricultural products.

NEWU employs the FAO Soil-Water Balance Model adapted for Zimbabwe in its crop production forecasting. The model evaluates weather, soils and crop data and calculates a crop "Water Satisfaction Requirement Index (WRSI)", in order to estimate yield potentials while the crop is still on the field. A WRSI value of 100% is interpreted as "no yield reduction due to water stress" whereas less than 50% is taken as an indication of crop failures. The concept is adopted from an empirical linear yield-reduction relation (Saney et al. 2001).

The Famine Early Warning System (FEWS), a REWU collaborator, developed a grid-based WRSI model for Southern Africa that estimates the water-limited yield potential based on start and end of crop season, precipitation and potential evapotranspiration according to the Penman-Monteith equation (Saney et al. 2001). In the monitoring mode, the WRSI is produced till the most recent dekad using observed data in the growing season. The alternative forecasting mode continues the WRSI calculation until the end of the growing season using long-term potential evapotranspiration and rainfall data. A linear correction factor was added to the model to account for yield reduction caused by water logging.

The model has been in use for a number of years notably for forecasting expected national productions of maize, sorghum, millet, groundnut and sunflower. However, although the model is tuned with observed yield data, the accuracy of predictions is as yet unclear. Forecasts of planted areas and production are made by agro-statisticians, using statistical surveys based on a combination of subjective and objective crop area and yield assessments and measurements.

Limitations of the WRSI model

Crop yield estimation based on the "Water Requirements Satisfaction Index (WRSI) model" as employed by NEWU has its limitations:

- WRSI applies index values quantified with empirical relationships that link rainfall and vegetation index data with (observed) yields. Significant limitations emerge from neglecting vital physiological and environmental mechanisms that control crop growth. Empirical relationships are not universally valid and hence results may be location-specific and not transportable.
- It is assumed that rain is the main factor determining yields; other yield constraints are not taken into consideration.
- The accuracy of crop area estimates is not known.
- It is assumed that rain gauge (point) data represent the area for which the estimate is made. In areas with sparse distribution of rainfall gauging stations such as the communal lands this is questionable,
- Where applied, the accuracy of satellite-derived rainfall estimates is unknown.

Required and achievable improvements

NEWU can improve the quality of crop yield estimates through the following:

- Assessment of the accuracy level of satellite-derived rainfall estimates that are applied in modelling the water-limited production potential,

- Deployments of mechanistic models in the establishment of reference production that take into account the physiological and environmental processes that control crop growth. Production situation models (PS-1, PS-2) calculate the biophysical and the water-limited production potentials in order to assess production levels that are possible under respective constraint-free and rain-fed conditions,
- Understanding of actual farmer's production and yield levels in order to determine yield-gap and the constraints that cause it,
- Application of the actual farmer's production situation model (PS-n) that estimates actual production and yields using the crop canopy-ambient air temperature difference concept.

Hypotheses

The accuracy of current crop yield estimation in Zimbabwe is impaired by:

- Using the empirical Water Requirement Satisfaction Index (WRSI) model in the current crop yield monitoring and forecasting; WRSI solely considers a soil-water balance, ignoring other yield limiting and yield reducing factors,
- The sparse network of meteorological and rain gauge stations,
- The unknown accuracy of crop area estimates,
- The unknown accuracy of satellite-derived rainfall estimates applied in the WRSI model,
- The lack of understanding of constraints, which cause the gap between theoretical yield potentials and the actual farmers' yields.

The accuracy of crop yield estimation can be improved by:

- Determining what is possible in terms of production levels under rain-fed conditions using the production-situation-based analytical approach instead of empirical methods,

- Understanding yield-limiting and yield-reducing factors that prevail under actual farmers conditions,
- Estimating actual production analytically by directly linking crop production to the difference between ambient air and remotely sensed canopy temperature (PS-n).

Assumptions

- Actual yield data can be obtained with good reliability from interviews with farmers,
- Water-limited potential production can be linked to rainfall data predicted from Meteosat satellite data and interpolation techniques,
- Actual crop yield can be linked to crop canopy parameters derived from NOAA AVHRR satellite data.

Improving Reference Production Modelling using Remotely Sensed Precipitation

Regional crop yield forecasting makes use of area-specific crop state variables derived from satellite remote sensing. Available methods apply index values quantified with empirical relationships that link satellite data with (observed) yields. Furthermore, the accuracy of the satellite-derived variable values is often not well established. Significant limitations emerge from neglecting vital physiological and environmental mechanisms that control crop growth. Mechanistic models including PS-1, PS-2 and PS-n, dynamically describe and apply these mechanisms but are applicable only at field scale. The solution proposed here builds on the possibility to derive crop state information from satellite data and adopt appropriate strategies for its use as substitute field-level variables. In this case study, the derivation methodology and accuracy of the Meteosat-derived Rainfall Estimates (RFEs) are assessed. Trial PS-2 model runs using Meteosat-derived RFEs are presented later on in the proceeding section.

Rain-Gauge related Limitations

Rainfall values measured with rain gauges have limitations that negatively affect their application in regional crop growth analyses. The most disturbing of these limitations are:

Point measurements: Rain gauges provide point measurements. Areal values are required for regional analyses. However, convective effects tend to be localized. A single gauge may be representative of only a very small area. Figure 2 (Flitcroft et al. 1989) presents the variation of rainfall over a 10km square on 20th July 1985 in Niger, West Africa. Although the average rainfall over this square is 50mm, there is wide variation in values over very short distances.

If only one of the gauges was chosen to represent the 10km square, any value between 30mm and 78mm could be obtained. Nonetheless, rain gauge values are often taken to be representative of very large areas or used in simple interpolation procedures to construct isohyetal maps.

Thin plate spline has been used to incorporate elevation as an independent variable in addition to latitude and longitude to interpolate monthly mean precipitation within a standard error of 10% (Hutchinson and Corbett 1993). However, precipitation error of interpolated surfaces is strongly related to gauge network density (Lebel et al. 1997; Hutchinson and Corbett 1993; Flitcroft et al. 1989). Large interpolation errors would be expected in such areas of sparse gauge network density as the communal lands of Northwestern Zimbabwe.

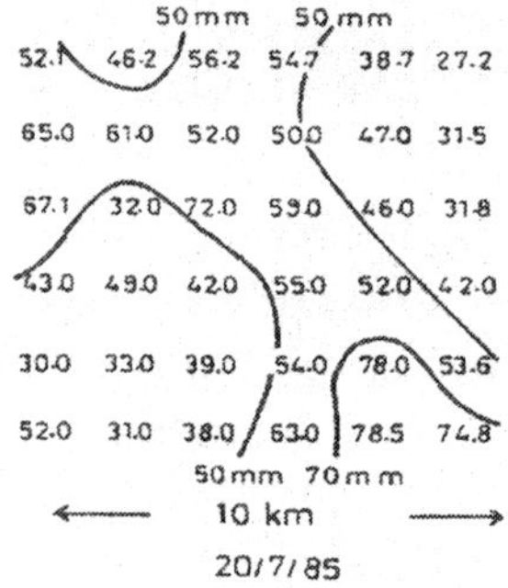

Figure 2. Rainfall variability within a 10km square on 20 July 1985 in Niger, West Africa (Source: Flitcroft et al. 1989).

- Rain gauge distribution. Rain gauge distribution is often corrupted by such deliberations as "ease of access" because recordings are made daily. This means that fewer gauges are placed in uninhabited areas and more are located along communication lines such as roads, railways and rivers. Figure 3 shows the distribution of meteorological and synoptic stations in Zimbabwe. Note the relative density of gauging stations along the major road network and the almost total absence in the communal areas.

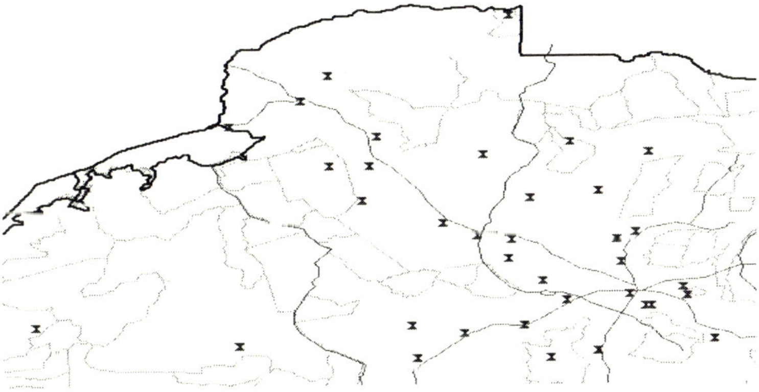

Figure 3: Rain Gauge Distribution in Northwestern Zimbabwe is Denser Along Major Roads.

- Measurement errors: Rainfall measured in strong winds may only be 60% of the actual rainfall because of turbulence created by the gauge itself (Grimes et al. 1998; Seed and Austin 1990). Other measurement errors include human errors such as omission of decimal points and assigning rainfall to the wrong day during the repetitive transcription of measurements before central archiving.
- Time delay: From the time of recording, rainfall data may take several weeks to months before becoming available for use.

Prospects for Satellite-Derived Rainfall Estimates

Meteorological satellites provide data, which can be used to make empirical estimates of rainfall based on cloud top temperatures. Data from meteorological satellites are available free of charge for non-commercial use. Initial costs incurred in the installation of data reception equipment and computer facilities for data processing are normally affordable. The data are received in near real-time by a relaying process from the satellite sensor through a receiver to the processing computer.

However, satellite-based rainfall estimates also have limitations. Sources of error in satellite-based rainfall estimations are discussed by Flitcroft et al. (1989) and by Grimes et al (1998). Errors stem mainly from the fact that data from geostationary satellites are only indirectly related to rainfall (e.g., cloud top temperature).

Sources of error include:

- Inability to monitor local variations in rain storm intensity
- Error commission on account of non-rain bearing cirrus clouds
- Error due to mismatch between gauge position and center of pixel (co-location error).

Published research on quantification of the above mentioned errors are not available yet.

The Cold Cloud Duration (CCD) Concept

Rainfall in the tropics can be related to clouds whose temperatures are colder than 235K (–38 °C). The GOES Precipitation Index (GPI) was subsequently introduced to convert hours of persistence of a cloud pixel at threshold temperature (235K) into an estimated rainfall value. This is the so-called Cold Cloud Duration (CCD) concept. A detailed description of various techniques for rainfall estimation from Meteosat satellite data is given by Snijders (1991); Grimes et al. (1998); Bonifacio et al. (1993) and Milford et al. (1994).

Two methods that are based on the CCD concept are in operation in Southern Africa. The SADC Regional Remote Sensing Project (RRSP) applies a method developed by the NOAA Climate Analysis Centre (CAC). The CAC technique uses daily rainfall data from the Global Telecommunication System (GTS), mainly oceanic floating buoys, to model a relationship with the CCD through statistical regression. The regression equation is then used to tune the GPI estimate. This approach works well over oceans but under-estimation is reported for land areas mainly on account of (ignored) land surface characteristics such as topography. A second method, developed by the TAMSAT group, University of Reading, is also based on the GEOS Precipitation Index. Grimes et al. (1998) and Flitcroft et al. (1989) provide a detailed description. This method was adopted in this study, to derive rainfall estimates for use in crop production simulations at PS-2 level.

Rainfall Estimation Methodology adopted in this Study

Half-hourly Thermal Infrared (TIR) image data at a (coarse) resolution of 5.5 to 7 km (smallest picture element, or pixel, 30 to 50 km^2) are captured by a Meteosat Primary Data User Station (PDUS) located at the Institute of Natural Resources, University of Natal, South Africa. The images are automatically processed using TAMSAT Rainfall Estimation Software (TRES). CCDs are derived at several temperature thresholds (usually 5 thresholds from –30°C to –70°C in steps of 10°) in the following way:

As TIR images come in every half-hour, new sets of cold cloud detections are added to the CCD images until a day is complete (24hrs). This image shows, for each pixel, the total number of time slots when cold cloud is detected. This is subsequently converted to hours and shows the relative persistence of cold cloud for each area. The process is also run for meteorological dekads to produce dekadal CCD images. From these dekadal CCD images, the software creates a dekadal rainfall estimate (RFE) image based on historical calibration rules pre-loaded into the system. The calibration rules are based on a number of past seasons of data and hence the quality of the rainfall estimate is influenced by the amount of data used in the calibration.

The calibrations are carried out by first determining which CCD threshold best discriminates rainfall occurrence. CCD data at this threshold is then used; pixel CCD values at gauged pixels are paired with corresponding gauge values and used in a robust regression scheme. Weighted least squares regression between median gauge rainfall within CCD classes (as narrow as data allows) and mid-class CCD, where the weights are proportional to the number of data pairs in each CCD class is then done. The final model is:

$$RFE = a*CCD+b \qquad \text{if } CCD > 0$$
$$RFE = 0 \text{ if } CCD = 0 \qquad \text{if } CCD = 0 \qquad \text{(Eq1)}$$

Where:

a and b are coefficients and CCD is at the determined temperature threshold.

The regression coefficients are derived on a spatially and temporally varying basis (Milford et al. (1994). These "TAMSAT method" calibrations and threshold temperatures for each month were developed in the mid-90s for (most) SADC countries, in workshops funded by FAO and USAID and are currently in use.

Comparison between Gauged Rainfall and Satellite-Derived Estimates

To make proper comparison between rainfall measured at a particular rain gauge and Meteosat-derived RFEs, it is necessary to determine the average rainfall in an area around the rain gauge that corresponds with the satellite pixel. Meteosat pixels are close to 36 km^2 in the situation discussed here.

The calculation of a pixel average rainfall value for the area around a rain gauge can be carried out using block-kriging procedures (Lebel et al. 1997). The procedure generates an estimate of rainfall over a predefined pixel size and an estimate of the error from a linear combination of gauge rainfall measurements.

The following equation describes the estimation of area average rainfall as a weighted mean of point values:

Where:

$$P_A = \sum_{j=1}^{n} \lambda_j p(x_j) \qquad \text{Eq2)}$$

P_A	is area average rainfall,
A	is pixel-sized area around a gauge,
$p(x_j)$	is observed rainfall for gauge j at position x_j,

λ is weight dependent on the degree of spatial variability of the data. The weights of these measurements are derived so as to give unbiased estimates of minimum variance (Curran, 1988).

Applying "*ordinary block-kriging*" of point estimates can produce estimates over finite areas (equivalent to satellite pixels). The derivation of the block-kriging system used in this study is provided hereafter. The method is used to generate estimates of pixel-size rainfall, but only for pixels that contain a gauge.

The semi-variance γ(h) is calculated as half the average squared difference between gauge values separated by a distance h according to the following equation:

$$\gamma(x_1,x_2)=\gamma(h)=\frac{1}{2}\frac{\sum_{N}(p(x_1)-p(x_2))^2}{N} \qquad \text{(Eq3)}$$

Where:

N is number of gauge pairs at locations x_1 and x_2 separated by distance h.

The set of values of γ(h) as a function of h is known as the experimental variogram.

Block-kriging requires that a model be fitted to this empirical semi-variogram; only a few are suitable for continuous variables. Basic types of variogram model are described in Journel and Huijbregts (1978). The three most commonly used are the spherical, exponential and linear models (Bogert et al. 1995, Journel and Huijbregts 1978, Curran 1988). The spherical and exponential models show a linear decline of spatial dependency of the variable with increasing distance until a sill value where it disappears. The linear model applies if spatial variability perpetually increases with increasing distance, never reaching the sill. The spherical model reaches the sill at a finite distance while the exponential model reaches its sill asymptotically according to the following mathematical formulations:

- Spherical:

$$\gamma(h)=\begin{cases} C_0+C_1\left[1.5\cdot\left(\frac{h}{a}\right)-0.5\cdot\left(-\frac{h}{a}\right)^3\right], h<a \\ C_0+C_1, h\geq a \end{cases} \qquad \text{(Eq.4)}$$

- Exponential:

$$\gamma(h)=C_0+C_1\left[1-\exp\left(-\frac{h}{a}\right)\right] \qquad \text{(Eq.5)}$$

The exact shape is determined by the values of the coefficients C_0, a and b, which characterize the spatial variability of a field. The coefficients are the theoretical nugget C_0, sill C_1 and range a. Model fitting can be done by simple visual adjustment, statistical methods of varying degrees of complexity (weighted least squares, maximum likelihood).

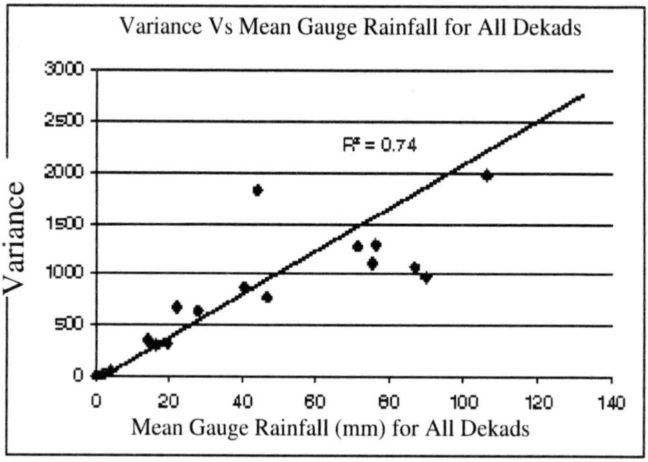

Figure 4: Increase of variance with increase in mean gauge rainfall indicating a seasonal effect.

Deriving a semi-variogram model for North-Western Zimbabwe

The usual practice in block-kriging interpolation would be to apply a variogram model from a single dataset. However, rainfall data are often characterized with several time steps with only a small number of data points in each step. The following limitations of the variogram model fitting are encountered:

- the work involved in fitting dozens of variogram models is enormous,
- often the rainfall field is sparsely sampled and the estimates of the variogram parameters may be poorly estimated,
- when many rainfall fields are available, treating each one separately makes only partial use of the global information contained in the whole data set.

Nevertheless, a single variogram would not be appropriate, as it would neglect seasonal and meteorological effects. Figure 4 shows increasing variance with increasing mean dekadal gauge rainfall. The high correlation shows the time variation in the mean rainfall confirming the seasonal effect. In Zimbabwe, the rains begin in October and fall until April with peak downfall in December and January. A practical solution suggested by Lebel et al. (1997) is to introduce a scale factor to the variogram as in the following:

$$\gamma(\mathrm{h},t)=\alpha(t)\cdot\gamma^*(\mathrm{h}) \qquad \text{(Eq.6)}$$

Where:

$\alpha(t)$ is a scale factor derived as the sample variance at each time step t

$\gamma^*(h)$ is a time-invariant scaled variogram

Ideally, a scaled variogram is required for each localized rainfall generating process. However, for practical applicability, one scaled variogram is derived for each calendar month. All dekadal rainfall data in a month are grouped together so as to have T fields of N data points each. A climatological variogram is then derived using all T*N data points in the following expression:

$$\gamma^*(h)=\frac{1}{2}\frac{\sum_{T}\sum_{N}(p^*(x_1,t)-p^*(x_2,t))^2}{N^*T} \qquad \text{(Eq.7)}$$

Note that in the climatological variogram formulation, rainfall data are standardized by their mean and variance by replacing p with p^*:

$$p^* = \frac{(p-\overline{p})}{\sigma_p} \qquad \text{(Eq.8)}$$

Where:

σ_p is dekadal rainfall data variance
p is observed rainfall
$\overline{p}$ is mean pixel area rainfall

This means using a variogram of fixed shape whose sill is 1.0 after variance normalization (Journel and Huijbregts 1978) and scaling it by the variance at each time step. The climatological variogram used for each time step t (dekad) takes the form:

$$\gamma(h,t)=\sigma^2(t)\cdot\gamma^*(h) \qquad \text{(Eq9)}$$

Note that this leaves the range unchanged, it simply scales the nugget and sill, so that the value at the sill becomes the sample variance. The climatological variogram achieves a higher accuracy by using a lot more data in its estimation. Another advantage is that the operators do not have to derive a variogram for each new time step. The fixed scaled variogram is simply scaled-up by the sample variance of the new incoming dataset. However, due consideration should be given to the possibility that the sample variance is poorly estimated by the available data.

A climatological rainfall variogram for the season 1997/98 was obtained from the dekadal rainfall data set of 37 stations in and around the study area bounded by the coordinates (27.7, -15.5) and (31.5, -18.5), Northwestern Zimbabwe. The derived variogram model in Figure 5 shows a high nugget indicating very significant short-range variability. The sill is reached at about 100km.

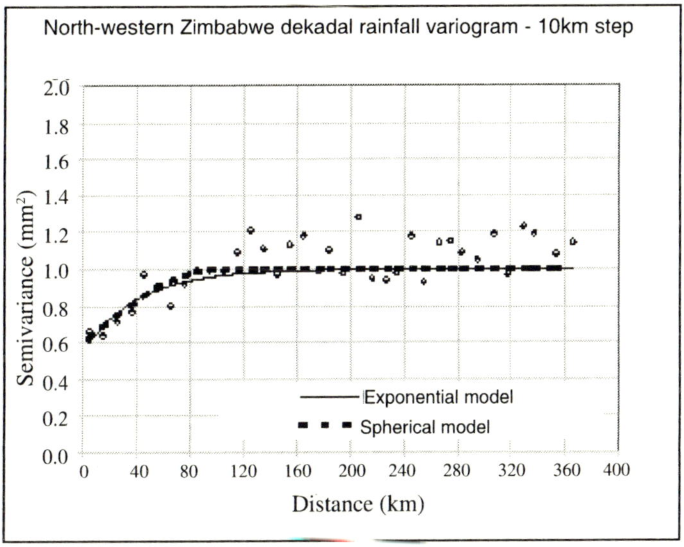

Figure 5: Mean seasonal rainfall variogram (1997-98) model for North-western Zimbabwe.

Estimating area average rainfall using block-kriging

Block-kriging assumes that an estimate (point or area) can be obtained from a linear combination of point measurements, i.e. the estimate (e.g. pixel average rainfall) is obtained as a *weighted mean* of point values (e.g. gauges in and around the pixel).

The block-kriging weights depend on the degree of spatial variability of the data as defined by means of the model semivariogram. Block-kriging weights are functions of the variogram model and depend on its type and on the value of its parameters.

Expressing the areal rainfall P_A as a linear combination of point values:

$$P_A = \sum_{j=1}^{n} \lambda_j p(x_j) \qquad \text{(Eq.10)}$$

The two basic conditions leading to the derivation of the block-kriging system are:

- the estimate is unbiased
- the estimate has minimum error (estimation variance)

These two conditions lead to the weights being calculated from a system of linear equations outlined below. The final result is:

$$\sum_{j=1} \lambda_j \gamma_p(x_i x_j) + \mu \bar{\gamma}_p(x_i, A) \qquad i = 1, \ldots, n \qquad \text{(Eq.11)}$$

With:

$$\sum_{j=1}^{n} \lambda_j = 1$$

The sets of equations above constitute the *block-kriging system.* The estimation variance is given by:

$$\sigma_A^2 = \sum_{i=1}^{n} \lambda_i \bar{\gamma}_p(x_i A) + \mu \bar{\gamma}_p(A, A) \qquad \text{(Eq.12)}$$

σ_A^2 is an estimate of the mean square error associated with P_A. Its square root will give an estimate of the uncertainty associated with the pixel rainfall estimate. In the above system of equations:

- μ is a dummy variable (Lagrange multiplier).
- $\gamma_p(x_i, x_j)$ is the rainfall variogram function between points x_j and x_i.
- $\bar{\gamma}_p(x_i, A)$ is the mean rainfall variogram function between point x_i and pixel A
- $\bar{\gamma}_p(A, A)$ is the mean rainfall variogram function within pixel A

For point block-kriging, to obtain estimates at a point location x_0 (rather than pixel A), the system is similar, only is replaced by $\gamma_p(x_i, x_0)$, the variogram function at the distance between a data point, *i,* and the estimate location. Also the term in the estimation variance vanishes as a point has infinitesimally small area.

The above system can be expressed more concisely in matrix form as:

$$K.\lambda = k \qquad \text{(Eq.13)}$$

To solve this equation for the weights , the k vector of point to pixel semivariances is multiplied by the inverse of matrix K, i.e. $\lambda = K^{-1}.k$ λ. The error variance in matrix form is:

$$\sigma^2 = \lambda^T.D + \lambda(A, A) \qquad \text{(Eq.14)}$$

Given a set of point data, the above system can be solved for the weights and hence suggests estimates of the areal average (or point value in the case of point block-kriging) and its estimation variance. The block-kriging system discussed above is solved with standard methods of matrix algebra. In the case of point estimation that is where the complexities end.

However, for areal (pixel) estimation there are areal mean variogram function terms in the system: — climatological rainfall variogram function between point and area/pixel *A* and — climatological rainfall variogram function within the area/pixel *A*. These terms are given by:

$$\bar{\gamma}_p(x_i, A) = \frac{1}{A}\int_A \gamma_p(x_i - u)\,du \qquad \text{(Eq.15)}$$

$$\bar{\gamma}_p(A, A) = \frac{1}{A^2}\int_A\int_A \gamma_p(u - v)\,du\,dv \qquad \text{(Eq.16)}$$

For rectangular areas (e.g., satellite pixels) these integrals can be derived by reduction to lower order integrals, to be solved by numerical integration (e.g., Gaussian quadrature). They can also be derived by simple summation over the discretized pixel. For irregular areas such as river catchments, discretization is the convenient approach.

Results and validation of Meteosat-derived RFEs

Figure 6 shows, for 524 dekads from 37 stations, Meteosat-derived RFEs against gauged rainfall (left scatter-plot) and Meteosat-derived RFEs against the block-kriged pixel average of gauged rainfall (right scatter-plot). Less scatter and stronger agreement between pixel-average rainfall and satellite estimates are apparent, with a 20% increase in R^2. However, notice the considerable reduction of gauge values of up to 100mm when averaged over a pixel-sized area in a block-kriging process. Global bias is illustrated by Figure 7; Table 2 shows Meteosat RFEs-gauge rainfall residuals versus Meteosat-derived RFEs and Meteosat RFEs-kriged pixel averages of gauged rainfall residuals versus Meteosat RFEs. Reduction in scatter (-1.3 to –0.9) is again evident in the residuals from kriged pixel averages of gauged rainfall.

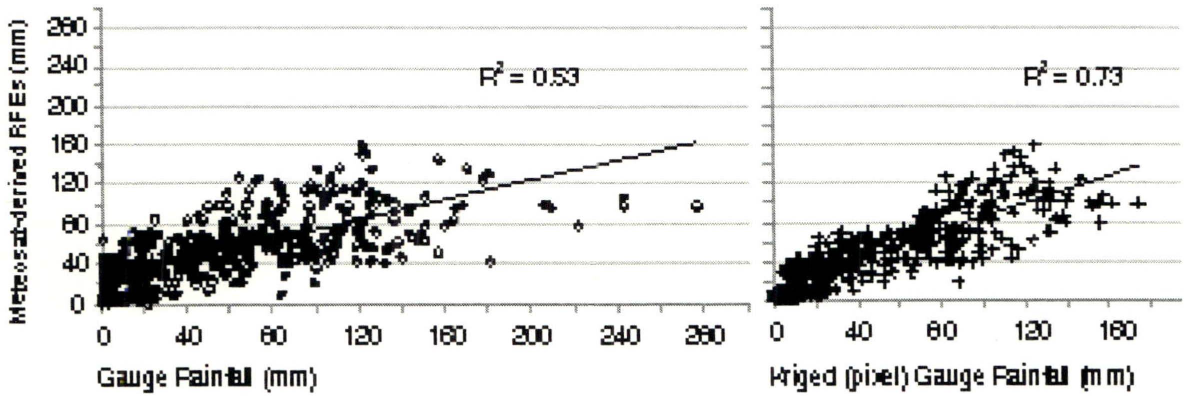

Figure 6: Scatter plots of gauge rainfall versus Meteosat-derived RFEs (left) and block-kriged (pixel) area mean rainfall versus Meteosat-derived RFEs (right). Note the 20% improvement in R^2 in the latter scatter plot due to reduced gauge rainfall variability brought about by block-kriging.

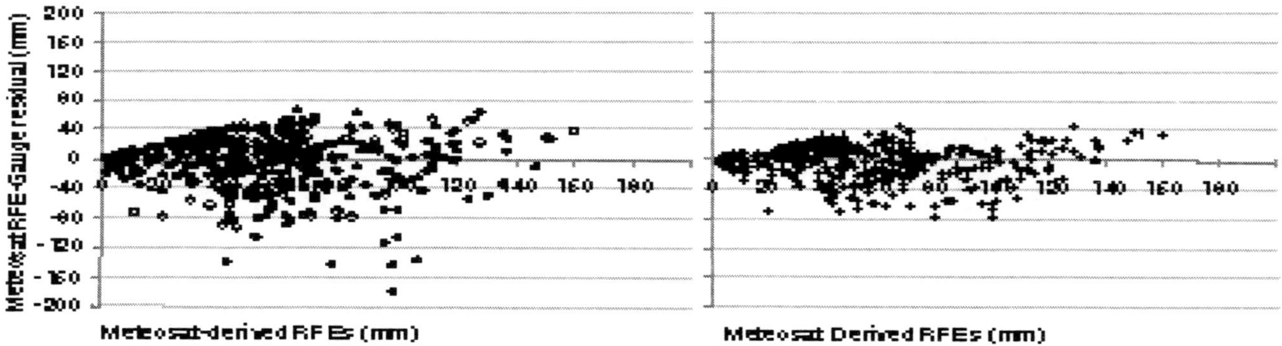

Figure 7: Scatter plots of Meteosat RFEs-gauge rainfall residuals versus Meteosat RFEs (left; bias = -1.3) and Meteosat RFEs-block-kriged gauge rainfall (pixel mean) residuals versus Meteosat RFEs (right; bias = -0.9). Note the reduced scatter and reduced tendency of residuals to increase with increase in RFE values.

Table 2: Global bias illustrated by Meteosat RFEs versus gauge rainfall (-1.3) compared with Meteosat RFEs versus block-kriged (pixel mean) rainfall (-0.9) and their respective rmse values

	Gauge (mm)	Meteosat-derived RFE (mm)
Bias	-1.3	-0.9
rmse	30.7	19.3

NE = 1.66

To better evaluate these differences, some simple statistical relations are used in both the satellite-point gauge and satellite-kriged gauge pixel comparisons. The statistics were:

$$bias = \frac{1}{N}\sum_{i=1}^{N}\delta_i \qquad \text{(Eq.17)}$$

$$rmse = \sqrt{\frac{1}{N}\sum_{i=1}^{N}\delta_i^2} \qquad \text{(Eq.18)}$$

Where:

$\delta_i = s_i - r_i$ is the residual of the satellite estimate and the gauge/pixel rainfall.

For satellite-pixel comparison a further parameter (normalized error) is of interest:

$$NE = \sqrt{\frac{1}{N}\sum_{i-1}^{N}\left(\frac{\delta_i}{\sigma_i}\right)^2} \qquad \text{(Eq.19)}$$

Where:

σ_i is block-kriging error estimate.

This parameter permits to evaluate how the satellite-pixel residual compares with the pixel error estimate. If the satellite-pixel residual is comparable to the gauge pixel block-kriging error, then NE tends to 1 and the satellite residual is on average of the same magnitude as the block-kriging standard deviation. The results show that both biases are close to 0, but as already obvious from the scatter plots using the pixel values leads to a much smaller *rmse* for the satellite data.

Finally some examples of comparisons for specific locations within the study area are presented. Two gauging sites, Karoi with a good agreement between the gauge rainfall and satellite derived estimates and Tengwe with a poor agreement are selected to illustrate the localized nature of convective rainfall as found in the study area. Figure 8 shows that both gauge rainfall and block-kriged pixel mean rainfall values are well correlated with Meteosat-derived RFEs.

However, a sharp contrast was observed between the correlations of gauge rainfall and block-kriged pixel mean rainfall with Meteosat-derived RFEs for the Tengwe site. Block-kriging of gauge rainfall improved the R^2 from 43% to 74%.

Figure 9 shows inter-comparison time-series plots of gauge rainfall, kriged gauge pixel rainfall and satellite-derived rainfall estimates for Tengwe synoptic station.

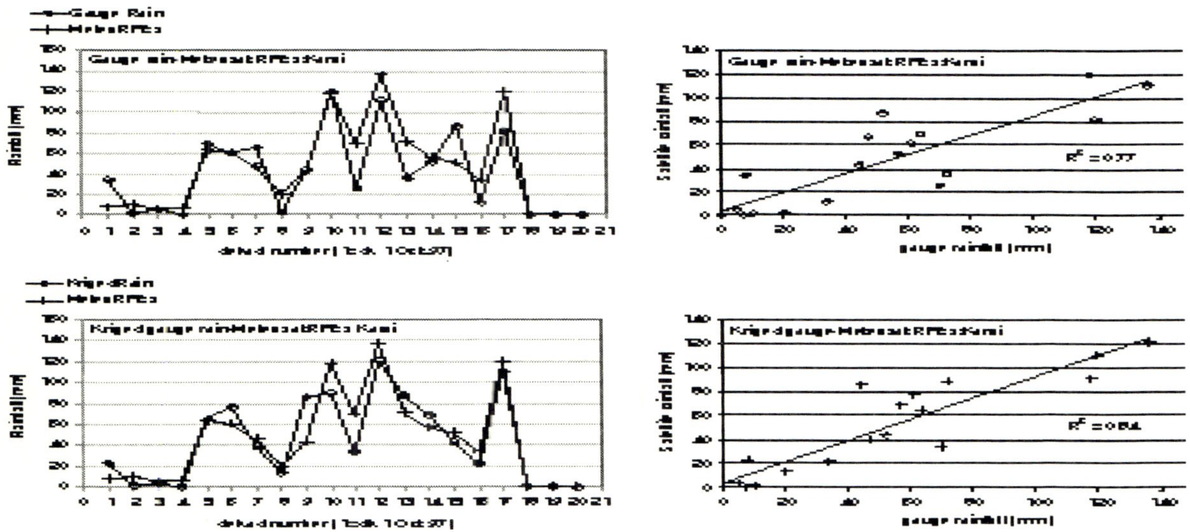

Figure 8: Karoi: Time-series comparison and R^2 of Meteosat-derived RFEs versus gauge rainfall (top) and Meteosat-derived RFEs versus block-kriged (pixel mean) rainfall (bottom). Although the R2 improvement was only 7% for this site, a better match between Meteosat-derived RFEs and block-kriged pixel means is evident throughout the time-series.

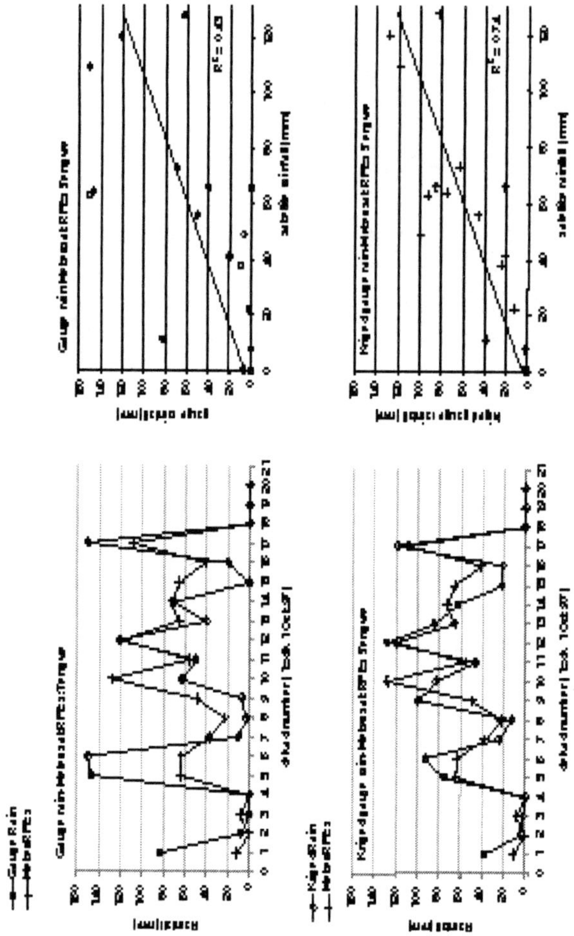

Figure 9: Tengwe: Time-series comparison and R^2 of Meteosat-derived RFEs versus gauge rainfall (top) and Meteosat-derived RFEs versus block-kriged (pixel mean) rainfall (bottom). Note the marked improvement in R2 from 43% to 74% when RFEs are plotted against block-kriged pixel mean rainfall.

Merits of Meteosat-Derived RFEs

The foregoing has shown that Meteosat-derived rainfall estimates are well correlated with area average (pixel) gauge rainfall. It is an additional merit of Meteosat-derived rainfall estimates based on CCD that they are spatially "consistent" whereas geostatistical analysis suggests that gauge-based area rainfall estimates degrade away from the gauge at distances shorter than the shortest inter-gauge distance. For areas with sparse rain gauge networks as found in the study area, Meteosat-derived rainfall provides an indication of rainfall distribution.

It is a further advantage of Meteosat-derived RFEs that the estimation procedure is automated and operational in the Southern African region and that such estimates are therefore easier and quicker to obtain. In the next section, the integration of Meteosat-derived RFEs in the modelling of water-limited production potentials is discussed.

Model Time-Step versus Available Data

Meteosat geo-stationary satellites provide meaningful rainfall estimates at best over 10 days and more reliably over a month (Dugdale et al. 1991 in Lebel et al. 1997). Available algorithms derive RFEs as dekadal values from Meteosat CCD data. However, dynamic crop models require daily precipitation values.

Disaggregation of 10-day rainfall estimates to daily values has been achieved using the third-order Markov rainfall generator. Disaggregated daily values showed no significant difference to the independent gauged daily rainfall values (p=0.51 for the 1997 rainy season and p=0.53 for the 1998 rainy season). Disaggregated Meteosat-derived rainfall values were used to update the value of the precipitation variable in simulations of the water-limited production potential (PS-2) of maize.

Disaggregating dekadal rainfall estimates to daily values

Methods for probabilistic disaggregation of *monthly* rainfall totals into *daily* time-series are commonly based on the notion that the probability of observing an event depends on the occurrence of previous events. Figure 10 presents a relational diagram of the final procedure followed in disaggregating RFEs events. Algorithms have been developed based on the Markov Chains (Groen, 2001; Jones and Thornton, 2000, 1999, 1997). The theoretical background and mathematical considerations involved are well explained and documented in Groen (2001), Jones and Thornton (1999) and Jones and Thornton (1997). In this study the software package MARKSIM (Jones and Thornton 2000) implementing the third order Markov weather generator was used to disaggregate Meteosat-derived dekadal RFE into daily values that are statistically similar to the observed values.

Climate normals derived from 9200 stations worldwide were grouped into 664 clusters. Rainfall model parameters for each of these clusters are derived from input monthly climate normals of rainfall, maximum and minimum air temperature, diurnal temperature range and station elevation and latitude. The program then assigns a cluster to a specified location using interpolated climate surfaces at a resolution of 10 minutes of arc (18km x 18km) and evaluates the parameter values for that point. The climate surfaces are derived from the NOAA dataset TGOP006 (Jones and Thornton, 2000) by spatial interpolation using a lapse rate to correct temperature for elevation effects.

The output weather file contains stochastically generated daily values of rainfall, solar radiation and extreme temperatures for a specified year and location (18km x 18km pixel centred around a specified latitude and longitude). Jones and Thornton (2000) acknowledge that inherent errors exist including wrongly interpolated surfaces caused by data gaps and interpolation method. Another error type described is due to local elevation changes within the 10 arc-minutes pixel resolution. This error could be minimized by applying an underlying, higher spatial resolution Digital Elevation Model (DEM) in the interpolation process (Jones and Thornton 2000). Higher quality DEMs can be generated using the method developed by Hutchinson (2000). Another serious source of error is the incorrect assignment of pixels to a particular climate cluster, e.g., because of gaps in the calibration dataset. A fourth error type is associated with the inaccurate estimation of parameter values within the climate cluster.

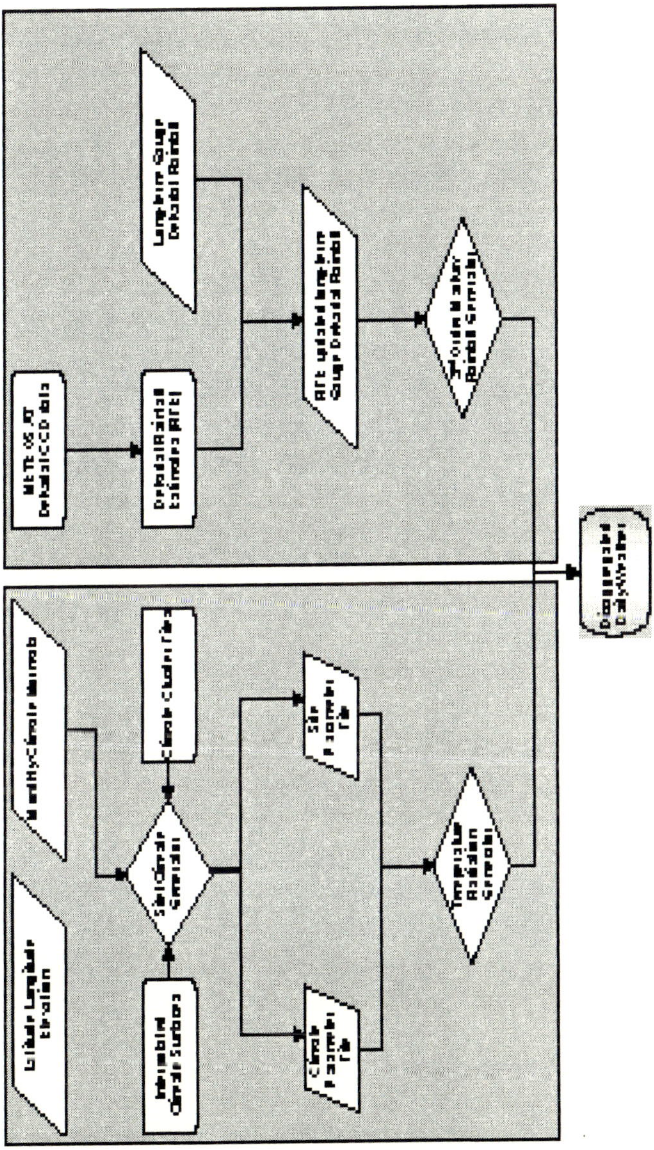

Figure 10: Relational diagram showing the disaggregation procedure of dekadal Meteosat-derived RFEs using the third order Markov weather generator after Jones and Thornton (2000).

Validation of the disaggregated rainfall estimates

The third-order Markov model was applied to Meteosat-derived dekadal RFEs over the pixel containing Karoi, the meteorological station nearest to the study area. Figure 11 shows a time-series comparison between gauged rainfall values and values obtained from stochastic disaggregation of Meteosat-derived RFEs for the 1997-98 rainy season at Karoi. The spikes visible on the disaggregated time series around day-of-year 21 and day-of year 70 reflect random error in disaggregation. MARKSIM is a prototype software that is still being refined. It is hoped that the interpolator will improve to acceptable output random error levels.

In order to show that the gauge rainfall values are statistically equivalent to disaggregated RFEs, the null hypothesis H_0 of no difference was tested. Denoting gauge rainfall as independent sample z and disaggregated RFEs as independent sample y, of sizes *m* and *n,* from possibly different probability distributions *F* and *G* respectively, the null hypothesis H_0: $F = G$ was tested. The test was based on the difference of means as a test statistic denoted by $t(\mathrm{x}) = \bar{z} - \bar{y}$, observed at 0.17 mm for the 1997/98 rainy season.

Neither the observed rain nor the generated rain showed normal distribution, symmetric distribution, or similarity of shape of the data. Conventional hypothesis testing could therefore not be applied. The bootstrap hypothesis test based on the method developed by Efron and Tibshirani (1993) was deemed more appropriate. The bootstrap re-sampling technique has the capability to randomly replicate a sample with replacement to create a proxy population from which it was spawned.

The theoretical background of this re-sampling technique is explained in Burt and Barber (1996), Efron and Tibshirani (1993) and Blank et al. (1999) as follows:

- *B* samples of size $m + n$ were iteratively drawn with replacement from the combined sample denoted x. The first *n* observations were denoted z* and the remaining *m* observations y*.

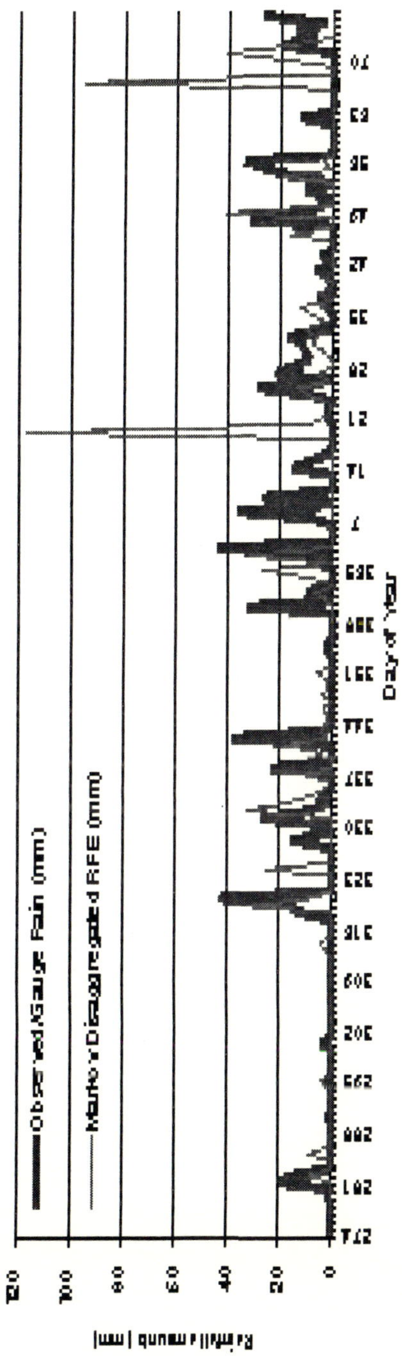

Figure 11: Time-series comparison of observed (gauged) rainfall versus disaggregated METEOSAT RFEs based on the third-order Markov rainfall generator (after Jones and Thornton 2000).

- *t(.)*, the mean difference was calculated on each of the *B* replicated samples,

$$t(\mathrm{x}^{*b}) = \mathrm{z}^{*} - \mathrm{y}^{*}, b = 1,2,\ldots B. \qquad \text{(Eq.20)}$$

- The *Achieved Significance Level* (ASL) or p-value was approximated, defined to be the probability of observing at least that large a value (of the observed mean difference) when the null hypothesis is true,

$$\mathrm{ASL} = \mathrm{Prob}_{H_0}\{t(\mathrm{x}^{*}) \geq t(\mathrm{x})\} \qquad \text{(Eq23)}$$

where $t(\mathrm{x})$ is fixed at its observed value and the random variable x* has an empirical distribution, specified by the null hypothesis H_0. The distribution constitutes a non-parametric estimate of the common population that produced both z and y.

$$\mathrm{ASL} = \#\{t(\mathrm{x}^{*b}) \geq t_{obs}\}/B \qquad \text{(Eq.24)}$$

where $t_{obs} = t(\mathrm{x})$ the observed value of the test statistic (observed mean difference)

Applying the above procedure, ten thousand bootstrap samples (B=10000) were drawn with replacement from the combined sample $(m + n)$, each time calculating and recording the difference between the gauge rainfall mean and the disaggregated RFEs mean. Both samples were the same size ($m = n = 365$) based on days of the year. Out of 10000 bootstrap samples generated, 4453 had $t(\mathrm{x}^*) \geq 0.17$, (see Figure 12) that is, ASL was 4453/10000 = 0.45. H_0 was accepted with ASL greater than the conventional significance level of 0.05, implying that there is no significant difference between gauge rainfall and disaggregated Meteosat derived RFEs.

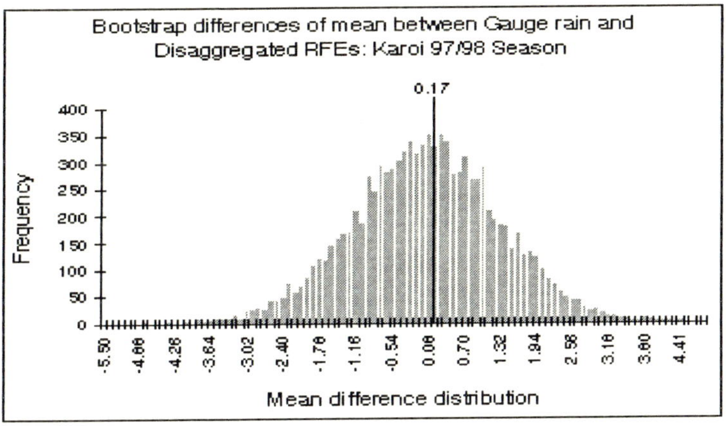

Figure 12: Histogram of bootstrap replications of the mean differences z–y for the test H_0: *F = G (Karoi 97/98 season)*. The observed value (t_{obs} = 0.17) is indicated by a vertical line through the distribution.

PS-2 Simulations of Maize-Based Land-Use Systems using Meteosat

Moulin et al. (1998) provide a detailed review of strategies employed in crop growth models that use satellite-derived data as input for a driving variable, and/or for updating a state variable, and/or as a reference for model re-initialization or for model re-calibration. In this study, the forcing strategy was adopted; Meteosat-derived dekadal rainfall estimates were used and only the rainfall state variable was updated.

Twelve PS-2 scenarios were run first updating the state variable with observed daily rainfall values then with disaggregated Meteosat-derived RFEs for Karoi in order to obtain two independent samples for equality of means hypothesis test H_0. Only rainfall data were updated as forcing variables in PS-2 runs. All other state variables were obtained from meteorological and agricultural extension sources. The twelve scenarios were based on the common planting dates (day of year) 274, 284 and 294, and on soil texture classes Loamy Sand (LS), Sandy Loam (SL), Loam (L) and Clay Loams (CL) that occur in the study area.

Table 3 shows the generated Storage Organ Mass (SOM) in kg/ha obtained by using gauged rainfall in comparison with maize SOM values obtained with disaggregated Meteosat-derived RFEs. A difference of 368 kg/ha was observed between the means of the SOM values obtained from PS-2 simulations using the observed rainfall and using the disaggregated RFEs.

Table 3: SOM (kg/ha) obtained by using gauged rainfall in comparison with SOM obtained using disaggregated Meteosat-derived RFEs

Planting Date	Texture	Gauge rain SOM (kg/ha)	Meteosat RFE SOM (kg/ha)
274	Sandy loam (SL)	8.781	6.435
	Loamy Sand (LS)	13.324	10.811
	Loam (L)	7.851	6.180
	Clay Loam (CL)	15.234	12.011
284	Sandy Loam (SL)	7.198	5.849
	Loamy Sand (LS)	12.185	9.988
	Loam (L)	6.838	5.619
	Clay Loam (CL)	14.282	10.466
294	Sandy Loam (SL)	2.136	9.867
	Loamy Sand (LS)	7.270	11.020
	Loam (L)	1.174	9.680
	Clay Loam (CL)	8.857	11.622
	Mean	8.761	9.129
	Difference of means		368

Although the observed difference of the means of 368 kg/ha was relatively small, there was evidence of bias (Figure 13) in the sense that SOM values from scenarios involving METEOSAT RFEs for planting dates 274 and 284 were underestimated for all soil texture classes. The opposite occurred for scenarios with planting date 294 and Sandy Loam or Loam soils.

However, consistency in growth trends in PS-2 simulations using rainfall data from the two different sources is evident in time-series plots as displayed in Figure 14. Time-series growth curves in Figure 14 show gauge scenarios on the left and disaggregated RFE scenarios on the right.

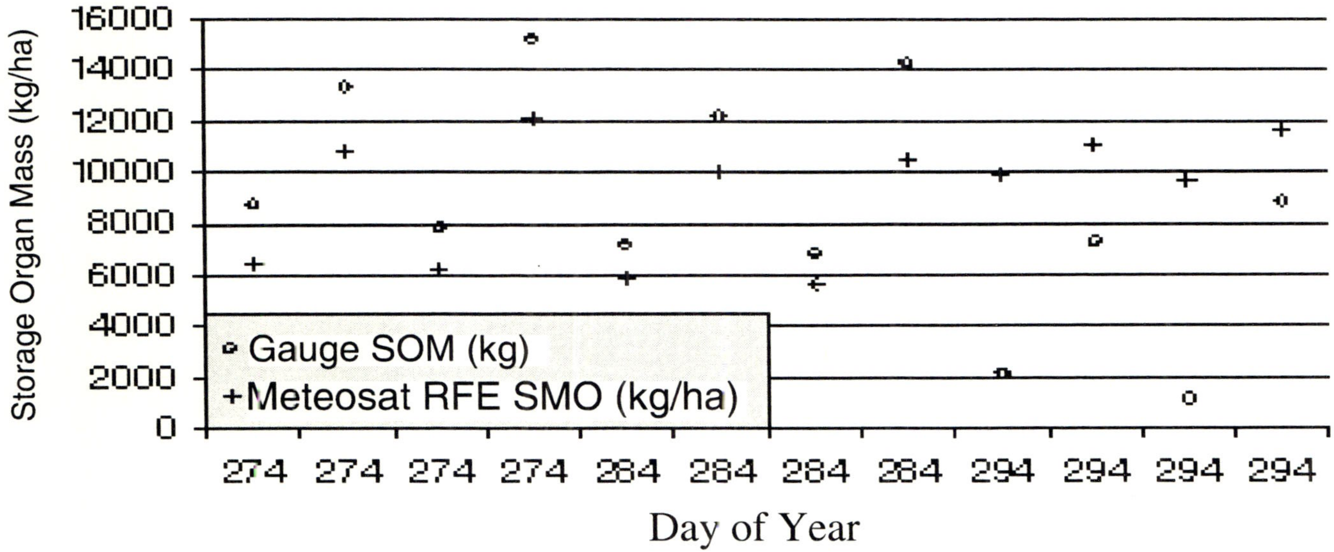

Figure 13: Graph showing underestimation of SOM based on disaggregated Meteosat-derived RFEs for planting dates 274 and 284 for all soil texture classes. Overestimations can be seen for planting date 294 and soil texture classes Sandy Loam and Loam.

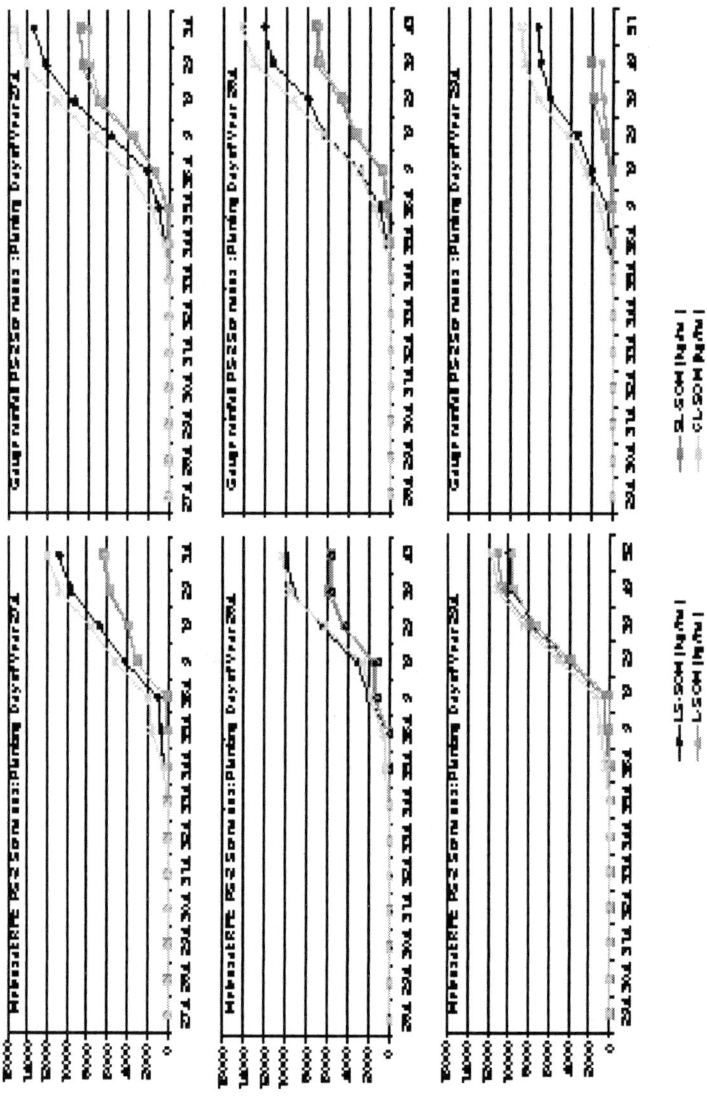

Figure 14: Paired time-series comparing PS-2 simulated SOM from gauged rainfall and Meteosat-derived RFEs for scenarios of four soil texture classes (LS, SL, L and CL) and for three planting dates (274, 284 and 294) within a 25-36 km² pixel size centred at Karoi.

A null hypothesis of no difference in means was again tested to show that no significant difference would exist between Storage Organ Mass (SOM) obtained from the PS-2 model runs using gauge rainfall and those obtained using disaggregated RFEs. The null hypothesis is H_0: $F = G$, where F and G are possibly different distributions of the two independent SOM samples simulated using gauge rainfall or using disaggregated RFEs. The bootstrap hypothesis test method was applied as described earlier in this chapter.

Ten thousand bootstrap replications were generated from the combined gauge and disaggregated RFEs SOM samples each time scoring the difference of the means. Figure 15 shows a histogram of the distribution of differences between mean SOM values simulated using gauge rainfall data and mean SOM values simulated using Meteosat derived RFEs. 5186 of the generated mean differences had values greater than or equal to the observed mean difference value of –368 kg/ha (ASL = 0.52). H_0: $F = G$ was accepted at 0.05 significance level suggesting that there was no significant difference between SOM predictions based on gauge rain and those based on Meteosat-derived RFEs.

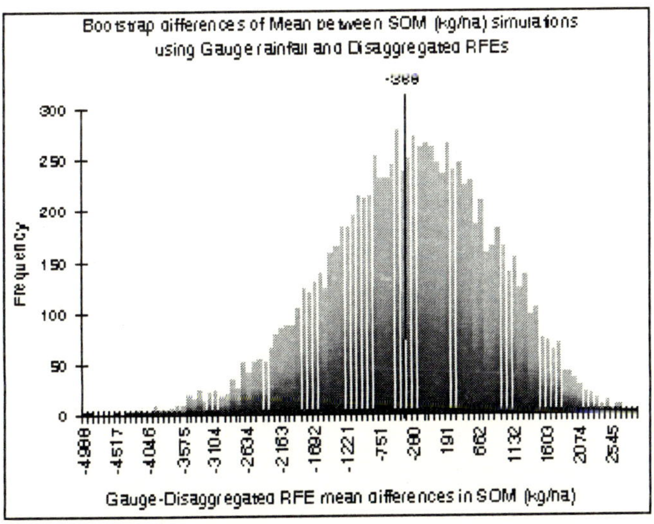

Figure 15: Bootstrap differences of the mean between SOM values simulated using gauged rainfall and disaggregated Meteosat-derived RFEs.

Experimental data for the validation of the predicted water-limited crop production potential are not available. However, re-sampling techniques can be applied on the simulated values, randomly replicating them into a proxy universe from which the distribution can be analyzed and confidence intervals of the mean determined.

The bootstrap method was used to generate a normal distribution of the predicted SOM values from which confidence intervals for the mean could be determined. The bootstrap percentile method was used to derive the confidence limits for the mean. The following bootstrap percentile confidence interval derivation is based on Burt and Barber (1996) and Efron and Tibshirani (1993):

For a random variable $\hat{\theta}^*$ drawn from the distribution $N(\hat{\theta}^* se^2)$,

it follows that:

$$\hat{\theta}^* \approx N(\hat{\theta}^* se^2) \qquad \text{(Eq.25)}$$

Where:

$\hat{\theta}$ is the sample estimate of the parameter θ
sê is the estimated standard error
N is the number of measurements making-up the entire population

Then, $\hat{\theta}_{lower} = \hat{\theta} - z^{(1-\alpha)}.\hat{se}$ and $\hat{\theta}_{upper} = \hat{\theta} - z^{(\alpha)}.\hat{se}$ are the 100. α^{th} and 100. $(1-\alpha)^{th}$ percentiles of .

Alternatively:

$\hat{\theta}_{lower} = \hat{\theta}^{*(\alpha)} = 100.\ \alpha^{th}$ percentile of $\hat{\theta}^*$distribution

$\hat{\theta}_{upper} = \hat{\theta}^{*(1-\alpha)} = 100.\ (1-\alpha)^{th}$ percentile of distribution

Figure 16 shows histograms of the bootstrap distributions for the respective SOM means. The distributions showed that 95% of SOM values predicted with gauged rainfall data lay between 6375 and 11102 kg/ha with a mean θ of 8761 kg/ha. 95% of the SOM predictions using disaggregated Meteosat RFEs were bound between 7823 and 10409 kg/ha with the mean θ = 9129 kg/ha. The lines flanking the distributions in Figure 16 indicate the 2.5th and 97.5th percentiles whereas the central lines show the calculated bootstrap mean

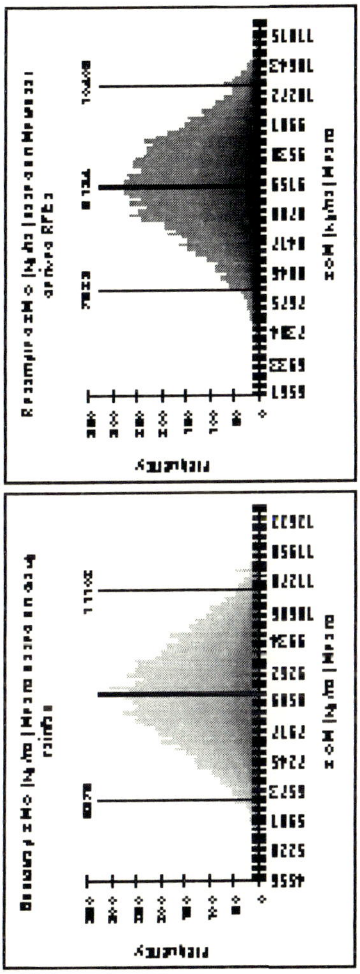

Figure 16: Histograms of 10000 bootstrap replications each of the mean SOM (kg/ha) predicated based on gauged rainfall (left) and based on disaggregated Meteosat-derived RFEs (right). The flanking vertical lines in each histogram indicate the 95% confidence interval values while the central lines indicate the respective means.

Discussion

The findings presented suggest that satellite-derived cumulative rainfall can be disaggregated to daily values and used for modelling of water-limited production potentials with sufficient accuracy based on universally valid physical, chemical and biological relationships.

One difficulty affecting the regionalization of reference crop production estimates is that in addition to precipitation, dynamic crop models require other meteorological data that include minimum/maximum air temperatures, relative air humidity and number of sunshine hours (or solar radiation data). The coarse spatial resolution of meteorological satellites does not allow the retrieval of these parameters or their equivalents with sufficient accuracy required for the crop growth models. Meteorological stations in Zimbabwe are not only very sparse but are almost completely absent in the communal lands. This makes interpolation of meteorological data for model input very risky. Recent research efforts have revealed that high spatial and temporal resolution climate surfaces can be generated (Hutchinson, 2000; Price et al, 2000) using Digital Elevation Models (DEM), hydrological maps and some basic meteorological station data. The applicability of this approach warrants further testing regionally and internationally.

REFERENCES

Adam, L. (1999) *African Connectivity, Problems, Solutions and Actions: Some Recommendations from Inet'96.* Addis Ababa: NRSC Report.

Adam, L. and F. Wood. (1999) "An Investigation of the Impact of Information and Communications Technologies in Sub-Saharan Africa." *Journal of Information Science* 25, 4: 239-247.

Adams, W.M. (1995) "Sustainable Development?" In R.J. Johnston, P.J. Taylor, and M.J. Watts (eds.), *Geographies of Global Change: Remapping the World in the Late Twentieth Century.* Oxford: Blackwell: 354-373.

Adamu, H. (2000) "We'll Feed Our People As We See Fit." *Washington Post* 11 September: A 23.

Adesina, S.K. (1976) "Phytochemical Studies of *T. danielli* (Benth) and *D. cumminsi* (Diels)." Ph.D. Thesis, University of Reading, Reading, England.

Adesina, S.K., E.A. Sofowora, and D. Higginbotham. (1979) "Sweet Tasting Fruits of Some Nigerian Plants." *Journal of African Medicinal Plants* 2: 17-31.

Adkins, S.L. and D.M. Bushnell. (1995) "Learning From Others in New World Order." *Forum for Applied Research and Public Policy* 10 (Summer): 124-127.

Adubifa, O. A. (2000) "Research Proposal for the Study of Science and Technology Assets in Sub-Saharan Africa." Unpublished Paper.

Aduda, D. (2000) "Computer Studies Getting Popular with Teachers." *Daily Nation* 4 September.

AFRE-FEM. (1998) Online. http://clients.tdg.ca/worldbank.

African Governors of the World Bank. (1998) "Partnership for Capacity Building in Africa: Strategy and Business Plan." Report of the African Governors of the World Bank to Mr. James D. Wolfensohn, September 1998.

Africa Policy Information Center (APIC) (2000) *Africa Policy Outlook 2000.* Washington, DC: APIC.

African Virtual University (AVU). (1997a) *A Project of the World Bank: Feasibility Study Report*. Washington, DC: The World Bank.

_____). (1997b) *Pilot Phase Operation: (1 July 1997 - June1998*. Washington, DC: The World Bank.

_____. (1997c) *Knowledge is Power*. Washington, DC: The World Bank.

Agor, W. (ed.). (1989) *Intuition in Organizations*. London: Sage.

Aiyepeku, W.O. (1982) "Information Utilization by Policy-Makers in Nigeria, Part I: Assessing Degrees of Information Consciousness." *Journal of Information Science* 4: 203-211.

Akotia, P. (2000) "Management of Public Sector Financial Records in Ghana." *African Journal of Library Archives and Information Science* 10, 2: 153-165.

Amanor, K.S. (1999) *Global Restructuring and Land Rights in Ghana: Forest Food Chains, Timber and Rural Livelihoods*. Uppsala, Sweden: Nordiska Afrikainstitutet.

Amosu, A. (2000) "People Power." *BBC Focus on Africa*. October-December 2000.

Amutabi, M.N. (1997a) "The Plight of Adult Education in Kenya." In Shirley Walters (ed.), *Globalization, Adult Education and Training: Impacts and Issues*. London: Zed Books: 196-201.

_____. (1997b) "Globalization and Its Impact on Higher Education in Africa: Problems and Prospects: The Case of Kenya." *Conference Proceedings for Southern African Society of Education held in Pretoria, 5-6 July 1997*; Johannesburg: Southern African Society of Education: 1-17.

Amutabi, M.N. et al. (1997) "Introduction." In Shirley Walters (ed.), *Globalization, Adult Education and Training: Impacts and Issues*. London: Zed Books: 1-14.

Amund, O.O. and C.O. Igiri. (1990) "Biodegradation of Petroleum Hydrocarbons Under Tropical Estuarine Conditions." *World Journal of*

Microbiology and Biotechnology 6: 255-262.

Anderson, D.M. (1984) "Depression, Dust Bowl, Demography, and Drought: The Colonial State and Soil Conservation in East Africa During the 1930's." *African Affairs* 83, 332: 321-343.

Anderson, D.M. and R. Grove (eds.). (1987) *Conservation in Africa: People, Policies and Practice*. Cambridge, UK: Cambridge University Press.

Anderson, P.J. (1993) *Physical Resource Inventory of the Communal Lands of Zimbabwe*. [Bulletin 60] Chatham, UK: Natural Resources Institute.

Asad, T. (1973) *Anthropology and the Colonial Encounter*. Atlantic Highlands, NJ: Humanities Press.

_____. (1987) "Are There Histories of People Without Europe." *Comparative Study in Society & History* 29: 594-607.

Atkinson, R. (1998) *The New Economy Index: Understanding America's Economic Transformation*. Washington, DC: Progressive Policy Institute.

Awobajo, A. (1981) "An Analysis of Oil Spillage Incidents in Nigeria, 1976-1980." *The Proceeding Industry and the Nigerian Environment*: 57-58.

Ayittey, G.B.N. (1998) *Africa in Chaos*. New York: St. Martins.

Baark, E. and R. Heeks. (1999) "Donor-Funded Information Technology Transfer Projects: Evaluating the Life Cycle Approach in Four Chinese Science and Technology Projects." *Information Technology for Development* 8, 4: 185-197.

Banders, U., B. Haverkort, and W. Hiemstra. (1996) *Biotechnology: Building of Farmer's Knowledge*. London, UK: Macmillan Education Ltd.

Bannock, G., R.E. Baxter, and E. Davis. (1992) *Dictionary of Economics*. London: Penguin.

Baranshamaje, E. (1995) *The African Virtual University (AVU): Concept Paper*. [Draft of 7 June 1995] Washington, DC: The World Bank.

Baranshamaje, E. et al. (1995) *Increasing Internet Connectivity in Sub-Saharan Africa: Issues, Options and World Bank Group Role.* [Draft of 29 March 1995] Washington, DC: The World Bank.

Barboza, D. (2000) "Industry Moves to Defend Biotechnology. (Public Relations Advertising Campaign to Ease Fears Over Genetically Altered Foods)" *New York Times* 4 April: C6(N) and C6(L).

_____. (2000) "Feminist Theory and Development: Implications for Policy, Research and Action." In J.L. Parpart, M.P. Connelly, and V.B. Barriteau (eds.), *Theoretical Perspectives on Gender and Development.* Ottawa, ON: International Development Research Center: 161-179.

Barriteau, E.V. (1995) "Postmodernist Feminist Theorizing and Development Policy and Practice in the Anglophone Caribbean: The Barbados Case." In M. Marchand and J. Parpart (eds.), *Feminism Postmodernism Development.* New York, NY: Routledge: 142-159.

Bassett, T.J. (1993) "Introduction: The Land Question and Agricultural Transformation in Sub-Saharan Africa." In T.J. Bassett and D.E. Crummey (eds.), *Land in African Agrarian Systems.* Madison, WI: The University of Wisconsin Press: 3-31.

Bates, A.W. (1995) *Technology, Open Learning and Distance Education.* London: Routledge.

Bates, R.H. (1981) *Markets and States in Tropical Africa: The Political Basis of Agricultural Policies.* Berkeley, CA: University of California Press.

Batsima, M. (1993) "Traditional Doctors' Beliefs, Knowledge and Practices Regarding AIDS in Botswana." Unpublished B.Ed. Nursing Thesis, University of Botswana.

Batterbury, S. and A.J. Bebbington. (1999) "Environmental Histories, Access to Resources and Landscape Change: An Introduction." *Land Degradation & Development* 10, 4: 279-289.

Batterbury, S., I.I.B.L. Turner, M. Thompson, W.M. Adams, I. Scoones, P. Woodhouse, O.T. Coomes, A. Bebbington, D. Preston, R.M. Auty, and D. Satterthwaite. (1997) "Environmental Transformations in Developing Countries." *Geographical Journal* 163, 2: 126-224.

Bebbington, A. (1996) "Movements, Modernizations, and Markets: Indigenous Organizations and Agrarian Strategies in Ecuador." In R. Peet and M.J. Watts (eds.), *Liberation Ecologies: Environment,*

Development, Social Movements. London: Routledge: 86-109.

Benjamin, P. and M. Dahms. (2001) *Socialize the Modern Production: The Role of Telecentres in Development*. Johannesburg: University of Witwatersrand, Link Centre.

Benner, C. (2000) "Building Community-Based Careers: Labor Markets and Intermediaries in Silicon Valley, Berkeley." Paper Presented at the Association of American Geographers held in Pittsburgh, PA, 4-8 April.

_____. (2002) *Flexible Work in the Information Economy: Labor Markets in Silicon Valley*. Oxford: Blackwell Press.

Benner, C. et al. (1999) *Walking the Lifelong Tightrope: Negotiating Work in the New Economy*. San Jose, CA: Working Partnerships USA with the Economic Policy Institute.

Bernard, B. (2000) "Local Loop Access Bottlenecks: A Tangled Web." *Global Telecoms Business* (February).

Berry, S. (1997) "Tomatoes, Land and Hearsay: Property and History in Asante in the Time of Structural Adjustment." *World Development* 25, 8: 1225-1241.

_____. (1993) *No Condition is Permanent: The Social Dynamics of Agrarian Change in Sub-Saharan Africa*. Madison, WI: The University of Wisconsin Press.

_____. (1986) "Macro-policy implications of research on farming systems and households." In J. Moock (ed.), *Understanding Africa's Rural Households and Farming Systems*. Boulder, CO: Westview: 199-216.

_____. (1984) "The Food Crisis and Agrarian Change in Africa: A Review Essay." *African Studies Review* 27, 2: 59-112.

Bessant, J. (1989) *Microelectronics and Change at Work*; Geneva: International Labor Office.

Bezuidenhout, A. (1996) "Centralization Through Decentralization: The Use of Subcontracted Labor in the Light of Intensification of Managerial Control." *South African Journal of Sociology* 27, 1: 1-10.

Bhagavan, M. (1997) *New Generic Technologies in Developing Countries*. New York: St. Martin's Press.

Bhagavan, M. (1990) *Technological Advance in the Third World: Strategies and Prospects*. Atlantic Highlands, NJ: Zed Books.

Bhorat, H. and J. Hodge. (1999) "Decomposing Shifts in Labor Demand in South Africa." *The South African Journal of Economics* 673: 348-80.

Biggs, S.D. (1995) "Farming Systems Research and Rural Poverty: Relationships Between Context and Content." *Agricultural Systems* 47: 161-174.

Binswanger, H. and P. Pingali. (1988) "Technological Priorities for Farming in Sub-Saharan Africa." *World Bank Research Observer* 3, 1: 81-98.

_____. (1987) "The Evolution of Farming systems in Sub-Saharan Africa." In V.W. Ruttan and C.E. Prey (eds.), *Policy for Agricultural Research*: Boulder, CO: Westview Press: 27-56.

_____. (1986) "The Evolution of Farming Systems and Agricultural Technology in Sub-Saharan Africa." In G.E. Schuh (ed.), *Technology, Human Capital, and the World Food Problem*. University of Minnesota Publication No. 37. Minneapolis, MN: University of Minnesota, Department of Agricultural and Applied Economics: 29-52

Binswanger, H. and V.W. Ruttan (eds.). (1978) *Induced Innovation: Technology, Institutions and Development*. Baltimore, MD: Johns Hopkins University Press.

Blalock, H.M. (1982) *Conceptualization and Measurement in the Social Sciences*. Beverly Hills, CA: Sage Publication.

Blank, S., C. Seiter, and P.Bruce. (1999) *Re-Sampling Stats in Excel Users Guide.* Arlington, VA: Resampling Stats, Inc.

BMI-TechKnowledge. (2001) *SA Telecommunications Services Overview and Opportunity Report*. Johannesburg: BMI-T Group.

_____. (1999) *Communication Technologies Handbook*. Johannesburg: BMI-Tech Group.

Bock, K.R. and R.D. Woods. (1983) "Etiology of African Mosaic Disease." *Plant Disease* 67: 994-995.

Bogert, P., P. Mahau, and F. Becker. (1995) *The Spatial Interpolation of Agro-Climatic Data.* Agrometeorology Working Series Paper No. 12. Rome, Italy: FAO.

Bonifacio, R., G. Dugdale, and J.R. Milford. (1993) "Sahelian Rangeland Production in Relation to Rainfall Estimates from METEOSAT." *International Journal of Remote Sensing* 14: 2695-2711.

Boostrom, E. et al. (1995) "Increasing Internet Connectivity in Sub-Saharan Africa: Issues, Options, and the World Bank Group Role." Online. http://www.un.org/Depts/eca/divis/disd/padis/telemat/africa03.htm.

Borlaug, N.E. and C.R. Dowswell. (1995) "Mobilising Science and Technology to Get Agriculture Moving in Africa." *Development Policy Review* 13, 2: 115-130.

Boserup, E. (1999) *My Professional Life and Publications 1929-1998.* Copenhagen, Denmark: Museum Tusculanum Press, University of Copenhagen.

_____. (1983) *Population & Technological Change: A Study of Long-Term Trends.* Chicago, IL: University of Chicago Press.

_____. (1970) *Women's Role in Economic Development.* London: St. Martin's Press.

_____. (1965) *The Conditions of Agricultural Growth: The Economics of Agrarian Change Under Population Pressure.* New York: Aldine Publishing Company.

Bourn, D. and W. Wint. (1994) *Livestock, Land Use and Agricultural Intensification in Sub-Saharan Africa.* London: Overseas Development Institute.

Boylaud, O. and G. Nicoletti. (2000) "Regulation, Market Structure and Performance in Telecommunications." OECD Economics Department Working Papers No. 237. Paris: OECD.

Braa, J., E. Monteiro, and E.S. Reinert. (1995) "Technology Transfer Versus Technology Learning: IT Infrastructure and Health Care in Developing Countries." *Information Technology for Development* 6: 15-23.

Brands, H. and E.T. Leo. (1999) *The Law and Regulation of Telecommunications Carriers*. Norwood: Artech House.

Brass, T. (1995) "Old Conservatism in 'New' Clothes." *Journal of Peasant Studies* 22, 3: 516-540.

Brink, C. (1996) "The Estimation of the Area Under Maize Using Area Frame Sampling in Remote Sensing." Conference Proceedings on the Applications of Remote Sensing and Geographic Information Systems in Resource Assessment in Africa, held in Harare, March: 242-244.

Brinkley, D. (1995) "Jimmy Carter's Modest Quest for Global Peace." *Foreign Affairs* 74, 6: 90-101.

Bryceson, D.F. (ed.). (1995) *Women Wielding the Hoe: Lessons From Rural Africa for Feminist Theory and Development Practice*. Oxford: Berg Publishers.

Bryceson, D. and V. Jamal (eds.). (1997) *Farewell to Farms: De-Agrarianisation and Employment in Africa*. Aldershot: Ashgate.

Bryceson, D., C. Kay, and J. Mooij. (2000) *Disappearing Peasantries? Rural Labour in Africa, Asia and Latin America*. London: Intermediate Publications.

Burt, J.E. and G.M. Barber. (1996) *Elementary Statistics for Geographers*. [2nd Ed.] NY: Guilford Press: 573-598.

Byerlee, D. and M. Collinson. (1980) *Planning Technologies Appropriate to Farmers: Concepts and Procedures*. Mexico, DF: CIMMYT.

Byerlee, D., L. Harrington, and D.L. Winkelman. (1982) "Farming Systems Research: Issues in Research Strategy and Technology Design." *American Journal of Agricultural Economics* 64, 5: 897-904.

Byerlee, D. and R. Tripp. (1988) "Strengthening Linkages in Agricultural Research Through a Farming Systems Perspective: The Role of Social Scientists." *Experimental Agriculture* 24: 137-151.

Callon, S. and Asia/Pacific Research Center. (1995) *Different Paths: The Rise of Taiwan and Singapore in the Global Personal Computer Industry*.

CA: Stanford University, Asia/Pacific Research Center Institute for International Studies.

Carson, R. (1965) *Silent Spring*. Harmondsworth: Penguin.

Chambers, R., A. Pacey, and L. Ann Thrupp (eds.). (1989) *Farmer First: Farmer Innovation and Agricultural Research*. London: Intermediate Technology Publications.

Chen, J.M. (2000) "IT Multinationals: Willing Partners to Repression in China?" *Digital Freedom Network* Online. http://dfn.org/focus/china/multinationals.htm. Retrieved 27 November 2000.

Chindah A.C. (1998) *The Effect of Industrial Activities on the Periphyton Communities At the Upper Reaches of New Calabar River, Niger Delta, Nigeria*. Port Harcourt, Nigeria.

Chowdry, G. (1995) "Engendering Development? Women in Development (WID) in International Development Theories." In M. Marchand and J. Parpart (eds.), *Feminism Postmodernism Development*. New York: Routledge: 23-26.

CIMMYT (International Maize and Wheat Improvement Center). (1992) *Enduring Designs For Change: An Account of CIMMYT's Research, Its Impact, and Its Future Directions*. Mexico, DF: CIMMYT.

Cleaver, K.M. and G.A. Schreiber. (1994) *Reversing the Spiral: the Population, Agriculture, and Environment Nexus in Sub-Saharan Africa*. Washington, DC: The World Bank.

Cogburn, D.L. and N.J. Hafkin (eds.). (1997) *Building the African Information Society with Public and Private Sector Cooperation. Global Information Infrastructure Commission Conference Report*. South Africa: GIIC and the Centre for Information Society Development in Africa (CISDA). Available online. http://www.giic.org/events/af970508.html.

Cohen, M.D., J.G. March, and J.P. Olsen. (1972) "A Garbage Can Model of Organizational Choice." *Administrative Science Quarterly* 17: 1-25.

Collinson, M.P. (ed.). (2000) *A History of Farming Systems Research*. Wallingford / Rome: CABI Publishing / FAO.

_____. (1982) *Farming Systems Research in Eastern Africa: The Experience of CIMMYT and Some National Agricultural Research Services, 1976-1981*. East Lansing, MI: Michigan State University, Department of Agricultural Economics.

_____. (1981) "A Low Cost Approach to Understanding Small Farmers." *Agricultural Administration* 8, 6: 434-50.

_____. (1980) "Farming System Research in the Context of an Agricultural Research Organization." In H. Ruthenberg (ed.), *Farming Systems in the Tropics*. Clarendon Press: Oxford: 381-389.

Comaroff, J.L. (1984) "The Closed Society and Its Critics: Historical Transformations in African Ethnography." *American Ethnologist* 11, 3: 571-583.

Conceicao, P., D.V. Gibson, M.V. Hector, and S. Shariq (eds.). (2000) *Science, Technology, and Innovation Policy*. Westport, CT: Quorum Books.

Convenor, K. (1994) *Science, Technology & Development: For Development in the SADC Region: Status, Needs, Prospects and Challenges*. Gaborone, Botswana: SADC.

Conway, G. (2000) "USDA Millennium Lecture." Online. http://www.ocav.usda.gov:8080/ramgen/secy/ml.rm.

_____. (1998) *The Doubly Green Revolution: Food for All in the 21st Century*. Ithaca, NY: Cornell University Press.

Conway, G.R. and E.B. Barbier. (1990) *After the Green Revolution: Sustainable Agriculture for Development*. London: Earthscan.

Conway, G. and S. Sechler. (2000) "Helping Africa Feed Itself." *Science* 289, 5485: 685.

Cooper, F. (1993) "Africa and the World Economy." In F. Cooper, F.E. Mallon, S.J. Stern, A.F. Isaacman, and W. Roseberry (eds.), *Confronting Historical Paradigms: Peasants, Labor, and the Capitalist World System in Africa and Latin America*. Madison, WI: University of Wisconsin Press: 84-204.

Crawford, E.W. (1982) *A Simulation Study of Constraints on Traditional Farming Systems in Northern Nigeria*. East Lansing, MI: Michigan State University, Department of Agricultural Economics.

Crehan, K. (1997) *The Fractured Community: Landscapes of Power and Gender in Rural Zambia*. Berkeley, CA: University of California Press.

Curran, P.J. (1988) "The Semivariogram in Remote." *Remote Sensing of Environment* 24: 493-507.

Cyberatlas. (2001) Online. http://www.cyberatlas.internet.com/big_picture/.

Department of Arts, Culture, Science and Technology (DACST). (1999) *Innovation Fund: Report on Round 1 (1998/1999)*. Pretoria, South Africa: DACST.

Department of Agricultural and Technical Services (DARTS). (1995) *A Handbook of Crop Varieties*. Liolongwe, Malawi: Ministry of Agriculture.

Day, B. and J. Miller. (2000) *National Information and Communications technology Policy and Planning Processes: Lessons From South Africa*. Cape Town, South Africa.

Degrande, A. and B. Duguma. (2000) *Adoption Potential of Rotational Hedgerow Intercropping in the Humid Lowlands of Cameroon*. London: Overseas Development Institute.

DeGrassi, A. (2000a) *Development Struggles: How Much Do Post-Development Accounts Reveal About a Classical Green Revolution Project in Africa in the Late Twentieth Century?* Unpublished Honors Thesis, International and Area Studies Department, University of California-Berkeley.

_____. (2000b) "(Mis)Understanding Agro-Environmental Change in Africa: Population Growth As the Primary Cause of Poverty and Environmental Degradation." *Berkeley Undergraduate Journal* 27 (Fall)

Denton Hall. (1995) *EC Telecommunications Law*. Colorado Springs, CO: John Wiley and Sons.

Developing an Information Infrastructure Agenda in Uganda (DIIAUP). (2000a) *First Quarter Report: DIIAUP. (April –30 June)*. Uganda: DIIAUP.

_____. (2000b) "Uganda Project – DIIAUP." Proceedings of the National Symposium on Developing an Information Infrastructure Agenda in Uganda, 5 December.

DeWalt, B.R. (1985) "Farming Systems Research." *Human Organization* 44, 2: 106-114.

DPR. (1991) *Standards and Guidelines for the Petroleum Industry*. Abuja: Department of Petroleum Resources.

Driver, H.E. (1962) *The Contribution of A.L. Kroeber to Culture Area Theory and Practice*. Baltimore, MD: Waverly Press.

Dugdale, G., V.D. McDougall and J.R. Milford. (1991) "Rainfall Estimates in the Sahel from Cold Cloud Statistics: Accuracy and Limitations of Operational Systems." In *Soil Water Balance in the Sudano-Sahelian Zone (Proceedings of the Niamey Workshop, February 1991). IAHS* 199: 65-74.

Earnshaw, G. (2001) "China Online." *Brill's Content* (February).

Easterbrook, G. (1997) "Forgotten Benefactor of Humanity." *The Atlantic Monthly* 279, 1: 75-82.

Eastman, J.R. (1996) "The Spatial Manifestation of ENSO Warm Phase Events in Southern Africa." Conference Proceedings on the Applications of Remote Sensing and Geographic Information Systems in Resource Assessment in Africa, held in Harare, Zimbabwe, March.

Ebeling, M. (2000) "State of Media in Nigeria." Unpublished Paper. 30 October.

Economic Commission for Africa (ECA). (2000) *Statistics 2000*. Addis Ababa: ECA.

_____. (1996) *African Information Society Initiative: An Action Framework to Build Africa's Information and Communication Infrastructure*. Addis Ababa: ECA. Available online. http://www.bellanet.org/partners/aisi/aisi.html.

_____. (1995a) "Communique of Addis Ababa." 7 April 1995. Online. www.sas.upenn.edu/African_Studies/Padis/telmatcs_communik.html.

_____. (1995b) "Final Report. Study of the Effectiveness of Informatics Policy Instruments in Africa." Online. www.bellanet.org/partners/aisi/policy/infopol/sumbrown.htm.

Economist Intelligence Unit. (1997) *Business Africa,* 1-5 May. London: The Economist.

Edwards, L. (2000) "Labor shedding output growth: Is trade the culprit?" *Trade and Industry Monitor* 14: 2-6.

Efron, B. and R. Tibshirani. (1993) *An Introduction to the Bootstrap.* New York: Chapman and Hall.

Egyptian Ministry of Education. (2000) *Annual Report.* Cairo: Ministry of Education.

Egyptian Ministry of Labor Force. (2000) *Annual Report.* Cairo: Ministry of Labor.

Ehrlich, P. (1968) *The Population Bomb.* New York: Ballantine.

Eicher, C.K. and D.C. Baker. (1982) *Research on Agricultural Development in Sub-Saharan Africa: A Critical Survey.* East Lansing, MI: Michigan State University, Department of Agricultural Economics.

Emdon, H. (2000) "Telecommunications, a Foundation for Socio-Economic Development." In BMI-TechKnowledge, *Communication Technologies Handbook 1999.* Johannesburg: BMI-Tech Group: 76-82.

Enos, J. (1995) *In Pursuit of Science and Technology in Sub-Saharan Africa: The Impact of Structural Adjustment Programmes.* New York: Routledge.

Erickson, P.A. (1998) *A History of Anthropological Theory.* New York: Broadview Press.

Ernst, D. and B.A. Lundvall. (1997) *Information Technology in the Learning Economy: Challenge for Developing Countries.* Aalborg, Denmark: Danish Research Unit for Industrial Dynamics.

Escobar, A. (1991) "Anthropology and the Development Encounter: The Making and Marketing of Development Anthropology." *American Ethnologist* 18: 658-682.

Fabian, J. (1983) *Time and the Other: How Anthropology Makes Its Object*. New York: Columbia University Press.

Fairhead, J. and M. Leach. (1998) *Reframing Deforestation: Global Analyses and Local Realities; Studies in West Africa*. London: Routledge.

____. (1996) *Misreading the African Landscape: Society and Ecology in a Forest-Savanna Mosaic*. Cambridge, UK: Cambridge University Press.

Federal Ministry of Agriculture and Natural Resources. (1994) *Guidelines on Biosafety for Nigeria*. (March) Abuja: Federal Ministry of Agriculture and Natural Resources.

Feldstein, H.S. and S.V. Poats (eds.). (1989) *Working Together: Gender Analysis in Agriculture*. West Hartford, CT: Kumarian Press.

Ferguson, J. (1999) *Expectations of Modernity: Myths and Meanings of Urban Life on the Zambian Copperbelt*. Berkeley, CA: University of California Press.

_____. (1997) "Anthropology and Its Evil Twin: 'Development' in the Constitution of a Discipline." In F. Cooper and R. Packard (eds.), *International Development and the Social Sciences: Essays on the History and Politics of Knowledge*. Berkeley, CA: University of California Press: 150-175.

_____. (1990) *The Anti-Politics Machine: "Development," Depoliticization, and Bureaucratic Power in Lesotho*, Minneapolis, MN: University of Minnesota Press.

FEPA. (1991) *Guidelines and Standards for Environmental Pollution Control in Nigeria*. Nigeria: FEPA.

Field, M. J. (1996) *Telemedicine: A Guide to Assessing Telecommunications in Health Care*. Washington, DC: National Academy Press.

Fine, B. (2001) *Social Capital Versus Social Theory*. London: Routledge.

Fineman, S. (ed.). (1993) *Emotion in Organizations*. London: Sage.

Flavell, R. (1995) "Plant Biotechnology R & D. The Next Ten Years." *TIBTECH* 13: 313-318.

Flitcroft, I.D., J.R. Milford, and G. Dugsdale. (1989) "Relating Point to Area Average Rainfall in Semiarid West Africa and the Implications for Rainfall Estimates Derived from Satellite Data." *Journal of Applied Meteorology* 28: 252-266.

Florida, R.L. (1999) "Other Countries' Money." *MIT's Technology Review* 101 (March/April): 29-32.

_____. (1995) "Technology Policy for a Global Economy." *Issues in Science and Technology* 11 (Spring): 49-56.

Food and Agriculture Organization (FAO). (2000) "The Electronic Forum on Bio-technology in Food and Agriculture." Online. http: //www.fao.org/biotech/forum.htm.

_____. (1999a) "State of the World's Forests 1999." *FAO Online*. www.fao.org/forestry/FO/SOFO/SOFO99/sofo99_e.stm.

_____. (1999b) *Production Yearbook: 1998 Statistics Series*. Rome: FAO.

_____. (1996) *Report on the World's Plant Genetic Resources*. (May) Rome: FAO.

_____. (1994) "Biotechnology Application in Agriculture." *World Agriculture* 1994: 85-88.

Forje, J. (1989) *Science and Technology in Africa*. London: Longman.

Frazer, J. (1890) *The Golden Bough*. NJ: Macmillan.

Freedom Forum. (2000) "Africa Media Forum: Press, Power, and Politics." Report from Conference held in Accra, Ghana, 27-28 September.

Frissen, V. (1992) "Trapped in Electronic Cages? Gender and New Information Technologies in Public and Private Domain." *Media, Culture and Society* 14: 31-49.

Fukuyama, F. (2002) *Our Post-human Society: Consequences of the Biotechnology Revolution.* New York: Farrar, Straus and Giroux.

Gaillard, J. (1991) *Scientists in the Third World.* Lexington, KY: University of Kentucky Press.

Galavotti, M.S. (2000) "What Makes Science Unique in the Experience of Mankind? The Specificity of the Scientific Approach." In UNESCO, *World Conference on Science: Science for the Twenty-First Century: A New Commitment.* Paris: UNESCO: 80-82.

Galliers, R.D., S. Madon, and R. Rashid. (1998). "Information Systems and Culture: Applying 'Stages of Growth' Concepts to Development Administration." *Information Technology for Development* 8: 89-100.

Galliers, R.D. and A.R. Sunderland. (1991) "Information Systems Management and Strategy Formation: The 'Stages of Growth' Model Revisited." *Journal of Information Systems* 1: 89-114.

Gaonkar, D. (ed.). (2001) *Alternative Modernities.* Durham/London: Duke University Press.

Geertz, C. (1963) *Agricultural Involution: The Process of Ecological Change in Indonesia.* Berkeley, CA: University of California Press.

Gibbons, J.H. (1995) "Science and Technology in Post-Cold War Era." *Forum for Applied Research and Public Policy* 10 (Spring): 119-122.

Gilbert, E.H., D.W. Norman, and F.E. Winch. (1980) *Farming Systems Research: A Critical Appraisal.* MSU Rural Development Paper No. 6. East Lansing, MI: Michigan State University, Department of Agricultural Economics.

Global Information Infrastructure Commission. (1996) *Information and Communications for Development: Nationalism, Regionalism and Globalism in Bulding the Information Society: GIIC Report on the Global Information Society and Development Forum.* Washington, DC: Global Information Infrastructure Commission. Available Online. http://www.cyberatlas.internet.com/big_picture/.

Global Witness. (1999) "A Crude Awakening." Online. www.globalwitness.org/campaigns/oil/downloads/cruderep.pdf.

Glover, L.A. (1994) "Environmental and Regulatory Aspects of Using Genetically Modified Microorganisms." In G. Marshall and O. Walters (eds.), *Molecular Biology in Crop Protection*. London: Chapman and Hall.

Gould, S.J. (1997) *Full House: The Spread of Excellence: From Plato to Darwin*. New York: Random House.

_____. (1994) *Hen's Teeth and Horse's Toes: Further Reflections in Natural History*. New York: W.W. Norton & Company.

_____. (1992) *The Panda's Thumb: More Reflections in Natural History*. New York: W.W. Norton & Company.

Gouvevitch, P., R. Bohn, and D. McKendrick. (2000) "Globalization of Production: Insights from the Hard Disk Drive Industry." *World Development* 28, 2: 301-317.

Government of Uganda. (1997) *The Uganda Communications Act, September 1997. Acts Supplement No. 6 to the Uganda Gazette No. 64, Volume XC, dated 26 September 1997.* Uganda: UPPC Entebbe (by order of the Government of Uganda).

Grant, G.G. (1999) *Implementing Electronic Networking in Africa: The Case of GOVERNET*. London: Report Commonwealth Secretariat.

Gray, L. (1999) "Is Land Being Degraded? A Multi-Scale Investigation of Landscape Change in Southwestern Burkina Faso." *Land Degradation & Development* 10, 4: 329-343.

Greene, J. (1959) *The Death of Adam: Evolution and Its Impact on Western Thought*. Ames, IA: Iowa State University Press.

Gressel, J. (1996) "Plant Biotechnology Can Quickly Offer Solutions to Hunger in Africa." *Scientist* 10, 19: 10-13.

Grimes, D.I.F., R. Bonifacio, and H.R.L. Loftie. (1998) *Rainfall Estimation Workbook*. Chatham, UK: Natural Resources Institute.

Groen, M.M. de. (2001) *Modeling Interception and Transpiration at Monthly Time Steps. Introducing Daily Variability Through Markov Chains*. Delft, The Netherlands: IHE Delft.

Grove, R.H. (1998) *Ecology, Climate and Empire: Colonialism and Global Environmental History, 1400-1940*. Washington, DC: The White House Press.

Gullish, J. (2000) "Credible and Transparent Regulation - Prerequisite for Successful Privatization and Liberalization." In BMI-TechKnowledge, *Communication Technologies Handbook, 1999*. Johannesburg: BMI-Tech Group: 60-74.

Gupta, A. (1998) *Post-Colonial Developments: Agriculture in the Making of Modern India*. Durham: Duke University Press.

Gupta, A. and J. Ferguson. (1997a) "Beyond Culture: Space, Identity, and the Politics of Difference." In A. Gupta and J. Ferguson (eds.), *Culture, Power, Place: Explorations in Critical Anthropology*. Durham: Duke University Press: 33-51.

_____. (1997b) "Culture, Power, Place: Ethnography at the End of an Era." In A. Gupta and J. Ferguson (eds.), *Culture, Power, Place: Explorations in Critical Anthropology*. Durham: Duke University Press: 1-32.

Gut, C.W. (1998) *National Science and Technology Policy, Situation Analysis*. Lilongwe, Malawi: Joint National/International Consultancy Report, UNESCO.

Guyer, J.I. (1997) *An African Niche Economy: Farming to Feed Ibadan, 1968-88*. London: Edinburgh University Press for the International African Institute.

_____. (1981) "Household and Community in African Studies." *African Studies Review* 24: 87-137.

Hafkin, N. (2000) "Convergence of Concepts: Gender and ICT in Africa." In E. Rathgeber and E. Adera (eds.), *Gender and the Information Revolution in Africa*. Ottawa, ON: International Development Research Center.

Hague Academy of International Law. (1999) *Legal Implications of Global Telecommunications*. The Hague: Centre for Studies in International Law and International Relations.

Haines, R.J. and B.E. Martin. (1997) *Biotechnology and the Sustainable Production of Tropical Timber*. Forest Genetic Resources No 25. Rome: FAO. Available Online. http://www.fao.org/forestry/FOR/FORM/FOGENRES/genresbu/125/125e/arte11.stm.

Hall, T. (1995) "Addis Hosts Historic `Superhighway' Conference, But Will Policymakers Go With the Vision?" Online. http://www.sas.upenn.edu/African_Studies/Padis/telmtcs_Hall.html.

Harcourt, W. (ed.). (1999) *Women @ Internet: Creating New Cultures in Cyberspace*. London/New York: Zed Books Ltd.

Harrison, B.D. (1987) "Properties and Geographical Variation of Geminivirus Isolates from Mosaic-Affected Cassava." In C. Fauquet and D. Fargette (eds.), *Proceedings of the International Seminar: African Cassava Mosaic Disease and Its Control, 4-8 May 1987*. Yamoussoukro, Côte d'Ivoire: D.CTA-ORSTOM: 36-42.

Harrison, B.D., X. Zhou, G.W. Otim-Nape, Y. Liu, and D.J. Robinson. (1997) "Role of a Novel Type of Double Infection in the Geminivirus-Induced Epidemic of Severe Cassava Mosaic in Uganda." *Annuals of Applied Biology* 131: 437-448.

Harrison, N.E. (1998) "Why Science and Technology Require Political Guidance to Sustain Development." *Politics and the Life Sciences* 17, 2: 179-188.

Harrison, P. (1996) *The Greening of Africa: Breaking Through in the Battle for Land and Food*. Nairobi, Kenya: Academy Science Publishers.

Hart, G. (forthcoming) *Disabling Globalization: Tales of Two Towns in South Africa*. Berkeley, CA: University of California Press.

_____. (1997) "Multiple Trajectories of Rural Industrialisation: An Agrarian Critique of Industrial Restructuring and the New Institutionalism." In D. Goodman and M.J. Watts (eds.), *Globalising Food: Agrarian Questions and Global Restructuring*. London: Routledge: 56-78.

_____. (1996) "The Agrarian Question and Industrial Dispersal in South Africa: Agro-Industrial Linkages Through Asian Lenses." *Journal of Peasant Studies* 23, 2/3: 245-277.

_____. (1986) *Power, Labor, and Livelihood: Processes of Change in Rural Java.* Berkeley, CA: University of California Press.

Hart, G., A. Turton, and B. White (eds.). (1989) *Agrarian Transformations: Local Processes and the State in Southeast Asia.* Berkeley, CA: University of California Press.

Hay, C. and D. Marsh. (2000) *Demystifying Globalization.* New York: Macmillan.

Hayami, Y. and V.W. Ruttan. (1971) *Agricultural Development: An International Perspective.* Baltimore, MD: The Johns Hopkins University Press.

Heeks, R. (1996) *India's Software Industry: State Policy, Liberalization and Industrial Development.* Thousand Oaks, CA: Sage Publications.

Hefner, R.W. (1998) "Multiple Modernities: Christianity, Islam, and Hinduism in a Globalizing Age." *Annual Review of Anthropology* 27: 83-104.

Henao, J. and C. Baanante. (1999) "Nutrient Depletion in the Agricultural Soils of Africa." *IFPRI 2020 Vision 62.* Washington, DC: International Food Policy Research Institute.

Herskovits, M. (1930) "The Culture Areas of Africa." *Africa* 3: 59-77.

Hildebrand, P.E. (1977) "Generating Small Farm Technology: An Integrated Multidisciplinary Approach." In *Annual West Indian Agricultural Economics Conference.*

Hilton, R.H. (1985) "Introduction." In T.H. Aston and C.H.E. Philton (eds.), *The Brenner Debate: Agrarian Class Structure and Economic Development in Pre-Industrial Europe.* Cambridge, UK: Cambridge University Press: 1-9.

Hodge, J. (1998) "The Midrand Area; An Emerging High-Technology Cluster." *Development Southern Africa* 15: 851-973.

Hodge, J. and A. Driver. (1999) *Understanding the High Tech Sector in the Cape Metropolitan Region: A Contribution to the Development of a Regional Strategy for High Tech Industry*. Cape Town: University of Cape Town, Development Policy Research Unit.

Hodge, J. and J. Miller. (1997). *Information Technology in South Africa: The State-of-the Art and Implications for National IT Policy*. Cape Town: University of Cape Town, Development Policy Research Unit.

Horvot, J. and J. Rosen. (1990) "Singlethink: Thoughts on the Havel Episode." *Gannett Center Journal* 4, 4: 31-43.

Howells, J. (1996) "Tacit Knowledge, Innovation and Technology Transfer." *Technology Analysis and Strategic Management* 82: 91-106.

Hoyningen-Huene, P. (2000) "The Nature of Science." In UNESCO, *World Conference on Science: Science for the Twenty-First Century: A New Commitment*. Paris: UNESCO: 52-56.

Hsu, J. Y. and A. Saxenian. (2000) "The Limits of Guanxi Capitalsim: Transnational Collaboration Between Taiwan and the USA." *Environment and Planning*. 3211: 1991-2005.

Hutchinson, M.F. (2000) *ANUSPLIN User's Guide*. Canberra: The Australian National University.

Hutchinson, M.F. and J.D. Corbett. (1993) *Spatial Interpolation of Climate Data Using Thin Plate Smoothing Splines*. FAO Agrometeorological Series Working Paper No. 13. Rome: FAO.

Huyer, S. (1997) "Supporting Women's Use of Information Technologies for Sustainable Development." Paper Submitted to the Gender and Sustainable Development Unit. Ontario, Canada: IDRC. February.

Hyden, G. (1983) *No Shortcuts to Progress: African Development Management in Perspective*. London: Heinemann.

International Development Research Centre (IDRC). "The Acacia Initiative." Online. http://www.idrc.ca/acacia/.

International Federation of Journalists (IFJ). (2001a) "Zimbabwe Government Gives No Safety Guarantee Amidst Restrictions on Foreign Media." Press Release. 2 March.

_____. (2001b) "Government Gives No Safety Guarantee to Journalists Amidst Restrictions on Foreign Media." Press Release. 5 March.

International Institute of Tropical Agriculture (IITA). (2000) "Knowledge Exchange and Learning Partnerships: Enhancing Sustainable Agriculture Education in Nigeria." A Proposal Prepared for USAID by the International Institute of Tropical Agriculture, Ibadan, Nigeria.

_____. (1992) *Sustainable Food Production in Sub-Saharan Africa: IITA's Contribution*. Ibadan, Nigeria: IITA.

_____. (1990). "Cassava in Tropical Africa." *A Reference Manual.* Ibadan, Nigeria: IITA: 61-63.

_____. (by William B. Ward). (1977) *Decade of Progress and Promise for Tomorrow: 10th Anniversary, International Institute of Tropical Agriculture.* Ibadan, Nigeria: International Institute of Tropical Agriculture.

International Labor Organization (ILO). (2001) *World Employment Report 2001: Life at Work in the Information Economy*. Geneva: ILO.

International Press Institute (IPI). (2000) "Resolutions Adopted by IPI's General Assembly." Presented at the 49th IPI General Assembly held in Boston, MA, 1 May.

International Telecommunications Union (ITU). (2000a) *Internet Indicators, 2000.* Geneva: ITU.

_____. (2000b) *World Telecommunication Development Report, 2000.* Geneva: ITU.

_____. (1998) *African Telecommunication Indicators, 1998.* Geneva: ITU.

_____. (1998b) *Role of Telecommunications in Economic, Social and Cultural Development.* Geneva: ITU.

_____. (1997) "Telecommunication Policies for Africa: 'The African Green Paper'." Online. http://www7.itu.int/treg/reform/Policy_Papers/Green/green_eng.htm.

_____. (1996) *African Telecommunication Indicators, 1996*. Geneva: ITU.

Irele, A.F. (1999) "The Political Kingdom: Toward Reconstruction in Africa." Online. http://www.africahome.com/scholar/stories/irele_reconstruct.shtml. Retrieved 19 June 1999.

Isaacman, A.F. (1993) "Peasants and Rural Social Protest in Africa." In F. Cooper, F.E. Mallon, S.J. Stern, A.F. Isaacman, and W. Roseberry (eds.), *Confronting Historical Paradigms: Peasants, Labor, and the Capitalist World System in Africa and Latin America*. Madison, WI: University of Wisconsin Press: 205-317.

Jennings, B.H. (1988) *Foundations of International Agricultural Research*. Boulder, CO: Westview Press.

Jensen, M.R. (2001) "African Internet Status." Online. http://www.cyberatlas.internet.com/big_picture/ and http://www2sn.apc.org/africa/countdet.CFM?countries_ISOCode=NG.

_____. (2000a) "African Internet Status 2000." Message to Global Knowledge for Development Discussion List. 28 November 2000. Online. http://www.globalknowledge.org/english/archives/mailarchives/gkd/home.html and http://www.africapolicy.org/docs00/inet0010.htm.

_____. (2000b) "African Internet Status November 2000." *Report of the African Information Society Initiative*. http://www3.sn.apc.org/africa

_____. (2000c) "The Internet in Africa - Status in 1999." In BMI-TechKnowledge, *Communication Technologies Handbook 1999*, Johannesburg: BMI-Tech Group: 48-57.

_____. (2000d) "Internet and Computer Infrastructure in Africa Information and Communication Technologies (ICTs) Telecommunications." Online. www2sn.apc.org.

_____. (1999) "Strengthening Africa's Information Infrastructure." Paper Prepared for United Nations Economic Commission for Africa African Development Forum. Available Online. http://www.bellanet.org/partners/aisi/adf99docs/infrastructure.htm.

_____. (1996) "Bridging the Gaps in Internet Development in Africa." International Development Research Centre Study.

_____. (1995) "Telematics for Development: Discussion Paper." Online. http://www.sas.upenn.edu/African_Studies/PADIS/telmatics_Jensen.htm.

Jeppe, J. (1994) "Cultural Dimensions of Development Policy Management in the new South Africa." *DPMN Bulletin* 2, 2: 8-10.

Jones, P.G. and P.K. Thornton. (2000) "MARKSIM: Software to Generate Daily Weather Data for Latin America and Africa." *Agronomy Journal* 92: 445-453.

_____. (1999) "Fitting a Third-Order Markov Rainfall Model to Interpolated Climate Surface." *Agricultural and Forest Meteorology* 97: 213-231.

_____. (1997) "Spatial and Temporal Variability of Rainfall Related to a Third-Order Markov Model." *Agricultural and Forest Meteorology* 86: 127-138.

Joint Venture Silicon Valley (JV:SVN). (1999) *Joint Venture's 1999 Index of Silicon Valley*. Palo Alto, CA: Joint Venture Silicon Valley.

Journel, A.G. and C.J. Huijbregts. (1978) *Mining Geostatistics*. New York: Academic Press.

Jugessur, S. (1998) "Science and Technology for African Development: Partnerships in a Global Economy." Paper Presented at the Conference on Science and Technology for African Development: Partnerships in a Global Economy, held in Harare, Zimbabwe, 5-18 March.

Juma, M.N. (2000) *The African Virtual University: The Case of Kenyatta University*. London: Commonwealth Secretariat.

Kanter, J. (1984) *Management Information Systems*. [3rd Ed.] Englewood Cliffs, NJ: Prentice Hall.

Kaplan, D. (1990) *The Crossed Line: The South African Telecommunications Industry in Transition*. Johannesburg: Wits University Press.

Kaplan, D. and R. Kaplinsky. (2000) *Information and Communication Technology in a Middle Income-Country: The Case of South Africa*. Geneva: International Labor Organization.

Kaplinsky, R. (1987) *Microelectronics and Work Revisited: A Review*. Geneva: International Labor Office.

Kaul, M. (1995) "Civil Service Reform: A Commonwealth Perspective." *DPMN Bulletin* 3, 1: 4-6.

Keeley, J. and I. Scoones. (2000) *Global Science, Global Policy: Local to Global Policy Processes for Soils in Africa*. IDS Working Paper No.115. Brighton: Institute for Development Studies.

Kenney, M. (ed.). (2000) *Understanding Silicon Valley: The Anatomy of An Entrepreneurial Region*. Palo Alto, CA: Stanford University Press.

Kenniston, K. and D. Kumar (eds.). (2000) *Bridging the Digital Divide: Lessons From India*. Banglore: National Institute of Advanced Study.

Kenny, B. and A. Bezuidenhout. (1999) "Contracting, Complexity and Control: An Overview of the Changing Nature of Subcontracting in the South African Mining Industry." *Journal of The South African Institute of Mining and Metallurgy* (July/August): 185-191.

Kenny, B. and E. Webster. (1998) "Eroding the Core: Flexibility and the Re-Segregation of the South African Labor Market." *Critical Sociology* 24, 3.

Kitching, G. (1982) *Development and Underdevelopment in Historical Perspective: Populism, Nationalism, and Industrialization*. New York: Metheun.

Kizza, I.N. (1999) *Africa's Indigenous Institutions in Nation Building*. Lampeter, UK: The Edwin Mellen Press, Ltd.

Klerck, G. (1994) "Industrial Restructuring and the Casualization of Labor: A Case Study of Subcontracted Labor in the Process Industries." *South African Sociological Review* 71: 32-62.

Kling, R. (1980) "Social Analysis of Computing: Theoretical Perspectives in Recent Empirical Research." *ACM Computing Surveys* 12, 1: 61-110.

Knight, P.T. (1996) "Destined to Leapfrog: Why a Revolution in Learning Will Occur in Brazil, Russia and South Africa." Paper Presented at the Second International Conference on Distance Education in Russia, 2-5 July.

_____. (1995) "The Telematics Revolution in Africa and the World Bank Group." Paper Presented at the AFRISTECH '95 Symposium on Information Superhighways, 'What Strategy for Africa?,' held in Dakar, 12-15 December.

Kole, E. (1999) "Connecting Women From Developing Countries to the Internet: Searching for an Appropriate Paradigm." Paper Presented at the 41st Annual Convention of the International Studies Association held in Los Angeles, California.

Koppel, B.M. (ed.). (1995) *Induced Innovation Theory and International Agricultural Development: A Reassessment.* Baltimore, MD: Johns Hopkins University Press.

Koppel, B.M. and E. Oasa. (1987) "Induced Innovation Theory and Asia's Green Revolution: A Case Study of an Ideology of Neutrality." *Development and Change* 18: 29-67.

Korang, K. (2002) "Colonial Intellectuals." In P.T. Zeleza and D. Eyoh (eds.), *Encyclopedia of Twentieth Century African History.* London/New York: Routledge: 266-269.

Kramer, K. and J. Dedrick. (1994) "Payoffs from Investment in Information Technology: Lessons from the Asia-Pacific Region." *World Development* 22, 22: 1921-1931.

Kramer, J. and C. Kramarae. (1996) "Women's Political Webs: Global Electronic Networks." In A. Sreberny and L. Van Zoonem (eds.), *Gender Politics and Communication.* New Jersey: Hampton Press: 205- 222.

Kripalani, M. (2001) "Hanging Up on Private Cell Companies." *Business Week* 5 March.

Kroeber, A.L. (1931) "The Culture-Area and Age-Area Concepts of Clark Wissler." In S. Rice (ed.), *Methods in Social Science.* Chicago, IL: University of Chicago Press: 248-265.

Kumbula, T.S. (1997) "Press Freedom in Zimbabwe." In F. Eribo and W. Jong-Ebot (eds.), *Press Freedom and Communication in Africa.* Trenton, NJ: Africa World Press.

Larouche, P. (2000) "The Basis of EC Telecommunications Law after Liberalisation." Unpublished Ph.D. Thesis, Maaastricht University.

Laws, E.A. (1993) *Aquatic Pollution: An Introductory Text*. New York: John Wiley and Sons, Inc.

Lawson, C. and W. Lorenz. (1999) "Collective Learning, Tacit Knowledge and Regional Innovative Capacity." *Regional Studies* 334: 305-317.

Leach, M. and R. Mearns. (1996) *The Lie of the Land: Challenging Received Wisdom on the African Environment*. Oxford, UK: James Curry.

Lebel, T. and L. le Barbe. (1997) "Rainfall Monitoring During the HAPEX-Sahel. 2 Point and Areal Estimation at the Event and Seasonal Scales." *Journal of Hydrology* 188-189: 97-122.

Lebel, T., J.D. Taupin, and N. d'Amato. (1997) "Rainfall Monitoring During the HAPEX-Sahel. 1. General Rainfall Conditions and Climatology." *Journal of Hydrology* 188-189: 74-96.

Leftwich, A. (1995) "Bringing Politics Back In: Towards a Model of the Developmental State." *Journal of Development Studies* 31, 3: 400-427.

Legg, J.P. (2000) "The Cassava Mosaic Virus Disease Pandemic in East and Central Africa: Past, Present and Future." Paper Presented at a Workshop on the Management of Cassava Virus Disease in Tanzania, held in Kibaha, Tanzania, 6-8 September.

Legg, J.P. and S. Ogwal. (1998) "Changes in the Incidence of African Cassava Mosaic Geminivirus and the Abundance of its Whitefly Vector Along South-North Transects in Uganda." *Journal of Applied Entomology* 122: 169-178.

Legg, J.P. and G. Okao-Okuja. (1999) "Progress in the diagnosis and epidemiological characterisation of cassava mosaic geminiviruses in East Africa." *Proceedings of the 7th International Plant Virus Epidemiology Symposium, 11-16 April 1999*. Aguadulce (Almeria), Spain: 74-75.

Legg, J.P., P. Sseruwagi, J. Kamau, S. Ajanga, S.C. Jeremiah, V. Aritua, G.W. Otim-Nape, A. Muimba-Kankolongo, R.W. Gibson, and J.M. Thresh. (1999) "The Pandemic of Severe Cassava Mosaic Disease in East Africa: Current Status and Future Threats." In: M.O. Akoroda and J.M. Teri (eds.), *Proceedings of the Scientific Workshop of the Southern*

African Root Crops Research Network (SARRNET), 17-19 August 1998. Lusaka, Zambia: 236-251.

Lele, U. and S. Stone. (1989) *Population Pressure, the Environment and Agricultural Intensification: Variations on the Boserup Hypothesis.* MADIA Discussion Paper No. 4. Washington, DC: The World Bank.

Lesetedi, G.N. (1999) "HIV/AIDS and the Status of Women." In K. Hope (ed.), *AIDS and Development in Africa.* New York: The Haworth Press: 47-57.

Lettington, R. (2000) "Re: Field Trials of GM Crops in the Developing World." In FAO, *The Electronic Forum on Biotechnology in Food and Agriculture.* 3 April 2000. Online. http://www.fao.org/biotech/forum.htm.

Levin, L. (1996) "Report to the IDRC on the Use of Information and Communication Technologies (ICT's) in Sub-Saharan Africa in the Area of Governance." Online. http://www.idrc.ca/acacia/studies/ir-lev.htm. Retrieved 1 January 2000.

Levine, C.S. (1998) "Reorienting for Sustainable Development: Support for a National Science and Technology Policy." *Journal of International Affairs* 51, 2: 675-688.

Levy, B. and P. Spiller. (1994) "The Institutional Foundations of Regulatory Commitment: A Comparative Analysis of Telecommunications Regulation." *The Journal of Law, Economics, and Organisation* 10, 2: 201-246.

Little, P.D. (1985) "Adding a Regional Perspective to Farming Systems Research: Concepts and Analysis." *Human Organization* 44, 4: 331-337.

Little, P.D. (1984) *Regional Analysis and Farming Systems Research.* Birmingham, NY: Institute for Development Anthropology.

Little, P.D. and M. Painter. (1995) "Discourse, Politics, and the Development Process: Reflections on Escobar's 'Anthropology and the Development Encounter'." *American Ethnologist* 22, 3: 602-616.

Lovelock, J.E. (1979) *Gaia: A New Look at Life on Earth.* Oxford: Oxford University Press.

Lush, D., H. Rushwayo, and F. Banda (eds.). (2000) *Into or Out of the Digital Divide? Perspectives on ICTs and Development in Southern Africa*. Lusaka: PANOS Southern Africa.

Lutta-Mukhebi, M.C. and M.N. Amutabi. (2000) "Globalization of Communication and Media Technology on Democracy in Kenya." Paper Presented at the 6th Congress of the Organization of Social Research in Eastern and Southern Africa (OSSREA), held in Dar-Es-Salaam, Tanzania, 23-28 April.

Mabawonku, A.O., N.N. Jibiri, and O.O. Fasunwon. (2001) "Cultural Challenges to the Development of Science and Technology in Africa." Paper Presented at the 27th Annual Symposium on Technology and Development in Africa, Center for African Studies and College of Engineering, University of Illinois at Urbana-Champaign.

MacDonald, B.A. (1986) "Information Management in the Public Service: Summary of Discussions." *Canadian Public Administration* 29, 1: 1-16.

Mackenzie, D. and J. Wajcman (eds.). (1985) *The Social Shaping of Technology*. Buckingham: Open University Press.

Madon, S. (1992) "Computer-Based Information Systems for Development Planning: the Significance of Cultural Factors." *Journal of Strategic Information Systems* 1, 5: 250-257.

Madu, C. and R. Jacob. (1991) "Multiple Perspectives and Cognitive Mapping to Technology Transfer Decisions." *Futures* 23, 9: 978-97.

Mahoney, R.J., S. Ostry, and E.M. Graham. (1995) "In Defense of Techno-Globalism." *Issues in Science and Technology* 12: 5-6.

Malawi Government. (2000) *Public Expenditure Review*. Lilongwe, Malawi: The World Bank.

Malki, L.H. (1997) "National Geographic: The Rooting of Peoples and Territorialization of National Identity among Scholars and Refugees." In A. Gupta and J. Ferguson (eds.), *Culture, Power, Place: Explorations in Critical Anthropology*. Durham: Duke University Press: 52-75.

Mamdani, M. (1996) *Citizen and Subject: Contemporary Africa and The Legacy of Late Colonialism*. Princeton, NJ: Princeton University Press.

Mandel, M. (1996) *The High-Risk Society: Peril and Promise in the New Economy*. New York: Times Books.

Margherio, L., D. Henry, S. Cooke, S. Montes, and K. Hughes (eds.). (1998) *The Emerging Digital Economy*. Washington, DC: Department of Commerce, Secretariat on Electronic Commerce.

Marx, K. (1847) *Wage Labor and Capital*. New York: W.W. Norton & Company.

Maturana, H. R. and F.J. Varela. (1987) *The Tree of Knowledge: The Biological Roots of Human Understanding*. Boston, MA: Shambhala.

Mayer, J. and L. Price (eds.). (1999) *The Emerging Digital Economy II*. Washington, DC: Department of Commerce, Secretariat on Electronic Commerce.

Mayor, F. (2000) "Opening Address." In UNESCO, *World Conference on Science: Science for the Twenty-First Century: A New Commitment*. Paris: UNESCO: 26-28.

Mazrui, A.A. (1999) "The African Renaissance: A Triple Legacy of Skills, Values and Gender." Online. http://africacentre.org.uk/renaissance.htm. Retrieved 23 November 1999.

_____. (1995) "The Polity and the University in African Perspective." In Senjo B-S. K. Adjibolosoo (ed.), *The Human factor in Developing Africa*. Wesport, CT: Praegar.

Mazur, B.D. (1995) "Plant Biotechnology." *TIBTECH* 13: 319 -323.

McCann, J.C. (1995) *People of the Plow: An Agricultural History of Ethiopia, 1800-1990*. Madison, WI: University of Wisconsin Press.

McCorkle, C.M. (2000) "Anthropology, Sociology and FSR." In M. Collinson (ed.), *A History of Farming Systems Research*. Wallingford / Rome: CABI Publishing/ FAO.

McGuinness, H. (1993) *Living Soils: Sustainable Alternatives to Chemical Fertilizers for Developing Countries*. Yonkers, NY: Consumer Policy Institute.

McIntire, J., D. Bourzat, and P. Pingali. (1992) *Crop-Livestock Interaction in Sub-Saharan Africa*. Washington, DC: The World Bank.

McItughen, A. (1994) "Assessment of the Impact and Performance of Transgenic Plants." In G. Marshall and O. Walters (eds.), *Molecular Biology in Crop Protection*. London: Chapman and Hall.

McMillan, D.E. (1990) "Curriculum Development in the Study of African Food Issues at a US Land Grant University: A Case Study." In R. Huss-Ashmore and S.H. Katz (eds.), *African Food Systems in Crisis: Coping With Change*. New York: Gordon and Breach Science Publishers: 139-158

_____. (1987) "Monitoring the Evolution of Household Economic Systems Over Time in Farming Systems Research." *Development and Change* 18, 2: 295-314.

McWhinney, W. (2000) "Structural Coupling." Personal Email Communication.

Meadows, D.H., D.L. Meadows, J. Randers, and W.W. Behrens III. (1972) *The Limits to Growth*. New York: Universe Books.

Media Institute of Southern Africa. (2000) "Public Broadcaster and City Authorities Allegedly Ban Company's Adverts." *Media Institute of Southern Africa* 10 November.

Media Rights Agenda. (2000) "The Campaign for a Freedom of Information Act." *Nigerian Media Rights Monitor*. Online. www.internews.org/mra/ewatch/ewMay.pdf. Retrieved 24 May 2000.

Melody, W.H. (1999) "Telecom Reform: Progress and Prospects." *Telecommunications Policy* 23: 7-34.

Menou, M. and N. Hafkin. (1997) "Connectivity in Africa: Use, Benefits and Constraints of Electronic Communications." *Synthesis Report Part 2: Overview of the Project Findings*. Online. www.bellanet.org/partners/aisi/prof/findings.htm and http://cneng.netvalue.com/news/index.htm.

Mensah, M.C. (2000) *Feeding More People and Better in West Africa: The Case of Benin*. Lecture Series No. 6. Washington, DC: IFPRI.

Merril-Sands, D. (1986) "Farming Systems Research: Clarification of Terms and Concepts." *Experimental Agriculture* 22: 87-104.

Merrill-Sands, D., P. Ewell, S. Biggs, R.J. Bingen, J. McAllister, and S. Poats. (1992) "Management of Key Institutional Linkages in On-Farm Client-oriented Research." In J.L. Moock and R.E. Rhoades (eds.), *Diversity, Farmer Knowledge, and Sustainability*. Ithaca: Cornell University Press: 113-146.

Migdal, J. (1974) *Peasants, Politics and Revolution: Pressures Toward Political and Social Change in the Third World.* Princeton, NJ: Princeton University Press.

Milford, J.R., V.D. McDougall, and G. Dugdal. (1994) "Rainfall Estimation from Cold Cloud Duration." Proceedings of Niamey Workshop held at ORSTOM, Niamey, December 1994.

Miller, H.I. (2000) "The Science of Biotechnology Meets the Politics of Global Regulation." *Issues in Science and Technology* 17, 1: 47-54.

Miller, J. (1999) "Entering the Information Age: Setting Priorities in South Africa." Paper Presented at the Second International Al Shaam Conference on Information Technology held in Damascus, Syria, 26-29 April.

Min, W. (2000) "Telecommunications Regulations: Institutional Structures and Responsibilities." *OECD Working Party on Telecommunication and Information Services Policies*. Paris: OECD.

Mingers, J. (1995) *Self-Producing Systems Implications and Applications of Autopoiesis*. New York: Plenum Press.

Minges, M. (2000) "The African Telecommunications Renaissance." In BMI-TechKnowledge, *Communication Technologies Handbook 1999*. Johannesburg: BMI-Tech Group: 38-44.

Ministry of Agriculture and Irrigation. (2000) *Agriculture Extension in the New Millennium: Towards Pluralistic and Demand-Driven Services in Malawi*. Lilongwe, Malawi: Agricultural Communication.

Ministry of Agriculture and Irrigation. (1999) *Review of Malawi Agricultural Policies and Strategies*. Lilongwe, Malawi: Department of Economic Planning.

Ministry of Finance and Economic Development. (2001) *Poverty Reduction Strategy Growth and Diversification.* Lilongwe, Malawi: Malawi Government.

Ministry of Transport and Communications. (1999) *Telecommunications and Postal Sector Policy Statement, April.* Nairobi, Kenya: Ministry of Transport and Communications.

Molutsi, P and M.N. Badade. (1999) "Societal Responses to HIV/AIDS in Botswana." In K. Hope (ed.), *AIDS and Development in Africa.* New York: Haworth Press: 143-151.

Montealegre, R. (1999) "A Case for More Case Study Research in the Implementation of Information Technology in Less Developed Countries." *Information Technology for Development* 8, 4: 199-207.

Moock, J. (ed.). (1986) *Understanding Africa's Rural Households and Farming Systems.* Boulder, CO: Westview Press.

Moodley, S. (2001a) *E-Business and Supply Chain Management in the Automotive Industry: Preliminary Findings From the Eastern Cape and Kwazulu-Natal Benchmarking Club Pilot Surveys.* Durban, South Africa: University of Natal, School of Development Studies.

_____. (200b) *E-Business in a Low-Tech Sector: The Case of the South African Wooden Furniture Value Chain.* Durban, South Africa: University of Natal, School of Development Studies.

Moore, H.L. and M. Vaughan. (1994) *Cutting Down Trees: Gender, Nutrition, and Agricultural Change in the Northern Province of Zambia 1890-1990.* Cambridge, UK: Cambridge University Press.

Moore, S.F. (1994) *Anthropology and Africa: Changing Perspectives on a Changing Scene.* Charlottesville, VA: The University Press of Virginia.

More E. and H. Irwin. (1995). "Technology Transfer: Technocultures, Power and Communication - the Australian Experience." *Journal of Information Science* 21, 3: 149-59.

Morna, C. And Z. Khan. (2000) *Net Gains: African Women Take Stock of Information and Communication Technologies.* Johannesburg, South Africa: Gender Links.

Moulin, S., A. Bondeau, and R. Delecolle. (1998) "Combining Agricultural Crop Models and Satellite Observations from Field to Regional Scale." *International Journal of Remote Sensing* 19: 1021-1036.

Mozley, R.E. and M.J. Blaine (eds.). (1996) *Information Technology Development and Policy*. Ipswich, Suffolk: Ipswich Book Co.

Mulira N. (1995) "IT Management in Uganda, Strategies and Policies." *Information Technology for Development* 6 (September): 95-105.

Munsanje, E.M. (2000) "Appropriate Biotechnologies for Developing Nations." In FAO, *The Electronic Forum on Biotechnology in food and Agriculture*. 27 March 2000. Available Online. http://www.fao.org/biotech/forum/htm.

Murphy, D.D. (1998) "A Revolution in Progress: Today's Proliferating Science and Technology." *The Journal of Social, Political and Economic Studies* 23, 1: 75-92.

Murray, D.L. and P.L. Taylor. (2000) "Claim No Easy Victories: Evaluating the Pesticide Industry's Global Safe Use Campaign." *World Development* 28, 10: 1735-1749.

NALEDI. (2000) *Highlights of Current Labor Market Conditions in South Africa*. Johannesburg, South Africa: NALEDI.

Nanda, P. (2000) *Health Sector Reforms in Zambia: Implications for Reproductive Health and Rights*. Takoma Park: Center for Health and Gender Equity.

National Cancer Institute. (1992) *Making Health Communication Programs Work*. Washington, DC: National Institutes of Health.

National Economic Council of Malawi, (2000) *Malawi Vision 2020 National Long-term Development Perspective for Malawi*. Lilongwe, Malawi: National Economic Council.

_____. (1999) *National Agricultural and Natural Resources Research Master Plan*. Lilongwe, Malawi: Agricultural Science Committee.

Neumann, R.P. (1998) *Imposing Wilderness: Struggles Over Livelihood and Nature Preservation in Africa*. Berkeley, CA: University of California Press.

Newsom, M. (1995) "The Earth as Output: Pollution." In R.J. Johnston, P.J. Taylor, and M.J. Watts (eds.), *Geographies of Global Change: Re-mapping the World in the Late Twentieth Century*. Oxford: Blackwell: 333-353.

Niemeijer, D. (1996) "The Dynamics of African Agricultural History - Is It Time for a New Development Paradigm?" *Development and Change* 27, 1: 87-110.

Nolan, R.L. (1979) "Managing the Crises in Data Processing." *Harvard Business Review* 52, 7: 115-26.

_____. (1973) "Managing the Computer Resource: A Stage Hypothesis." *Communications of the ACM* 16, 7: 399-405.

Nonaka, I. and H. Takeuchi. (1995) *The Knowledge-Creating Company: How Japanese Companies Create the Dynamics of Innovation*. New York: Oxford University Press.

Norman, D.W. (2000) "Farm Management and the Farming Systems Approach." In M. Collinson (ed.), *A History of Farming Systems Research*. Wallingford / Rome: CABI Publishing / FAO: 293-299.

Norman, D.W. and D.C. Baker. (1986) "Components of Farming Systems Research, FSR Credibility, and Experiences in Botswana." In J. Moock (ed.), *Understanding Africa's Rural Households and Farming Systems*. Boulder, CO: Westview Press: 36-57.

Norman, D.W. and E. Gilbert. (1982) "A General Overview of Farming Systems Research." In W.W. Shaner, P.F. Philipp, and W.R. Schmehl (eds.), *Readings in Farming Systems Research and Development*. Boulder, CO: Westview Press: 17-30.

Norman, D.W. and P.J. Matlon. (2000) "Agricultural Systems Research and Technical Change." In J.-P. Colin and E.W. Crawford (eds.), *Research on Agricultural Systems: Accomplishments, Perspectives and Issues*. Huntington, NY: Nova Science Publishers, Inc: 17-47.

Nthoi, J. (2001) "Traditional Concepts of HIV/AIDS: Implications for the Control of the Affliction in Botswana." Paper Presented at the Theology and Religious Studies Department Seminar held at the University of Botswana, Gaborone.

NUA. (2001) "NUA Internet Surveys." Online. http://www.nua.com/surveys/how_many_online/index.html.

Nzomo, M. (1995) "Women and Democratization Struggles in Africa: What Relevance to Postmodernist Discourse?" In M. Marchard and J. Parpart (eds.), *Feminism Postmodernism Development.* New York, NY: Routledge: 127-130.

Oasa, E.K. (1987) "The Political Economy of the International Agricultural Research: A Review of CGIAR's Response to Criticisms of the 'Green Revolution'." In B. Glaeser (ed.), *The Green Revolution Revisited.* London: Allen and Unwin.

_____. (1985) "Farming Systems Research: A Change in Form But Not Content." *Human Organization* 44, 3: 219-227.

Odedra, M. (1992) "Is Information Technology Really Being Transferred to the African Countries?" In G. Cyranek and S.C. Bhatnagar (eds.), *Technology Transfer for Development.* New Delhi: Tata McGraw Hill: 47-58.

Ogunseitan, O., B. Durojaije, B. Denloye, and O. Hanciles. (1991) "Africa Leads the Neglect of the Technology of the Future." *African Science Monitor* 1, 1: 27-37.

Ojo, S.O. (1992) "Socio-cultural and Organizational Issues in IT Applications in Nigeria: Key Research Issues." In S.C. Bhatnagar and M. Odedra (eds.), *Social Implications of Computers in Developing Countries*. New Delhi: Tat McGraw Hill: 99-109.

Okafor, N. (1994) "Biotechnology and Sustainable Development in Sub-Saharan Africa: Role of Biotechnology in Developing Country Development." *World Journal of Microbiology and Biotechnology* 10, 3: 243-248.

Oliver, A. and J. Smith. (1999) "Implausible Deniability." *Brill's Content* February.

Olivier de Sardan, J.-P. (2000) "Social Anthropology, Peasantry, and Agricultural Systems Analysis." In J.-P. Colin and E.W. Crawford (eds.), *Research on Agricultural Systems: Accomplishments, Perspectives and Issues*. Huntington, NY: Nova Science Publishers, Inc: 223-242.

Ong, A. (1996) "Anthropology, China and Modernities: The Geopolitics of Cultural Knowledge." In H.L. Moore (ed.), *The Future of Anthropological Knowledge*. London: Routledge: 60-93.

Organization for Economic Cooperation and Development (OECD). (1996) *The Knowledge-Based Economy*. Paris: OECD.

_____. (1991) *Managing Technological Change in Less-Advanced Developing Countries*. Paris: OECD.

_____. (1992) *Information Computer Communications Policy 27: Telecommunications Type Approval: Policies and Procedures for Market Access*. Paris: OECD.

O'Riain, S. (1997) "The Birth of Celtic Tiger." *Communications of the ACM* 403: 11-16.

_____. (2000) "The Flexible Developmental State: Globalization, Information Technology and the 'Celtic Tiger'." *Politics and Society* 28, 2: 157-193.

Osiro, D.A. (1999) "Investing in ICT: Legal Perspectives." Unpublished LL.B. Dissertation, University of Nairobi, Kenya.

Otim-Nape, G.W., A. Bua, and Y. Baguma. (1994) "Accelerating the Transfer of Improved Production Technologies: Controlling Cassava Mosaic Virus Disease Epidemics in Uganda." *African Crop Science Journal* 2: 479-495.

Otim-Nape, G.W., A. Bua, J.M. Thresh, Y. Baguma, S. Ogwal, G.N. Semakula, G. Acola, B. Byabakama, and A. Martin. (1997) *Cassava Mosaic Virus Disease in Uganda: The Current Pandemic and Approaches to Control*. Chatham, UK: Natural Resources Institute.

Overseas Development Institute (ODI). (1994) *Comments on 'Livestock, Land Use and Agricultural Intensification in Sub-Saharan Africa'*. London: Overseas Development Institute.

Owino, W. (2000) "Cultural Factors Hamper Botswana's AIDS Campaign." *Pan-African News Agency Report* 1 December 2000. Online. http://allafrica.com/stories/200012010344.html.

Paige, J.M. (1975) *Agrarian Revolution: Social Movements and Export Agriculture in the Underdeveloped World.* New York: Free Press.

Parker, E. and J. Rogers. (1995) *The Wisconsin Regional Training Partnership: Lessons for National Policy.* Berkeley, CA: National Center for the Workplace, IIR.

Parpart, J. (1995) "Deconstructing the Development 'Expert': Gender, Development and the Vulnerable Groups." In M. Marchard and J. Parpart (eds.), *Feminism Postmodernism Development.* New York, NY: Routledge: 221-143.

Parpart, J. and M. Marchard. (1995) "Exploding the Canon: An Introduction/Conclusion." In M. Marchard and J. Parpart (eds.), *Feminism Postmodernism Development.* New York, NY: Routledge: 23-41.

Parpart, J., P. Connelly, and V. Barriteau. (2000) *Theoretical Perspectives on Gender and Development.* Ottawa, ON: International Development Research Center.

Parpart, L.J. (1995) "Postmodernism, Gender and development." In J. Crush (ed.), *Power of Development.* New York, NY: Routledge: 252-265.

Parthasarathy, B. (2000) "Globalization and Agglomeration in Newly Industrializating Countries: The State and the Information Technology Industry in Bangalore, India." *City and Regional Planning.* Berkeley, CA: University of California.

Patterson, R. (2001) "Strategic Policy Benchmarking for Technological Development: The IT Cases of Zimbabwe and Botswana." *International Journal of Comparative Sociology* 41: 2.

Patterson, R. and J. Bozeman. (1999) "Comparativist Study of State Promotion of Science and Technology: Cases of Botswana and Singapore." In R. Patterson (ed.), *Science and Technology in Southern Africa and East and South Asia.* Boston, MA: Brill Academic Publishers.

_____. (1997) "Relative Prospects for Foreign Direct Investment and Foreign State Investment into Southern Africa." *Journal of Developing Societies* 13: 3-4.

Paye, J-C. (1995) "Making Life-Long Learning a Reality for All." *OECD Observer* 193: 4-5.

Peled, A. (2000) "First Class Technology, Third Rate Bureaucracy - The Case of Israel." *Information Technology for Development* 9, 45-58.

Pfeffer, J. (1992) *Managing with Power: Politics and Influence in Organizations.* Boston, MA: Harvard Business School Press.

Pincus. (1996) *Class, Power and Agrarian Change: Land and Labor in Rural West Java.* London: St. Martin's Press.

Pingali, P., Y. Bigot, and H.P. Binswanger. (1987) *Mechanization and the Evolution of Farming Systems in Sub-Saharan Africa.* Baltimore, MD: Johns Hopkins for the World Bank.

Pingali, P. and H. Binswanger. (1988) "Population Density and Farming Systems: The Changing Locus of Innovations and Technical Change." In R.D. Lee, W.B. Arthur, A.C. Kelley, G. Rogers, and T.N. Srinivasan (eds.), *Population, Food and Rural Development.* Oxford: Clarendon Press: 51-76.

_____. (1987) "Population Density and Agricultural Intensification: A Study of the Evolution of Technologies in Tropical Agriculture." In D.G. Johnson and R.D. Lee (eds.), *Population Growth and Economic Development: Issues and Evidence.* Madison, WI: University of Wisconsin Press: 283-318.

_____. (1984) *Population Density and Agricultural Intensification: A Study of the Evolution of Technologies in Tropical Agriculture.* Agriculture and Rural Development Department, Research Unit, Report No. ARU 22. Washington, DC: The World Bank.

Pinstrup-Andersen, P. (1999) "Modern Biotechnology and Small Farmers in Developing Countries." *IFPRI Research Perspectives* 21, 2.

Poats, S.V., M. Schmink, and A. Spring. (1988) "Linking FSR/E and Gender: An Introduction." In S.V. Poats, M. Schmink, and A. Spring (eds.), *Gender Issues in Farming Systems Research and Extension..* Boulder, CO: Westview Press: 1-18

Popkin, S.L. (1979) *The Rational Peasant: The Political Economy of Rural Society in Vietnam.* Berkeley, CA: University of California Press.

Pretty, J. (1998) "On Feeding the World." *The Splice of Life: The Genetics Forum's Magazine* 4: 6.

Prewitt, K. (ed.). (1998) *Networks in International Capacity Building: Cases from Sub-Saharan Africa.* SSRC Working Paper 2. New York: SSRC.

Price, D.T., D.W. McKenny, I.A. Nalder, M.F. Hutchinson, and J.L. Kesteven. (2000) "A Comparison of Two Statistical Methods for Interpolation of Canadian Monthly Mean Climate Data." *Agricultural Forest Meteorology* 101: 81-94.

Priotrow, P.T., D.L. Kincaid, J.G. Rimon, and W. Rinehart. (1997) *Health Communication: Lessons From Family Planning and Reproductive Health.* Westport, CT: Praeger.

Radcliffe-Brown, A.R. (1952) *Structure and Function in Primitive Society.* London: Cohen & West.

Radcliffe-Brown, A.R. and D. Forde (eds.). (1950) *African Systems of Kinship and Marriage.* London: Oxford University Press for the International African Institute.

Ranger, T. (1983) "The Invention of Tradition in Colonial Africa." In E. Hobsbawm and T. Ranger (eds.), *The Invention of Tradition.* Cambridge, UK: Cambridge University Press: 211-262.

Rasheed, S. (1995) "Ethics and Accountability in the African Civil Service." *DPMN Bulletin III* 1: 12-14.

Rasiah, R. and M. Best. (2000) *Industrial Transition in the Malaysian Electronics Industry.* Vienna: United Nations Industrial Development Organization.

Rathgeber, E. and E.O. Adera. (2000) *Gender and the Information Revolution in Africa.* Ottawa, ON: International Development Research Center.

Ratzan, S.C., G.L. Filerman, and J.W. LeSar. (2000) *The Population Bulletin.* Washington, DC.

Reardon, T. (1998) "African Agriculture: Productivity and Sustainability Issues." In C.K. Eicher and J.M. Statz (eds.), *International Agricultural Development.* Baltimore, MD: The Johns Hopkins University Press: 444-457.

Reid, D. (1995) *Sustainable Development.* London: Earthscan Publications Ltd.

Reporters Without Borders. (2001) *The Enemies of the Internet: Obstacle to the Free Flow of Information on the Internet.* Paris: Reporters Without Borders.

Republic of Botswana. (1998) *Botswana National Policy on HIV/AIDS.* Gaborone: Ministry of Health.

_____. (1997) *Botswana HIV and AIDS: Second Medium Term Plan.* Gaborone: Ministry of Health.

Riaño, P. (ed.). (1994) *Women In Grassroots Communication: Furthering Social Change.* Thousand Oaks, London/ New Delhi: SAGE Publications.

Ribot, J.C. (1999) "Decentralization, Participation, and Accountability in Sahelian Forestry." *Africa* 69,1: 23-65.

Richards, P. (1996) *Fighting for the Rainforest: War, Youth & Resources in Sierra Leone.* Oxford: James Currey.

_____. (1985) *Indigenous Agricultural Revolution: Ecology and Food Production in West Africa.* London: Hutchinson.

_____. (1983) "Ecological Change and the Politics of African Land Use." *African Studies Review* 25, 2: 1-83.

Richardson, D. and L. Paisley (eds.). (1998) *The First Mile of Connectivity: Advancing Telecommunications for Rural Development Through a Participatory Communication Approach,* Rome: FAO.

Roberts, B. (2000) "Freedom of the Press and Freedom of the Internet." *Broadcast Dialogue,* North American Broadcasters Association. May 2000. Online. http://www.nabanet.com/news/docs/freedom.doc.

Robotham, D. (2000) "Postcolonialities: The Challenge of New Modernities." *International Social Science Journal* 52: 357-371.

Roche, M.E. and M.J. Blaine (eds.). (1996) *Information Technology Development and Policy.* Ipswich, Suffolk: Ipswich Book Co.

Rogers, E.M. (1995) *Diffusion of Innovations*. New York: Free Press.

Rogerson, C.M. (2001) "Knowledge-Based or Smart Regions in South Africa." *South African Geographical Journal* 83: 34-47.

_____. (1998) "High-Technology Clusters and Infrastructure Development: International and South African Experiences." *Development Southern Africa* 15: 875-905.

Rukuni, C. (2000) "Boycott Sends Circulation of Bulawayo Papers Tumbling." *Insider Online*. http://users.harare.iafrica.com/~insider/. Retrieved 28 September 2000.

Rukuni, M. and C.K. Eicher. (1994) *Zimbabwe's Agricultural Revolution*. Harare: University of Zimbabwe.

Ruthenberg, H. (1980) *Farming Systems in the Tropics*. London: Oxford University Press.

Ruttan, V.W. and Y. Hayami. (1972) "Induced Innovation Model of Agricultural Development." In C.K. Eicher and J.M. Staatz (eds.), *International Agricultural Development*. [3rd Ed.] Baltimore, MD: The Johns Hopkins University Press: 163-178.

Ruttan, V.W. and Y. Hayami. (1995) "Induced Innovation Theory and Agricultural Development." In B.M. Koppel (ed.), *Induced Innovation Theory and International Agricultural Development: A Reassessment*. Baltimore, MD: Johns Hopkins University Press: 22-38.

Ruttan, V.W. and C. Thirtle. (1989) "Induced Technical and Institutional Change in African Agriculture." *Journal of International Development* 1, 1: 1-45.

Sagan, C. (1990) "Science and Technology in the 20th Century: Good and Bad." *New Perspectives Quarterly* 16, 2: 25-30.

Sahay, S. and G. Walsham. (1995) "Information Technology in Developing Countries: a Need for Theory Building." *Information Technology for Development* 6: 111-124.

Salam, A. (1987) "Science Transfer for Development and Global Problems of Science and Technology." In *A Man of Science: Abdus Salam Nobel Laureate in Physics, 1979*. Ljubljana: Regional Centre for Cooperation with Developing Countries: 35-55.

Salih, M.A.M. (1999) *Environmental Politics and Liberation in Contemporary Africa*. Boston, MA: Kluwer Academic Publishers.

Saney, G., J. Verdin, and R. Smith. (2001) "Crop Yield Estimate Comparison Between Two Methods: Energy Balance EWBMS and Water Balance WRSI." *FEWS NET*.USGS/EROS Data Centre.

Sanghvi, A.P. (2000) "African Rural and Renewable Energy Literature (AFRREL)." Kisiizi Telecentre, Pilot Project. 1 November.

Sarker, A. (2001) "The Internet: Bridging the Digital Divide." *India: Dataquest*. Online. http://www.dqindia.com/content/Trends /201032102.asp.

Sawyer C.N. and P.L. McCarty. (1978) *Chemistry for Environmental Engineering*. [3rd Ed.] New York: McGraw-Hill Book Company.

Saxenian, A. (1994) *Regional Advantage: Culture and Competition in Silicon Valley and Route 128*. Cambridge, MA: Harvard University Press.

Scherf, B.D. (1995) *World Watch List for Domestic Animal Diversity*. Rome: FAO.

Schiere, J.B., J. Lyklema, J. Schakel, and K.G. Rickert. (1999) "Evolution of Farming Systems and Systems Philosophy." *Systems Research and Behavioral Science* 16, 4: 375.

Schultz, T. (1964) *Transforming Traditional Agriculture*. New Haven, CT: Yale University Press.

Schumacher, E.F. (1973) *Small Is Beautiful: a Study of Economics As If People Mattered*. London: Blond and Briggs.

Scoones, I. (1994) *Living With Uncertainty: New Directions in Pastoral Development in Africa*. London. Intermediate Technology Publications.

Sconnes, I. with C. Chibudu, S. Chikura, and P. Jeranyama. (1996) *Hazards and Opportunities: Farming Livelihoods in Dryland Africa. Lessons From Zimbabwe*. London: IIED.

Scoones, I. and J. Thompson (eds.). (1994) *Beyond Farmer First: Rural People's Knowledge, Agricultural Research and Extension Practice*. London: Intermediate Technology Publications.

Scoones, I. and C. Toulmin. (1999) *Policies for Soil Fertility Management in Africa*. Sussex, UK: IDS and IIED.

Scott, J.C. (1976) *The Moral Economy of the Peasant: Rebellion and Subsistence in Southeast Asia*. New Haven, CT: Yale University Press.

_____. (1972) "Patron-Client Politics and Political Change." *American Political Science Review* (March): 91-113.

Seed, A.W. and G.L. Austin. (1990) "Sampling Errors for Raingauge-Derived Mean Areal Daily and Monthly Rainfall." *Journal of Hydrology* 118: 163-173.

Serageldin, I. and J. Martin-Brown (eds.). (1999) *Culture in Sustainable Development: Investing in Cultural and National Endowments*. Washington, DC: The World Bank.

Silberfein, M. (ed.). (1998) *Rural Settlement Structure and African Development*. Boulder, CO: Westview Press.

Simon, H. (1960) *The New Science of Management*. Euglewood Cliffs, NJ: Prentice Hall.

Singh, I.J., L. Squire, and J. Strauss (eds.). (1986) *Agricultural Household Models: Extensions, Applications, and Policy*. Washington, DC: The World Bank.

Singh, J.P. (1998) "Evaluating Telecommunication Privatization and Liberalization in India and the Far East." Paper Presented at the 'Has Privatization Worked? The International Experience' Conference of the Columbia Institute of Tele-Information, held at Columbia University, New York.

Singhal, A. and E. Rogers. (20001) *India's Communication Revolution: From Bullock Carts to Cyber Marts*. Thousand Oaks, CA: Sage Publications.

Skocpol, T. (1982) "What Makes Peasants Revolutionary." In T. Skocpol (ed.), *Social Revolutions in the Modern World*. Cambridge, UK: Cambridge University Press: 213-239.

Snijders, F.L. (1991) "Rainfall Monitoring Based on Meteostat Data: A Comparison of Techniques Applied to the Western Sahel." *International Journal of Remote Sensing* 12: 1331-1347.

Sofowora, E.A. and W.A. Isaacs. (1971) "Anti-Sickling Activity of *F. xanthoxyloides*." *Lloydia* 24: 383.

Sonaiya, E B. (1999) "International Network for Family Poultry Development: Origins, Activities, Objectives and Visions." Proceedings of the Development Workers' Course: 'Poultry as a Tool in Poverty Eradication and Promotion of Gender Equality' held at the Course Center Tune Landboskole in Greve, Denmark, 22-26 March. Available Online at: http://www.husdyr.kvl.dk/htm/php/tune99/index.htm

_____. (1997) "African Network on Rural Poultry Development: Progress Report, November 1989 - June 1995." In E. B. Sonaiya (ed.), *Sustainable Rural Poultry Production in Africa. Proceedings ANRPD Workshop, Addis Ababa, Ethiopia, 13-16 June 1995*. Ife-Ife, Nigeria: ANRPD.

Sonaiya, E.B. and O.G. Omitogun. (2001) "A Proposal and Budget for the Establishment of the Nigerian Agricultural Biotechnology Education and Resource Program." Submitted to the National Genetic Engineering and Biotechnology Program, Department of Agricultural and Natural Sciences, Federal Ministry of Science and Technology, Abuja, Nigeria.

South African Information Technology Industry Strategy Project (SAITIS). (2000a) *Baseline Study: A Survey of the IT Industry and Related Jobs and Skills in South Africa*. Pretoria: South African Information Technology Industry Strategy Project.

_____. (2000b) *South African ICT Sector Development Framework*. Pretoria: South African Information Technology Industry Strategy Project.

Sseruwagi, P., J.P. Legg, and G.W. Otim-Nape. (1999) "An Overview of the Incidence of Cassava Mosaic Disease in East Africa, 1998 Update." *Proceedings of the Seventh International Plant Virus Epidemiology Symposium, 11-16 April 1999*. Aguadulce (Almeria), Spain: 87-88.

Stamp, P. (1989) *Technology, Gender, and Power in Africa*. Canada: International Development Research Center.

Statistics South Africa. (2001) *Bulletin of Statistics*. Pretoria: Statistics South Africa.

Stavrou, A., P. Benjamin, P. Burton, and C. McCarthy. (2000) *TeleCentre 2000: The Way Forward*. Johannesburg, South Africa: Vodadom Link Center.

Stebbing, E.P. (1938) "The Man-Made Desert in Africa: Erosion and Drought." *Journal of the Royal African Society Supplement* (January): 3-40.

Stein, H. and E.J. Wilson III. (1993) "Robert Bates, Rational Choice and the Political Economy of Development in Africa." *World Development* 21, 6: 1033-1081.

Stringer, E. T. (1996) *Action Research: A Handbook for Practitioners*. Thousand Oaks, CA: Sage Publications.

Susman, G. I. and R.D. Evered. (1978) "An Assessment of the Scientific Merits of Action Research." *Administrative Science Quarterly* 23: 582-603.

Swaminathan, M.S. (2000) "Science in Response to Basic Human Needs." In UNESCO, *World Conference on Science: Science for the Twenty-First Century: A New Commitment*. Paris: UNESCO: 33-40.

Swannel, R.P.J., K. Lee, and M. McDonagh. (1996) "Field Evaluation of Marine Oil Spill Bioremediation." *Microbiological Reviews* 60: 324-365.

Swanson, M.M. and B.D. Harrison. (1994) "Properties, Relationships and Distribution of Cassava Mosaic Geminiviruses." *Tropical Science* 34:15-25.

Sweeney, P. (1999) *The Celtic Tiger: Ireland's Continuing Economic Miracle*. Dublin: Oak Tree Press.

Swift, J. (1996) "Desertification: Narratives, Winners & Losers." In M. Leach and R. Mearns (eds.), *The Lie of the Land: Challenging Recieved Wisdom on the African Environment*. Oxford: James Currey: 73-90.

TAC. (1988) *Sustainable Agricultural Production: Implications for International Agricultural Research*. Rome: FAO.

_____. (1978) *Farming Systems Research at the International Agricultural Research Centers*. Rome: TAC Secretariat.

Takigawa, T. (1998) "The Impact of WTO Telecommunications Agreement on US and Japanese Telecommunications Regulations." *Journal of World Trade* 32, 6: 33-55.

Taylor, C.S. (1995) "Kenaf as a new annual fiber source for industrial uses: general economic and environmental aspects." Kenaf International, Ltd., McAllen, TX, KI Pub. #13, 4 pgs.

Telkom. (2000a) *Creating Value Through People and Technology: Telkom Annual Report 1999/2000*. Pretoria, South Africa: Telkom.

_____. (2000b) *National Telecommunications Policy Colloquium Submission*. Pretoria, South Africa: Telkom.

Thottappilly, G. and O.G. Omitogun. (1997) "Application of Molecular Strategies for Crop Improvement in Nigeria." In *International Workshop on Biotechnology: Towards Sustainable Agricultural Development in Africa, 10-15 August*. Umuahia, Nigeria: Federal University of Agriculture.

Thresh, J.M. and M.W. Mbwana. (1998) "Cassava Mosaic and Cassava Brown Streak Disease in Zanzibar." *Roots* 5: 6-9.

Tiamiyu, M.A. (2000a) "Availability, Accessibility and Use of Information Technologies in Nigerian Federal Agencies: a Preliminary Survey." *Information Technology for Development* 9, 2: 91-104.

Tiamiyu, M.A. (2000b) "Information Technology in Nigerian Federal Agencies: Problems, Impact and Strategies." *Journal of Information Science* 26, 4: 227-37.

Tiamiyu, M.A. (1993) "The Realities of Developing Modern Information Resources Management Systems in Government Organizations in Developing Countries with Particular Reference to Nigeria." *Journal of Information Science* 19, 3: 189-98.

Tiffen, M., M. Mortimore, and F. Gichuki. (1994) *More People, Less Erosion: Environmental Recovery in Kenya*. New York: Wiley.

Tiffin, S. and F. Osotimehin. (1992) *New Technologies and Enterprise Development in Africa*. Paris: OECD.

Trigo, E.J. and W.R. Jaffe. (1994) "Agricultural Biotechnology Development in Latin America and the Caribbean." *FAO World Agriculture* 1994:89-92.

Tripp, R. (ed.). (1991) *Planned Change in Farming Systems: Progress in On-Farm Research.* New York: John Wiley & Sons.

Tripp, R. (1985) "Anthropology and On-Farm Research." *Human Organization* 44, 2: 114-124.

Tripp, R., P. Anandajayasekeram, D. Byerlee, and L. Harrington. (1990) "Farming Systems Research Revisited." In C.K. Eicher and J.M. Staatz (eds.), *Agricultural Development in the Third World.* [2nd Ed.] Baltimore / London: Johns Hopkins University Press: 384-99.

Turner, B., G. Hyden, and R.W. Kates (eds.). (1993) *Population Growth and Agricultural Change in Africa.* Gainesville, FL: University of Florida Press.

Turner, M. (1999) "Merging Local and Regional Analyses of Land-Use Change: The Case of Livestock in the Sahel." *Annals of the Association of American Geographers* 89, 2: 191-219.

Tusubira, F.F. "Integration of DIIAUP in Uganda 's Emerging ICT Infrastructure." Proceedings from the DIIAUP Validation Workshop.

TWNSO. (1998) *Profiles of Institutions for Scientific Exchange and Training in the South.* Trieste: TWNSO.

_____. (1994) *Profiles of Institutions for Scientific Exchange and Training in the South.* Trieste: TWNSO.

Uganda Communications Commission Database. (2001) "Policies and Strategies for Rural Communications in Uganda, Including A Plan for the Rural Communications Development Fund and its Administration." Final Report Submitted to the Uganda Communications Commission by Intelcon Research and Consultancy, 1 March.

UN Advisory Committee on the Application of Science and Technology to Development. (1971) *World Plan of Action for the Application of Science and Technology to Development.* New York: UN.

UNAIDS/WHO. (2000) *Botswana: Epidemiological Fact Sheet on HIV/ AIDS and Sexually Transmitted Infections*. Geneva: UNAIDS/WHO. Available Online. http://www.unaids.org/hivaidsinfo/statistics/june00/ fact-sheets/pdf/botswana.pdf.

UN Commission on Trade and Development (UNCTAD). (2000) *World Investment Report 2000*. New York: UN.

UNCTAD Secretariat (1999a) *A Common Vision on Science and Technology for Development*. Geneva: UN Commission on Science and Technology for Development.

_____. (1999b) *Expert Group Meeting On A Common Vision For The Future Contribution Of Science and Technology For Development*. Geneva: UN Commission on Science and Technology for Development.

_____. (1999c) *Working Group on Science and Technology Partnerships and Networking for National Capacity-Building*. Geneva: UN Commission on Science and Technology for Development.

_____. (1998) *New Approaches to Science and Technology Cooperation and Capacity Building*. New York: UN.

_____. (1997) *Consideration of Ways and Means of Commemorating The Twentieth Anniversary of the Vienna Conference on Science and Technology for Development*. Geneva: UN Commission on Science and Technology for Development.

UN Economic Commission for Africa (UNECA). (1999a) *Information and Communications Technologies for Improved Governance in Africa*. Addis Ababa: UNECA.

_____. (1999b) "Africa on the Internet: An Annotated Guide to African Web Sites." *Bellanet Online*. http://www.bellanet.org/partners/aisi/nici/ index.htm.

_____. (1999c) "African Country Profiles: National Information and Communication Infrastructure." *Bellanet Online*. http://www. bellante.org/partners/aisi/nici.htm.

_____. (1999d) "Democratizing Access to the Information Society." *Bellanet Online*. http://www.bellanet.org/partners/aisi/adf99docs/ democratising.htm.

_____. (1999e) "Globalization and the Information Economy: Challenges and Opportunities for Africa." *Bellanet Online*. http://www.bellanet.org/partners/aisi/adf99docs/infoeconomy.htm.

_____. (1999f) "Information and Communication Technologies for Improved Governance in Africa." *Bellanet Online*. http://www.bellanet.org/parnters/aisi/adf99docs/governance.htm.

_____. (1999g) "The Process of Developing National Information and Communications Infrastructure (NICI) Policies in Africa." E/ECA/ADF/99/9. *Bellanet Online*. http://www.bellanet.org/partners/aisi/adf99docs/nici.htm, http://www.bellanet.org/partners/aisi/new.htm, and http://www.bellanet.org/partners/aisi/activities.htm.

UN Education, Scientific and Cultural Organization (UNESCO). (2000) *World Conference on Science: Science for the Twenty-First Century: A New Commitment*. Paris: UNESCO.

_____. (1999a) *Women, Science and Technology: Towards a New Development?* Paris: UNESCO.

_____. (1999b) *Science and Technology in the SADC Region for the 21st Century*. Paris: UNESCO.

_____. (1999c) *UNESCO Statistical Yearbook*. New York: United Nations.

_____. (1998a) *World Education Report 1998*. Paris: UNESCO.

_____. (1998b) *World Science Report 1998*. Paris: UNESCO.

_____. (1996) "African Information Society Initiative: An Action Framework to Build Africa's Information and Communication Infrastructure." *Bellanet Online*. http://www.bellanet.org/partners/aisi/aisi.htm.

_____. (1995a) "Communique of Addis Ababa." 7 April 1995. Online. http://www.sas.upenn.edu/African_Studies/Padis/telmatcs_communik.html.

_____. (1995b) "Final Report: Study of the Effectiveness of Informatics Policy Instruments in Africa." *Bellanet Online*. http://www.bellanet.org/partners/aisi/policy/infopol/sumbrown.htm.

_____. (1994) *Symposium on Science and Technology in Africa.* Nairobi, Kenya: UNESCO Regional Office for Science and Technology for Africa.

_____. (1974) *Science and Technology in African Development.* Paris: UNESCO.

USTR. (2000) "Annual Review of Telecommunication Trade Agreements: Highlights Concerns in Columbia, Mexico, South Africa and Taiwan." Online. http://www.ustr.gov/releases/2001/04/01-20html. Retrieved 6 April 2001.

Uwachukwa, A.N. (1997) "Internet Regulations in Nigeria." In *Telematics for Development: Infotech '96.* Ife-Ife, Nigeria: National Centre for Technology Management.

Uzo, E. and C. Akalezi. (2000) "Sustainable Bio-Remediation: A New Approach to Oil Spill Response." Workshop Proceedings.

Vail, L. (ed.). (1988) *The Creation of Tribalism in Southern Africa: Studies in the Political Economy of an Ideology.* Berkeley, CA: University of California Press.

van Ameringen, M. (1995) *Science and Technology Policy: Building a New South Africa.* Ottawa, ON: IDRC.

van Audenhove, L.J., J-C. Burgelman, G. Nulens and B. Cammaerts. (2001) "Telecommunication and Information Society Policy in Africa. A Critical Assessment of the Dominant Scenario." In N. Hafkin, L.J. van Audenhove and G. Nulens (eds.), *The African Information Society.* Brussels: The Free University of Brussles Press.

van Eijk, T. (2000) "Holism and FSR." In M. Collinson (ed.), *A History of Farming Systems Research.* Wallingford / Rome: CABI Publishing / FAO: 323-334.

van Horen, C. and A. Eberhard. (1995) "Energy, Environment and the Rural Poor in South Africa." *Development South Africa* 12, 2: 197-211.

van Zoonen, L. (1992) "Feminist Theory and Information Technology." *Media, Culture and Society* 14: 9-29.

Vayda, A.P. (ed.). (1969) *Environment and Cultural Behavior: Ecological Studies in Cultural Anthropology.* Garden City, NY: Natural History Press.

Vincent, V. and R.G. Thomas. (1960) *An Agricultural Survey of Southern Rhodesia, Part I. The Agro-Ecological Survey*. Salisbury: Government Printer.

von Bertalanffy, L. (1968) *General Systems Theory: Foundations, Development Applications*. London: Allen Lane.

Wachira, N. (2001) "One Man's Fight to Wire Africa." *Wired Magazine* 15 March.

Wade, R. (1997) "Greening the Bank: The Struggle over the Environment, 1970-1995 (v.2)." In D. Kapur, J.P. Lewis, and R. Webb (eds.), *The World Bank: Its First Half Century*. Washington, DC: Brookings Institution Press: 611-734.

Wajcman, J. (2000) "Reflections on Gender and Technology Studies: In What State is the Art?" *Social Studies of Science* 30, 3: 447-464.

_____. (1992) "Feminist Theory and Information Technology." *Media, Culture, and Society* 14: 9-29.

_____. (1991) *Feminism Confronts Technology*. State College, PA: The Pennsylvania State University Press.

Walters, P.J.C. (1989) *Famine Early Warning Systems*. London: Earthscan Publications, Ltd.

Walters, S. (1997) *Globalization, Adult Education and Training: Impacts and Issues*. London: Zed Books.

Wambugu, F. (1999) "Why Africa Needs Agricultural Biotech: There Is an Urgent Need for the Development and Use of Agricultural Biotechnology in Africa to Help Counter Famine, Environmental Degradation and Poverty." *Nature* 400, 6739: 15-17.

_____. (1996) "Control of African Sweet Potato Virus Diseases Through Biotechnology and Technology Transfer: Turning Priorities into Feasible Programs." In J. Komen, J.I. Cohen, and Z. Ofir (eds.), *Proceedings of a Policy Seminar on Agricultural Biotechnology for East and Southern Africa*. Agricultural Biotechnology Policy Seminars No. 2: 75-81.

Watts, M.J. (1983a) "Good Try, Mr. Paul: Populism and the Politics of African Land Use." *African Studies Review* 26, 2: 73-84.

Watts, M.J. (1983b) *Silent Violence: Food, Famine & Peasantry in Northern Nigeria*. Berkeley, CA: University of California Press.

Weight, D. and V. Kelley. (1999) *Fertilizer Impacts on Soils and Crops of Sub-Saharan Africa*. International Development Paper No. 21. East Lansing, MI: Michigan State University, Department of Agricultural Economics.

Whingwiri, E.E., K. Mashingaidze, and M. Rukini (eds.). (1992) *Small Scale Agriculture in Zimbabwe. Book One: Farming Systems, Policy and Infrastructural Development*. Harare, Zimbabwe: Rockwood Publishers.

White, W. (ed.). (1996) *Bridge Builders: African Experiences With Information and Communication Technology*. Washington, DC: National Academy Press.

Whiteman, A. and C. Brown. (1999) "The Potential Role of Forest Plantations in Meeting Future Demands for Industrial Wood Products." *International Forestry Review* 1,3: (September): 143-152.

Wilmsen, E.N. (1989) *Land Filled With Flies: A Political Economy of the Kalahari*. Chicago, IL: The University of Chicago Press.

Wilson, E. II and K. Wong. (1999) "African Information Revolution: A Balance Sheet." Unpublished Paper.

Wimmers, K., S. Ponsuksili, F. Schmoll, T. Hardge, E.B. Sonaiya, K. Schellander, and P. Horst. (1999) "Application of Microsatellite Analysis to Group Chicken According to Their Genetic Similarity." *Archiv fuer tierzucht (Archives for Animal Breeding)* 42, 6: 629-639.

Wingfield, B. (2000) "Biotechnology by Africans for Africa." 3 April 2000. In FAO, *The Electronic Forum on Biotechnology in Food and Agriculture*. Online. http://www.fao.org/biotech/forum.htm.

Winkleman, D. and E. Moscardi. (1982) "Aiming Agricultural Research at the Needs of Farmers." In W.W. Shaner, P.F. Philipp, and W.R. Schmehl (eds.), *Readings in Farming Systems Research and Development*. Boulder, CO: Westview Press: 31-43.

Woherem, E.E. (1995) "Towards A Culture of Management of Software Systems Maintenance in Africa." *Information Technology for Development* 6: 5-14.

Wolf, E. (1982) *Europe and the People Without History*. Berkeley, CA: University of California Press.

_____. (1969) *Peasant Wars of the Twentieth Century*. New York: Harper & Row.

Wood, E. (1999) *Thwate Consulting (Pty) Ltd.: A Case Study*. Cape Town: University of Cape Town, Graduate School of Business.

Working Party on Telecommunications and Information Services Policies (WPTISP). (2000) *Telecommunications Regulations: Institutional Structures and Responsibilities*. [DSTI/TISP(99)5/FINAL] Paris: OECD.

_____. (1999) *A Review of Market Openness and Trade in Telecommunications*. [DSTI/TISP(99)5/FINAL] Paris: OECD.

World Bank. (2001) *World Development Report 2000/2001*. Washington, DC: The World Bank.

_____. (2000) *Sub-Saharan Africa Data Profile*. [July] Washington, DC: The World Bank.

_____. (1999a) *World Development Report: Knowledge for Development*. Washington, DC: The World Bank.

_____. (1999b) *Culture in Sustainable Development: Investing in Cultural and Natural Endowments*. Washington, DC: The World Bank.

_____. (1998) *World Development Report Background Papers: Knowledge for Africa*. Washington, DC: The World Bank.

_____. (1995) *Increasing Internet Connectivity in Sub-Saharan Africa: Issues, Options and World Bank Group Role*. [Draft of 29 March 1995] Washington, DC: The World Bank.

World Bank. (1993) *Vetiver Grass: The Hedge Against Erosion*. Washington, DC: The World Bank.

World Trade Organization (WTO). (1998) *The Results of the Uruguay Round of Multilateral Trade Negotiations: Legal Texts*. Geneva: WTO Secretariat.

World Wildlife Fund (WWF). (1999) "An Overview of GM Technology in the Forest Sector. _ A Scoping Study for WWF-UK and WWF International." (by Rachel Asante Owusu), October 1999. Online. http: //www.wwf_uk.org/news/news108.htm.

Zeleza, P.T. (2002a) *Rethinking Africa's Globalization. Vol.1: The Intellectual Challenges*. Trenton, NJ: Africa World Press.

_____. (2002b) "Contemporary African Migrations in a Global Context." *African Issues* (forthcoming).

Zongo, G. and P. Benjamin. (2001) *Telecentre Models World-Wide: The Situation in Sub-Saharan Africa.* Johannesburg: University of Witwatersrand, Link Center.

Zwizwai, B. et al. (1997) *The State of S and T Policy, Dialogue, and Institutions in Zimbabwe.* Harare, Zimbabwe: n.p.

ABBREVIATIONS

AIRS	Academic Records Information Service
AAS	African Academy of Sciences
AAU	Association of African Universities
ABERP	Agricultural Biotechnology Education and Resource Program
ADDs	Agricultural Development Divisions
ADB	African Development Bank
ADF	African Development Forum
AERC	African Economics Research Consortium
AFPREPREN	African Energy Policy Research Network
AGR	Agriculture and Agri-Food Canada
AIHTTR	African Institute for Higher Technical Training and Research
AISI	African Information Society Initiative
ALDSAP	Agriculture and Livestock Development Action Plan
AMIS	Agriculture Management Information System
APEC	Asia–Pacific Economic Cooperation
AR	Action Research
ARCEDEM	African Regional Centre for Engineering Design and Manufacturing
ARCIS	Africa Regional Centre for Information Science
ARCT	African Regional Centre for Technology
ARET	Agricultural Research and Extension Trust
ARIPO	African Regional Industrial Property Organization
ARSO	African Regional Standards Organization
ARSTD	African Regional Symposium on Telematics for Development.
ASARECA	Association for Strengthening Agricultural Research in Eastern and Central Africa
ATM	Automatic Teller Machine
ATPS	African Technology Policy Studies Network
AU	African Unity, previously OAU

AVU	African Virtual University
BANDWD	International bandwidth
BDP	Botswana Democratic Party
BoTECH	Botswana Technology Center
Bt	Bacillus thuringiensis
BTA	Basic Telecommunication Annex
CABECA	Capacity Building for Electronic Communication in Africa
CALLCOST	Cost of utilization
CAPAS	Coordinated African Program of Assistance in Services
CCK	Communications Commission of Kenya
CCSI	Client Consultation System Infrastructure
CERSE	Center for Environmental Resources and Sustainable Ecosystem
CGIAR	Consultative Group for International Agricultural Research
CIDA	Canadian International Development Agency
CIMMYT	International Maize and Wheat Improvement Center
CITIPOP	Number of cities with points-of-presence
CMD	Cassava Mosaic Virus Disease
CNN	Cable News Network
COHRED	Council on Health Research for Development
COMPINE	Combating Paedophile information networks in Europe
CORAF	West and Central African Council for Agricultural Research and Development
COSATU	Congress of South African Trade Unions
CSIR	Council for Scientific and Industrial Research and Development
CZI	Confederation of Zimbabwe Industries
DAHI	Department of Animal Health and Industry, Malawi
DARTS	Department of Agricultural research and Technical Services
DIIAUP	Developing Information Infrastructure Agenda in Uganda Project

DNA	Deoxyribonucleic acid amplification, see also PCR
DRC	Democratic Republic of the Congo
DTI	South African Department of Trade and Industry
EC	European Commission
ECA	United Nations Economic Commission for Africa
EMIS	Educational Management Information System
EPR	Electronic Patient Records
ERT	Energy for Rural Transformation
ESAP	Economic Structural Adjustment Program
FADIB	Foundation for African Development through International Biotechnology
FAO	Food and Agriculture Organization
FDI	Foreign Direct Investment
FEI	Financial Executives Research Foundation
FIDO	Field Integrated Design & Operations
FINIS	Financial Management Information System
FRIM	Forestry Research Institute of Malawi
FS	Farming systems
FSR	Farming systems Research
G-7	Group of Seven Highly Industrialized Nations
GATS	General Agreement on Trade in Services
GCB	Government Computer Bureau
GDP	Gross Domestic Product
GEAR	Growth, Employment and Redistribution
GIIC	Global Information Infrastructure Commission
GIS	Global Information Society
GMO	Genetically modified organisms
GOUEXECNET	Uganda presidential internal/public office network
GTZ	German Technical Cooperation
HEN	Health and Education Network
HIPC	Highly Indebted Poor Countries
HMIS	Health Management Information System
HSRC	Human Science Research Council

HURIS	Human Resource Information System
IAPS	Internet Access Providers
IARC	International Agricultural Research Center
IBPGR	International Board for Plant Genetic Resources
ICLAM	International Center for Living Aquatic Resource Management
ICPD	International Conference on Population and Development
ICRAF	International Center for Research in Agro-forestry
ICRISAT	International Center for Research in Semi-Arid Tropics
ICS	Institute of Computer Science
ICT	Information and communication technology
IDAD	Internet Design & Development
IDEP	Institut pour de development economique et la planification
IDRC	International Development Research Centre (of Canada)
IHPP	International Health Policy Program
IIR	Institute of Industrial Relations
IITA	International Institute of Tropical Agriculture
IMF	International Monetary Fund
INFPD	International Network for Family Poultry Development
INTECU	Information Technology Coordinating Unit
ISETT	Information Systems, Electronics and Telecommunication Technologies
Isis-WICCE	Isis Women's International Cross Cultural Exchange
ISP	Internet Service Point or Provider
ICT	Information and communication technology
IT	Information technology
ITU	International Telecommunication Union
IXP	Internet Exchange Point
JICST	Japan Information Center of Science and Technology
KCA	Kenya Communications Act

KP&TC	Kenya Post and Telecommunication Corporation
LAN	Network neighborhood
LDC	Least developed countries
LIBIS	Library Information System
LITRATE	Literacy rate
LSTOCK	Local stock exchange
MDC	Movement for Democratic Change, Zimbabwe
MFEE	Monthly fee paid by providers
MIRTDC	Malawi Industrial Research and Technology Development Center
MISA	Media Institute of Southern Africa
MMPZ	Media Monitoring Project Zimbabwe
MOE	Ministry of Education
MPS	Ministry of Public Service
MSI	Millennium Science Initiative
MSU	Michigan State University
MTN	Mobile Telecommunication Network
NACGRAB	National Center for Genetic Resources and Biotechnology
NAS	National Academy of Sciences
NCC	Nigerian Communications Commission
NCDC	National Curriculum Development Centre
NEAP	National Environmental Action Plan
NGOs	Non–governmental organizations
NICI	National Information and Communication Infrastructure
NICs	Newsgroup Information Centers
NIIA	National Information Infrastructure Agenda
NORAD	Norwegian Agency for Development
NRC	National Research Council
NSF	National Science Foundation
NSTRDC	National Science & Technology Research and Development Council
NTSAB	Nigerian Technical Subcommittee on Agricultural Biotechnology
NUA	Internet survey services

NUAfrica	Interactive discussion group on African affairs
NUMSA	National Union of Metalworkers of South Africa
OAU	Organization of African Unity see also AU
OECD	Organization for Economic Cooperation and Development
PADIS	Pan African Development Information System
PANA	Pan-African News Agency
PCR	Polymerase Chain Reaction
PICTA	Partnership for Information and Communication Technologies in Africa
PINCOME	Personal income
PIWH	Pacific Institute for Women's Health
PTO	Public telephone operators
R&D	Research and development
RCZ	Research Council of Zimbabwe
RDCF	Rural Development Communication Fund
RDPs	Rural Development Projects
RENACO	West and Central Africa Cowpea Research Network
RFLP	Restriction fragment length polymorphisms
RTO	Research and Technology Agency of NATO
RTPMPCA	Restrictive Trade Practices, Monopolies and Price Control Act
S&T	Science and Technology
SADC	Southern African Development Community
SAITIS	South African Information Technology Industry Strategy
SAPs	Structural adjustment programs
SAPT	South African Posts and Telecommunications
SIRDC	Scientific and Industrial Research Development Centre
SWIFT	Society for Worldwide International Financial Telecommunications
TEEAL	The essential electronic agriculture library
TNCs	Transnational corporations
TRF	The Rotary Foundation
TRIM	Trade Related Investment Measures

TWAS	Third World Academy of Sciences
TWNSO	Third World Network of Scientific Organizations
UBOS	Uganda Bureau of Statistics
UCB	Uganda Commercial Bank
UCC	Uganda Communication Commission
UCS	Uganda Computer Services
UNCST	Uganda National Council of Science and Technology
UNCTAD	United Nations Conference on Trade and Development
UNDP	United Nations Development Program
UNECA	United Nations Economic Commission for Africa
UNESCO	United Nations Educational, Scientific and Cultural Organization
UNU	United Nations University
UPMA	Uganda Private Midwives Association
USAID	United States Agency for International Development
USDA	United States Department of Agriculture
USERATE	Internet utilization rate
USHEPiA	Universities Science, Humanities, and Engineering Partnerships in Africa
USTR	United States Trade Representative (office of)
UTL	Uganda Telecom Limited
UWONET	Uganda Women's Network
WACAF	West and central Africa
WAG	Women's Action Group
WB	World Bank
WCS	World Conference on Science
WET	Whole Earth Telescope
WFP	World Food Program
WHO	World Health Organization
WPTISP	Working Party on Telecommunication and Information Services Policies
WSP	Workplace Skills Plan

WTO	World Trade Organization
ZANU-PF	Zimbabwe African National Union-Patriotic Front
ZIDS	Zimbabwe Institute of Development Studies
ZNADWO	Zambia National Association of Disabled Women
ZWB	Zimbabwe Women's Bureau

Index

B

C

D

E

F

G

H

I

J

K

L

M

N

T

U

V

W

Z